LIVERPOOL JMU LIBRARY

VOLUME FOUR HUNDRED AND SEVENTY-FOUR

METHODS IN ENZYMOLOGY

Thiol Redox Transitions in Cell Signaling, Part B

Cellular Localization and Signaling

METHODS IN ENZYMOLOGY

Editors-in-Chief

JOHN N. ABELSON AND MELVIN I. SIMON

Division of Biology
California Institute of Technology
Pasadena, California, USA

Founding Editors

SIDNEY P. COLOWICK AND NATHAN O. KAPLAN

VOLUME FOUR HUNDRED AND SEVENTY-FOUR

METHODS IN ENZYMOLOGY

Thiol Redox Transitions in Cell Signaling, Part B

Cellular Localization and Signaling

EDITED BY

ENRIQUE CADENAS AND **LESTER PACKER**
*Department of Pharmacology & Pharmaceutical Sciences
School of Pharmacy
University of Southern California
Los Angeles, CA, USA*

AMSTERDAM • BOSTON • HEIDELBERG • LONDON
NEW YORK • OXFORD • PARIS • SAN DIEGO
SAN FRANCISCO • SINGAPORE • SYDNEY • TOKYO
Academic Press is an imprint of Elsevier

Academic Press is an imprint of Elsevier
525 B Street, Suite 1900, San Diego, CA 92101-4495, USA
30 Corporate Drive, Suite 400, Burlington, MA 01803, USA
32 Jamestown Road, London NW1 7BY, UK

First edition 2010

Copyright © 2010, Elsevier Inc. All Rights Reserved.

No part of this publication may be reproduced, stored in a retrieval system or transmitted in any form or by any means electronic, mechanical, photocopying, recording or otherwise without the prior written permission of the publisher

Permissions may be sought directly from Elsevier's Science & Technology Rights Department in Oxford, UK: phone (+44) (0) 1865 843830; fax (+44) (0) 1865 853333; email: permissions@elsevier.com. Alternatively you can submit your request online by visiting the Elsevier web site at http://elsevier.com/locate/permissions, and selecting *Obtaining permission to use Elsevier material*

Notice
No responsibility is assumed by the publisher for any injury and/or damage to persons or property as a matter of products liability, negligence or otherwise, or from any use or operation of any methods, products, instructions or ideas contained in the material herein. Because of rapid advances in the medical sciences, in particular, independent verification of diagnoses and drug dosages should be made

For information on all Academic Press publications
visit our website at elsevierdirect.com

ISBN: 978-0-12-381003-8
ISSN: 0076-6879

Printed and bound in United States of America
10 11 12   10 9 8 7 6 5 4 3 2 1

Working together to grow
libraries in developing countries

www.elsevier.com | www.bookaid.org | www.sabre.org

ELSEVIER   BOOK AID International   Sabre Foundation

Contents

Contributors	xiii
Preface	xix
Volumes in Series	xxi

1. Engineering of Fluorescent Reporters into Redox Domains to Monitor Electron Transfers — 1
Derek Parsonage, Stacy A. Reeves, P. Andrew Karplus, and Leslie B. Poole

1. Introduction	2
2. The Problem: Low Sensitivity and Improperly Rate-Limited Assays for Redox Functions of Bacterial Peroxiredoxin Systems	3
3. The Solution: Engineering of Fluorescent Redox Reporters into the N-Terminal Domain of AhpF and *E. coli* Grx1	4
4. Engineering of Disulfide-Containing Electron Acceptor Domains to Detect Electron Transfers via Fluorescence Changes; Linkage of Fluorescein to Bacterial AhpC via a Reducible Disulfide Bond	7
5. Materials	8
6. Methods	10
7. Summary	19
References	19

2. Blot-Based Detection of Dehydroalanine-Containing Glutathione Peroxidase with the Use of Biotin-Conjugated Cysteamine — 23
Sue Goo Rhee and Chun-Seok Cho

1. Introduction	24
2. Oxidative Inactivation of Glutathione Peroxidase and the Conversion of Its Active Site Sec to DHA	25
3. Preparation of Biotin-Conjugated Cysteamine	27
4. Blot-Based Detection of DHA–GPx1 in RBCs	28
5. Effects of Oxidative Stress on the Formation of DHA–GPx1 in RBCs	29
6. Concluding Remarks	30
Acknowledgment	32
References	32

3. Analysis of the Redox Regulation of Protein Tyrosine Phosphatase Superfamily Members Utilizing a Cysteinyl-Labeling Assay — 35

Benoit Boivin and Nicholas K. Tonks

1. Introduction — 36
2. Active-Site Structure, Catalysis, and Oxidation — 37
3. Detection Methods — 39
4. General Principle of the Assay — 40
5. Solutions — 42
6. Preparation of the Lysis Buffer — 42
7. Preparation of the Hypoxic Glove Box — 43
8. Preparation of Cell Lysates — 43
9. Cysteinyl-Labeling Assay — 44
10. Acute Stimulus-Induced Reversible Oxidation of PTPs — 44
11. Perspectives — 45
12. Conclusion — 48
References — 48

4. Measuring the Redox State of Cellular Peroxiredoxins by Immunoblotting — 51

Andrew G. Cox, Christine C. Winterbourn, and Mark B. Hampton

1. Introduction — 52
2. Measurement of Prx Dimerization — 54
3. Measurement of Prx Hyperoxidation — 59
4. Discussion — 61
Acknowledgments — 64
References — 64

5. Thiol Redox Transitions by Thioredoxin and Thioredoxin-Binding Protein-2 in Cell Signaling — 67

Eiji Yoshihara, Zhe Chen, Yoshiyuki Matsuo, Hiroshi Masutani, and Junji Yodoi

1. Functional Regulation of Redox-Sensitive Proteins by Thiol Modification — 68
2. Thiol Reduction by the Thioredoxin Redox System — 70
3. Thioredoxin Superfamily — 73
4. Reversible Redox and Signal Regulation by Thioredoxin and Thioredoxin-binding Protein-2 (TBP-2) — 74
5. Conclusion — 76
References — 76

6. **Detection of Protein Thiols in Mitochondrial Oxidative Phosphorylation Complexes and Associated Proteins** — 83

Kelly K. Andringa and Shannon M. Bailey

1. Introduction — 84
2. Mitochondria Isolation and Protein Thiol Labeling — 86
3. Application of Blue Native-PAGE for the Isolation of Oxidative Phosphorylation Protein Subunits and Other Proteins Associated with the Complexes — 89
4. Detection of IBTP-Labeled Protein Thiols in Protein Complexes — 94
5. Analysis and Mass Spectrometry Identification of Protein — 95
6. Other Considerations — 104
7. Conclusion — 105
Acknowledgments — 106
References — 106

7. **Mitochondrial Thioredoxin Reductase: Purification, Inhibitor Studies, and Role in Cell Signaling** — 109

Maria Pia Rigobello and Alberto Bindoli

1. Introduction — 110
2. Purification of Thioredoxin Reductase from Isolated Mitochondria, Cultured Cells, and Whole Organs — 111
3. Estimation of Thioredoxin Reductase Activity — 117
4. Inhibitor Studies of Thioredoxin Reductase — 118
5. Role in Cell Signaling — 118
References — 120

8. **Measuring Mitochondrial Protein Thiol Redox State** — 123

Raquel Requejo, Edward T. Chouchani, Thomas R. Hurd, Katja E. Menger, Mark B. Hampton, and Michael P. Murphy

1. Introduction — 124
2. Quantification of Mitochondrial Protein Thiols — 127
3. Quantification of Glutathionylation of Mitochondrial Proteins — 130
4. Assessment of S-Nitrosated Protein Thiols — 135
5. Measurement of the Thioredoxin and Peroxiredoxin Redox States — 139
6. Conclusions — 143
Acknowledgments — 144
References — 144

9. **Measurement of Extracellular (Exofacial) Versus Intracellular Protein Thiols** — 149

Jolanta Skalska, Steven Bernstein, and Paul Brookes

1. Measurement of Mitochondrial Thiol Status — 150
2. Measurement of Cytosolic Thiol Status — 156
3. Measurement of Exofacial Thiols — 157
4. Exofacial Thiol Status and Cancer — 160
References — 162

10. **Redox Clamp Model for Study of Extracellular Thiols and Disulfides in Redox Signaling** — 165

Young-Mi Go and Dean P. Jones

1. Introduction — 166
2. Key Concepts for Use — 166
3. Principles for Experimental Design — 167
4. Summary of Available Redox Clamp Studies — 171
5. Perspectives and Conclusion — 177
Acknowledgments — 178
References — 178

11. **Redox State of Human Serum Albumin in Terms of Cysteine-34 in Health and Disease** — 181

Karl Oettl and Gunther Marsche

1. Background — 182
2. HPLC Analysis — 182
3. Albumin Thiol State and Exercise — 187
4. Influence of Supplementation — 188
5. Albumin Oxidation in Disease — 188
6. Albumin Thiol State During Aging — 191
7. Summary — 193
References — 193

12. **Methods for Studying Redox Cycling of Thioredoxin in Mediating Preconditioning-Induced Survival Genes and Proteins** — 197

Chuang C. Chiueh

1. Hormetic Mechanism: Role of Redox Cycling of Thioredoxin — 198
2. Redox Functioning of Thioredoxin — 199

3. Implications of Preconditioning Protection
   from Preclinical and Clinical Studies ... 200
4. Drugs Mimic Thioredoxin-Medicated Preconditioning-Induced
   Signaling and Protection in Cells ... 202
5. Methods and Materials ... 203
References ... 209

13. Oxidative Stress, Thiol Redox Signaling Methods in Epigenetics ... 213

Isaac K. Sundar, Samuel Caito, Hongwei Yao, and Irfan Rahman

1. Introduction ... 216
2. Histone Acetylation Assays Using [^{3}H]-Acetate Incorporation ... 219
3. Histone Acetylation by Immunoblotting ... 221
4. HAT Activity Assay ... 222
5. HDAC Activity Assay Using [^{3}H]-Labeled Histones ... 223
6. HDAC Activity Assay ... 225
7. HDACs Levels by Immunoblotting ... 227
8. Posttranslational Modifications of HDACs and SIRTs (Sirtuins 1–7) by Immunoprecipitation ... 228
9. Preparation of Whole Cell Lysate ... 229
10. Preparation of Cytoplasmic and Nuclear Proteins ... 229
11. Redox-Mediated Posttranslational Modification Assays ... 230
12. Chromatin Immunoprecipitation (ChIP) Assay ... 233
13. Conclusions ... 238
Acknowledgments ... 239
References ... 239

14. Characterization of Protein Targets of Mammalian Thioredoxin Reductases ... 245

Anton A. Turanov, Dolph L. Hatfield, and Vadim N. Gladyshev

1. Introduction ... 246
2. Preparation of TR-immobilized Affinity Resins ... 247
3. Identification of Targets of Mammalian TRs in Cell Lysates ... 250
4. Concluding Remarks and Future Perspectives ... 252
Acknowledgments ... 253
References ... 253

15. Alteration of Thioredoxin Reductase 1 Levels in Elucidating Cancer Etiology — 255

Min-Hyuk Yoo, Bradley A. Carlson, Petra Tsuji, Robert Irons, Vadim N. Gladyshev, and Dolph L. Hatfield

1. Introduction — 256
2. Materials and Methods — 257
3. Results and Discussion — 263
4. Conclusions and Future Perspectives — 271
Acknowledgments — 272
References — 272

16. Regulation of Apoptosis Signal-Regulating Kinase 1 in Redox Signaling — 277

Kazumi Katagiri, Atsushi Matsuzawa, and Hidenori Ichijo

1. Overview — 278
2. Materials — 281
3. Methods — 283
4. Comment — 286
References — 287

17. Protocols for the Detection of S-Glutathionylated and S-Nitrosylated Proteins In Situ — 289

Scott W. Aesif, Yvonne M. W. Janssen-Heininger, and Niki L. Reynaert

1. Introduction — 290
2. Protein S-Glutathionylation — 290
3. Protein S-Nitrosylation — 293
4. Summary — 295
References — 295

18. Synthesis, Quantification, Characterization, and Signaling Properties of Glutathionyl Conjugates of Enals — 297

Sanjay Srivastava, Kota V. Ramana, Aruni Bhatnagar, and Satish K. Srivastava

1. Introduction — 298
2. Synthesis, Quantification, and Characterization of Reagent Glutathionyl Conjugates of HNE — 300
3. Metabolism of HNE — 305

4. Signaling Properties of Glutathionyl Conjugates of HNE	309
5. Conclusions	310
Acknowledgments	311
References	311

19. Thioredoxin and Redox Signaling in Vasculature—Studies Using Trx2 Endothelium-Specific Transgenic Mice   315

Wang Min, Luyao (Kevin) Xu, Huanjiao (Jenny) Zhou, Qunhua Huang, Haifeng Zhang, Yun He, Xu Zhe, and Yan Luo

1. Introduction	316
2. Methods	318
References	323

Author Index	*325*
Subject Index	*333*

Contributors

Scott W. Aesif
Department of Pathology, University of Vermont College of Medicine, Burlington, Vermont, USA

Kelly K. Andringa
Department of Environmental Health Sciences, Center for Free Radical Biology, University of Alabama at Birmingham, Birmingham, Alabama, USA

Shannon M. Bailey
Department of Environmental Health Sciences, Center for Free Radical Biology, University of Alabama at Birmingham, Birmingham, Alabama, USA

Steven Bernstein
James P. Wilmot Cancer Center, University of Rochester Medical Center, Rochester, New York, USA

Aruni Bhatnagar
Diabetes and Obesity Center, University of Louisville, Louisville, Kentucky, USA

Alberto Bindoli
Institute of Neuroscience (CNR), Section of Padova, c/o Department of Biological Chemistry, Padova, Italy

Benoit Boivin
Cold Spring Harbor Laboratory, Cold Spring Harbor, New York, USA

Paul Brookes
Department of Anesthesiology, University of Rochester Medical Center, Rochester, New York, USA

Samuel Caito
Lung Biology and Disease Program, Department of Environmental Medicine, University of Rochester Medical Center, Rochester, New York, USA

Bradley A. Carlson
Molecular Biology of Selenium Section, Laboratory of Cancer Prevention, Center for Cancer Research, National Cancer Institute, National Institutes of Health, Bethesda, Maryland, USA

Zhe Chen
Department of Biological Responses, Institute for Virus Research, Kyoto University, Kyoto, Japan

Chuang C. Chiueh
Division of Clinical Pharmacy, School of Pharmacy and Taipei Medical University—Shuang Ho Hospital, Taipei, Taiwan

Chun-Seok Cho
Division of Life and Pharmaceutical Sciences, Ewha Womans University, Seodaemun-gu, Seoul, Korea

Edward T. Chouchani
Medical Research Council Mitochondrial Biology Unit, Wellcome Trust/MRC Building, Cambridge, UK

Andrew G. Cox
Free Radical Research Group, Department of Pathology, and National Research Centre for Growth and Development, University of Otago, Christchurch, New Zealand

Vadim N. Gladyshev
Division of Genetics, Department of Medicine, Brigham & Women's Hospital and Harvard Medical School, Boston, Massachusetts, USA

Young-Mi Go
Department of Medicine, Division of Pulmonary, Allergy and Critical Care Medicine, Emory University, Atlanta, Georgia, USA

Mark B. Hampton
Free Radical Research Group, Department of Pathology, and National Research Centre for Growth and Development, University of Otago, Christchurch, New Zealand

Dolph L. Hatfield
Molecular Biology of Selenium Section, Laboratory of Cancer Prevention, Center for Cancer Research, National Cancer Institute, National Institutes of Health, Bethesda, Maryland, USA

Yun He
Interdepartmental Program in Vascular Biology and Therapeutics, Yale University School of Medicine, New Haven, Connecticut, USA

Qunhua Huang
Interdepartmental Program in Vascular Biology and Therapeutics, Yale University School of Medicine, New Haven, Connecticut, USA

Thomas R. Hurd
Medical Research Council Mitochondrial Biology Unit, Wellcome Trust/MRC Building, Cambridge, UK

Hidenori Ichijo
Laboratory of Cell Signaling, Graduate School of Pharmaceutical Sciences, The University of Tokyo, Hongo, Bunkyo-ku, Tokyo, Japan

Robert Irons
Molecular Biology of Selenium Section, Laboratory of Cancer Prevention, Center for Cancer Research, National Cancer Institute, National Institutes of Health, Bethesda, Maryland, USA

Yvonne M. W. Janssen-Heininger
Department of Pathology, University of Vermont College of Medicine, Burlington, Vermont, USA

Dean P. Jones
Department of Medicine, Division of Pulmonary, Allergy and Critical Care Medicine, Emory University, Atlanta, Georgia, USA

P. Andrew Karplus
Department of Biochemistry and Biophysics, Oregon State University, Corvallis, Oregon, USA

Kazumi Katagiri
Laboratory of Cell Signaling, Graduate School of Pharmaceutical Sciences, The University of Tokyo, Hongo, Bunkyo-ku, Tokyo, Japan

Yan Luo
State Key Laboratory of Ophthalmology, Zhongshan Ophthalmic Center, Sun Yat-Sen University, Guangzhou, China

Gunther Marsche
Institute of Experimental and Clinical Pharmacology, Medical University of Graz, Universitätsplatz, Graz, Austria

Hiroshi Masutani
Department of Biological Responses, Institute for Virus Research, Kyoto University, Kyoto, Japan

Yoshiyuki Matsuo
Department of Biological Responses, Institute for Virus Research, Kyoto University, Kyoto, Japan

Atsushi Matsuzawa
Laboratory of Cell Signaling, Graduate School of Pharmaceutical Sciences, The University of Tokyo, Hongo, Bunkyo-ku, Tokyo, Japan

Katja E. Menger
Medical Research Council Mitochondrial Biology Unit, Wellcome Trust/MRC Building, Cambridge, UK

Wang Min
Interdepartmental Program in Vascular Biology and Therapeutics, Yale University School of Medicine, New Haven, Connecticut, USA, and State Key Laboratory of Ophthalmology, Zhongshan Ophthalmic Center, Sun Yat-Sen University, Guangzhou, China

Michael P. Murphy
Medical Research Council Mitochondrial Biology Unit, Wellcome Trust/MRC Building, Cambridge, UK

Karl Oettl
Institute of Physiological Chemistry, Medical University of Graz, Harrachgasse, Graz, Austria

Derek Parsonage
Department of Biochemistry, Wake Forest University School of Medicine, Winston-Salem, North Carolina, USA

Leslie B. Poole
Department of Biochemistry, Wake Forest University School of Medicine, Winston-Salem, North Carolina, USA

Irfan Rahman
Lung Biology and Disease Program, Department of Environmental Medicine, University of Rochester Medical Center, Rochester, New York, USA

Kota V. Ramana
Department of Biochemistry and Molecular Biology, University of Texas Medical Branch, Galveston, Texas, USA

Stacy A. Reeves
Department of Biochemistry, Wake Forest University School of Medicine, Winston-Salem, North Carolina, USA

Raquel Requejo
Medical Research Council Mitochondrial Biology Unit, Wellcome Trust/MRC Building, Cambridge, UK

Niki L. Reynaert
Department of Respiratory Medicine, Maastricht University, Maastricht, The Netherlands

Sue Goo Rhee
Division of Life and Pharmaceutical Sciences, Ewha Womans University, Seodaemun-gu, Seoul, Korea

Maria Pia Rigobello
Department of Biological Chemistry, University of Padova, Padova, Italy

Jolanta Skalska
James P. Wilmot Cancer Center, University of Rochester Medical Center, Rochester, New York, USA

Sanjay Srivastava
Diabetes and Obesity Center, University of Louisville, Louisville, Kentucky, USA

Satish K. Srivastava
Department of Biochemistry and Molecular Biology, University of Texas Medical Branch, Galveston, Texas, USA

Isaac K. Sundar
Lung Biology and Disease Program, Department of Environmental Medicine, University of Rochester Medical Center, Rochester, New York, USA

Nicholas K. Tonks
Cold Spring Harbor Laboratory, Cold Spring Harbor, New York, USA

Petra Tsuji
Molecular Biology of Selenium Section, Laboratory of Cancer Prevention, Center for Cancer Research, and Cancer Prevention Fellowship Program; Nutritional Science Research Group, Division of Cancer Prevention, National Cancer Institute, National Institutes of Health, Bethesda, Maryland, USA

Anton A. Turanov
Division of Genetics, Department of Medicine, Brigham & Women's Hospital and Harvard Medical School, Boston, Massachusetts, USA

Christine C. Winterbourn
Free Radical Research Group, Department of Pathology, and National Research Centre for Growth and Development, University of Otago, Christchurch, New Zealand

Luyao (Kevin) Xu
Interdepartmental Program in Vascular Biology and Therapeutics, Yale University School of Medicine, New Haven, Connecticut, USA

Hongwei Yao
Lung Biology and Disease Program, Department of Environmental Medicine, University of Rochester Medical Center, Rochester, New York, USA

Junji Yodoi
Department of Biological Responses, Institute for Virus Research, Kyoto University, Kyoto, Japan

Min-Hyuk Yoo
Molecular Biology of Selenium Section, Laboratory of Cancer Prevention, Center for Cancer Research, National Cancer Institute, National Institutes of Health, Bethesda, Maryland, USA

Eiji Yoshihara
Department of Biological Responses, Institute for Virus Research, and Division of Systemic Life Science, Graduate School of Biostudies, Kyoto University, Kyoto, Japan

Huanjiao (Jenny) Zhou
State Key Laboratory of Ophthalmology, Zhongshan Ophthalmic Center, Sun Yat-Sen University, Guangzhou, China

Haifeng Zhang
Interdepartmental Program in Vascular Biology and Therapeutics, Yale University School of Medicine, New Haven, Connecticut, USA

Xu Zhe
Interdepartmental Program in Vascular Biology and Therapeutics, Yale University School of Medicine, New Haven, Connecticut, USA

Preface

Signaling by reactive oxygen- and nitrogen species, mainly non-radical species such as hydrogen peroxide, lipid peroxides, and peroxynitrite—which may be viewed as second messengers—has emerged as a major regulatory process of cell function. Signaling targets are redox-sensitive protein cysteines and the large pool of low-molecular thiols, mainly glutathione. These volumes of *Methods in Enzymology* on Thiols Redox Transitions in Cell Signaling address two large topics Chemistry and Biochemistry of Low Molecular Weight and Protein Thiols (Part A, Volume 473) and Cellular Localization and Signaling (Part B, Volume 474). Both volumes serve to bring together current methods and concepts in the field of cell signaling driven by thiol redox modifications by techniques such as fluorescence-based proteomics, mass spectrometry approaches, and fluorescence reporters.

The editors thank all the contributors, whose thorough and innovative work is the basis of these two *Methods in Enzymology* volumes. The editors give special thanks to Leopold Flohé for providing the introductory chapter "Changing paradigms in thiology: from antioxidant defense toward redox regulation, whose critical thinking educated us on the major concepts by which metabolic regulation and adaptation are transduced via thiol modifications.

Enrique Cadenas
Lester Packer
March 2010

Methods in Enzymology

Volume I. Preparation and Assay of Enzymes
Edited by Sidney P. Colowick and Nathan O. Kaplan

Volume II. Preparation and Assay of Enzymes
Edited by Sidney P. Colowick and Nathan O. Kaplan

Volume III. Preparation and Assay of Substrates
Edited by Sidney P. Colowick and Nathan O. Kaplan

Volume IV. Special Techniques for the Enzymologist
Edited by Sidney P. Colowick and Nathan O. Kaplan

Volume V. Preparation and Assay of Enzymes
Edited by Sidney P. Colowick and Nathan O. Kaplan

Volume VI. Preparation and Assay of Enzymes *(Continued)*
Preparation and Assay of Substrates
Special Techniques
Edited by Sidney P. Colowick and Nathan O. Kaplan

Volume VII. Cumulative Subject Index
Edited by Sidney P. Colowick and Nathan O. Kaplan

Volume VIII. Complex Carbohydrates
Edited by Elizabeth F. Neufeld and Victor Ginsburg

Volume IX. Carbohydrate Metabolism
Edited by Willis A. Wood

Volume X. Oxidation and Phosphorylation
Edited by Ronald W. Estabrook and Maynard E. Pullman

Volume XI. Enzyme Structure
Edited by C. H. W. Hirs

Volume XII. Nucleic Acids (Parts A and B)
Edited by Lawrence Grossman and Kivie Moldave

Volume XIII. Citric Acid Cycle
Edited by J. M. Lowenstein

Volume XIV. Lipids
Edited by J. M. Lowenstein

Volume XV. Steroids and Terpenoids
Edited by Raymond B. Clayton

VOLUME XVI. Fast Reactions
Edited by KENNETH KUSTIN

VOLUME XVII. Metabolism of Amino Acids and Amines (Parts A and B)
Edited by HERBERT TABOR AND CELIA WHITE TABOR

VOLUME XVIII. Vitamins and Coenzymes (Parts A, B, and C)
Edited by DONALD B. MCCORMICK AND LEMUEL D. WRIGHT

VOLUME XIX. Proteolytic Enzymes
Edited by GERTRUDE E. PERLMANN AND LASZLO LORAND

VOLUME XX. Nucleic Acids and Protein Synthesis (Part C)
Edited by KIVIE MOLDAVE AND LAWRENCE GROSSMAN

VOLUME XXI. Nucleic Acids (Part D)
Edited by LAWRENCE GROSSMAN AND KIVIE MOLDAVE

VOLUME XXII. Enzyme Purification and Related Techniques
Edited by WILLIAM B. JAKOBY

VOLUME XXIII. Photosynthesis (Part A)
Edited by ANTHONY SAN PIETRO

VOLUME XXIV. Photosynthesis and Nitrogen Fixation (Part B)
Edited by ANTHONY SAN PIETRO

VOLUME XXV. Enzyme Structure (Part B)
Edited by C. H. W. HIRS AND SERGE N. TIMASHEFF

VOLUME XXVI. Enzyme Structure (Part C)
Edited by C. H. W. HIRS AND SERGE N. TIMASHEFF

VOLUME XXVII. Enzyme Structure (Part D)
Edited by C. H. W. HIRS AND SERGE N. TIMASHEFF

VOLUME XXVIII. Complex Carbohydrates (Part B)
Edited by VICTOR GINSBURG

VOLUME XXIX. Nucleic Acids and Protein Synthesis (Part E)
Edited by LAWRENCE GROSSMAN AND KIVIE MOLDAVE

VOLUME XXX. Nucleic Acids and Protein Synthesis (Part F)
Edited by KIVIE MOLDAVE AND LAWRENCE GROSSMAN

VOLUME XXXI. Biomembranes (Part A)
Edited by SIDNEY FLEISCHER AND LESTER PACKER

VOLUME XXXII. Biomembranes (Part B)
Edited by SIDNEY FLEISCHER AND LESTER PACKER

VOLUME XXXIII. Cumulative Subject Index Volumes I–XXX
Edited by MARTHA G. DENNIS AND EDWARD A. DENNIS

VOLUME XXXIV. Affinity Techniques (Enzyme Purification: Part B)
Edited by WILLIAM B. JAKOBY AND MEIR WILCHEK

VOLUME XXXV. Lipids (Part B)
Edited by JOHN M. LOWENSTEIN

VOLUME XXXVI. Hormone Action (Part A: Steroid Hormones)
Edited by BERT W. O'MALLEY AND JOEL G. HARDMAN

VOLUME XXXVII. Hormone Action (Part B: Peptide Hormones)
Edited by BERT W. O'MALLEY AND JOEL G. HARDMAN

VOLUME XXXVIII. Hormone Action (Part C: Cyclic Nucleotides)
Edited by JOEL G. HARDMAN AND BERT W. O'MALLEY

VOLUME XXXIX. Hormone Action (Part D: Isolated Cells, Tissues, and Organ Systems)
Edited by JOEL G. HARDMAN AND BERT W. O'MALLEY

VOLUME XL. Hormone Action (Part E: Nuclear Structure and Function)
Edited by BERT W. O'MALLEY AND JOEL G. HARDMAN

VOLUME XLI. Carbohydrate Metabolism (Part B)
Edited by W. A. WOOD

VOLUME XLII. Carbohydrate Metabolism (Part C)
Edited by W. A. WOOD

VOLUME XLIII. Antibiotics
Edited by JOHN H. HASH

VOLUME XLIV. Immobilized Enzymes
Edited by KLAUS MOSBACH

VOLUME XLV. Proteolytic Enzymes (Part B)
Edited by LASZLO LORAND

VOLUME XLVI. Affinity Labeling
Edited by WILLIAM B. JAKOBY AND MEIR WILCHEK

VOLUME XLVII. Enzyme Structure (Part E)
Edited by C. H. W. HIRS AND SERGE N. TIMASHEFF

VOLUME XLVIII. Enzyme Structure (Part F)
Edited by C. H. W. HIRS AND SERGE N. TIMASHEFF

VOLUME XLIX. Enzyme Structure (Part G)
Edited by C. H. W. HIRS AND SERGE N. TIMASHEFF

VOLUME L. Complex Carbohydrates (Part C)
Edited by VICTOR GINSBURG

VOLUME LI. Purine and Pyrimidine Nucleotide Metabolism
Edited by PATRICIA A. HOFFEE AND MARY ELLEN JONES

VOLUME LII. Biomembranes (Part C: Biological Oxidations)
Edited by SIDNEY FLEISCHER AND LESTER PACKER

VOLUME LIII. Biomembranes (Part D: Biological Oxidations)
Edited by SIDNEY FLEISCHER AND LESTER PACKER

VOLUME LIV. Biomembranes (Part E: Biological Oxidations)
Edited by SIDNEY FLEISCHER AND LESTER PACKER

VOLUME LV. Biomembranes (Part F: Bioenergetics)
Edited by SIDNEY FLEISCHER AND LESTER PACKER

VOLUME LVI. Biomembranes (Part G: Bioenergetics)
Edited by SIDNEY FLEISCHER AND LESTER PACKER

VOLUME LVII. Bioluminescence and Chemiluminescence
Edited by MARLENE A. DELUCA

VOLUME LVIII. Cell Culture
Edited by WILLIAM B. JAKOBY AND IRA PASTAN

VOLUME LIX. Nucleic Acids and Protein Synthesis (Part G)
Edited by KIVIE MOLDAVE AND LAWRENCE GROSSMAN

VOLUME LX. Nucleic Acids and Protein Synthesis (Part H)
Edited by KIVIE MOLDAVE AND LAWRENCE GROSSMAN

VOLUME 61. Enzyme Structure (Part H)
Edited by C. H. W. HIRS AND SERGE N. TIMASHEFF

VOLUME 62. Vitamins and Coenzymes (Part D)
Edited by DONALD B. MCCORMICK AND LEMUEL D. WRIGHT

VOLUME 63. Enzyme Kinetics and Mechanism (Part A: Initial Rate and Inhibitor Methods)
Edited by DANIEL L. PURICH

VOLUME 64. Enzyme Kinetics and Mechanism
(Part B: Isotopic Probes and Complex Enzyme Systems)
Edited by DANIEL L. PURICH

VOLUME 65. Nucleic Acids (Part I)
Edited by LAWRENCE GROSSMAN AND KIVIE MOLDAVE

VOLUME 66. Vitamins and Coenzymes (Part E)
Edited by DONALD B. MCCORMICK AND LEMUEL D. WRIGHT

VOLUME 67. Vitamins and Coenzymes (Part F)
Edited by DONALD B. MCCORMICK AND LEMUEL D. WRIGHT

VOLUME 68. Recombinant DNA
Edited by RAY WU

VOLUME 69. Photosynthesis and Nitrogen Fixation (Part C)
Edited by ANTHONY SAN PIETRO

VOLUME 70. Immunochemical Techniques (Part A)
Edited by HELEN VAN VUNAKIS AND JOHN J. LANGONE

VOLUME 71. Lipids (Part C)
Edited by JOHN M. LOWENSTEIN

VOLUME 72. Lipids (Part D)
Edited by JOHN M. LOWENSTEIN

VOLUME 73. Immunochemical Techniques (Part B)
Edited by JOHN J. LANGONE AND HELEN VAN VUNAKIS

VOLUME 74. Immunochemical Techniques (Part C)
Edited by JOHN J. LANGONE AND HELEN VAN VUNAKIS

VOLUME 75. Cumulative Subject Index Volumes XXXI, XXXII, XXXIV–LX
Edited by EDWARD A. DENNIS AND MARTHA G. DENNIS

VOLUME 76. Hemoglobins
Edited by ERALDO ANTONINI, LUIGI ROSSI-BERNARDI, AND EMILIA CHIANCONE

VOLUME 77. Detoxication and Drug Metabolism
Edited by WILLIAM B. JAKOBY

VOLUME 78. Interferons (Part A)
Edited by SIDNEY PESTKA

VOLUME 79. Interferons (Part B)
Edited by SIDNEY PESTKA

VOLUME 80. Proteolytic Enzymes (Part C)
Edited by LASZLO LORAND

VOLUME 81. Biomembranes (Part H: Visual Pigments and Purple Membranes, I)
Edited by LESTER PACKER

VOLUME 82. Structural and Contractile Proteins (Part A: Extracellular Matrix)
Edited by LEON W. CUNNINGHAM AND DIXIE W. FREDERIKSEN

VOLUME 83. Complex Carbohydrates (Part D)
Edited by VICTOR GINSBURG

VOLUME 84. Immunochemical Techniques (Part D: Selected Immunoassays)
Edited by JOHN J. LANGONE AND HELEN VAN VUNAKIS

VOLUME 85. Structural and Contractile Proteins (Part B: The Contractile Apparatus and the Cytoskeleton)
Edited by DIXIE W. FREDERIKSEN AND LEON W. CUNNINGHAM

VOLUME 86. Prostaglandins and Arachidonate Metabolites
Edited by WILLIAM E. M. LANDS AND WILLIAM L. SMITH

VOLUME 87. Enzyme Kinetics and Mechanism (Part C: Intermediates, Stereo-chemistry, and Rate Studies)
Edited by DANIEL L. PURICH

VOLUME 88. Biomembranes (Part I: Visual Pigments and Purple Membranes, II)
Edited by LESTER PACKER

VOLUME 89. Carbohydrate Metabolism (Part D)
Edited by WILLIS A. WOOD

VOLUME 90. Carbohydrate Metabolism (Part E)
Edited by WILLIS A. WOOD

VOLUME 91. Enzyme Structure (Part I)
Edited by C. H. W. HIRS AND SERGE N. TIMASHEFF

VOLUME 92. Immunochemical Techniques (Part E: Monoclonal Antibodies and General Immunoassay Methods)
Edited by JOHN J. LANGONE AND HELEN VAN VUNAKIS

VOLUME 93. Immunochemical Techniques (Part F: Conventional Antibodies, Fc Receptors, and Cytotoxicity)
Edited by JOHN J. LANGONE AND HELEN VAN VUNAKIS

VOLUME 94. Polyamines
Edited by HERBERT TABOR AND CELIA WHITE TABOR

VOLUME 95. Cumulative Subject Index Volumes 61–74, 76–80
Edited by EDWARD A. DENNIS AND MARTHA G. DENNIS

VOLUME 96. Biomembranes [Part J: Membrane Biogenesis: Assembly and Targeting (General Methods; Eukaryotes)]
Edited by SIDNEY FLEISCHER AND BECCA FLEISCHER

VOLUME 97. Biomembranes [Part K: Membrane Biogenesis: Assembly and Targeting (Prokaryotes, Mitochondria, and Chloroplasts)]
Edited by SIDNEY FLEISCHER AND BECCA FLEISCHER

VOLUME 98. Biomembranes (Part L: Membrane Biogenesis: Processing and Recycling)
Edited by SIDNEY FLEISCHER AND BECCA FLEISCHER

VOLUME 99. Hormone Action (Part F: Protein Kinases)
Edited by JACKIE D. CORBIN AND JOEL G. HARDMAN

VOLUME 100. Recombinant DNA (Part B)
Edited by RAY WU, LAWRENCE GROSSMAN, AND KIVIE MOLDAVE

VOLUME 101. Recombinant DNA (Part C)
Edited by RAY WU, LAWRENCE GROSSMAN, AND KIVIE MOLDAVE

VOLUME 102. Hormone Action (Part G: Calmodulin and Calcium-Binding Proteins)
Edited by ANTHONY R. MEANS AND BERT W. O'MALLEY

VOLUME 103. Hormone Action (Part H: Neuroendocrine Peptides)
Edited by P. MICHAEL CONN

VOLUME 104. Enzyme Purification and Related Techniques (Part C)
Edited by WILLIAM B. JAKOBY

VOLUME 105. Oxygen Radicals in Biological Systems
Edited by LESTER PACKER

VOLUME 106. Posttranslational Modifications (Part A)
Edited by FINN WOLD AND KIVIE MOLDAVE

VOLUME 107. Posttranslational Modifications (Part B)
Edited by FINN WOLD AND KIVIE MOLDAVE

VOLUME 108. Immunochemical Techniques (Part G: Separation and Characterization of Lymphoid Cells)
Edited by GIOVANNI DI SABATO, JOHN J. LANGONE, AND HELEN VAN VUNAKIS

VOLUME 109. Hormone Action (Part I: Peptide Hormones)
Edited by LUTZ BIRNBAUMER AND BERT W. O'MALLEY

VOLUME 110. Steroids and Isoprenoids (Part A)
Edited by JOHN H. LAW AND HANS C. RILLING

VOLUME 111. Steroids and Isoprenoids (Part B)
Edited by JOHN H. LAW AND HANS C. RILLING

VOLUME 112. Drug and Enzyme Targeting (Part A)
Edited by KENNETH J. WIDDER AND RALPH GREEN

VOLUME 113. Glutamate, Glutamine, Glutathione, and Related Compounds
Edited by ALTON MEISTER

VOLUME 114. Diffraction Methods for Biological Macromolecules (Part A)
Edited by HAROLD W. WYCKOFF, C. H. W. HIRS, AND SERGE N. TIMASHEFF

VOLUME 115. Diffraction Methods for Biological Macromolecules (Part B)
Edited by HAROLD W. WYCKOFF, C. H. W. HIRS, AND SERGE N. TIMASHEFF

VOLUME 116. Immunochemical Techniques (Part H: Effectors and Mediators of Lymphoid Cell Functions)
Edited by GIOVANNI DI SABATO, JOHN J. LANGONE, AND HELEN VAN VUNAKIS

VOLUME 117. Enzyme Structure (Part J)
Edited by C. H. W. HIRS AND SERGE N. TIMASHEFF

VOLUME 118. Plant Molecular Biology
Edited by ARTHUR WEISSBACH AND HERBERT WEISSBACH

VOLUME 119. Interferons (Part C)
Edited by SIDNEY PESTKA

VOLUME 120. Cumulative Subject Index Volumes 81–94, 96–101

VOLUME 121. Immunochemical Techniques (Part I: Hybridoma Technology and Monoclonal Antibodies)
Edited by JOHN J. LANGONE AND HELEN VAN VUNAKIS

VOLUME 122. Vitamins and Coenzymes (Part G)
Edited by FRANK CHYTIL AND DONALD B. MCCORMICK

VOLUME 123. Vitamins and Coenzymes (Part H)
Edited by FRANK CHYTIL AND DONALD B. MCCORMICK

VOLUME 124. Hormone Action (Part J: Neuroendocrine Peptides)
Edited by P. MICHAEL CONN

VOLUME 125. Biomembranes (Part M: Transport in Bacteria, Mitochondria, and Chloroplasts: General Approaches and Transport Systems)
Edited by SIDNEY FLEISCHER AND BECCA FLEISCHER

VOLUME 126. Biomembranes (Part N: Transport in Bacteria, Mitochondria, and Chloroplasts: Protonmotive Force)
Edited by SIDNEY FLEISCHER AND BECCA FLEISCHER

VOLUME 127. Biomembranes (Part O: Protons and Water: Structure and Translocation)
Edited by LESTER PACKER

VOLUME 128. Plasma Lipoproteins (Part A: Preparation, Structure, and Molecular Biology)
Edited by JERE P. SEGREST AND JOHN J. ALBERS

VOLUME 129. Plasma Lipoproteins (Part B: Characterization, Cell Biology, and Metabolism)
Edited by JOHN J. ALBERS AND JERE P. SEGREST

VOLUME 130. Enzyme Structure (Part K)
Edited by C. H. W. HIRS AND SERGE N. TIMASHEFF

VOLUME 131. Enzyme Structure (Part L)
Edited by C. H. W. HIRS AND SERGE N. TIMASHEFF

VOLUME 132. Immunochemical Techniques (Part J: Phagocytosis and Cell-Mediated Cytotoxicity)
Edited by GIOVANNI DI SABATO AND JOHANNES EVERSE

VOLUME 133. Bioluminescence and Chemiluminescence (Part B)
Edited by MARLENE DELUCA AND WILLIAM D. MCELROY

VOLUME 134. Structural and Contractile Proteins (Part C: The Contractile Apparatus and the Cytoskeleton)
Edited by RICHARD B. VALLEE

VOLUME 135. Immobilized Enzymes and Cells (Part B)
Edited by KLAUS MOSBACH

VOLUME 136. Immobilized Enzymes and Cells (Part C)
Edited by KLAUS MOSBACH

VOLUME 137. Immobilized Enzymes and Cells (Part D)
Edited by KLAUS MOSBACH

VOLUME 138. Complex Carbohydrates (Part E)
Edited by VICTOR GINSBURG

VOLUME 139. Cellular Regulators (Part A: Calcium- and
Calmodulin-Binding Proteins)
Edited by ANTHONY R. MEANS AND P. MICHAEL CONN

VOLUME 140. Cumulative Subject Index Volumes 102–119, 121–134

VOLUME 141. Cellular Regulators (Part B: Calcium and Lipids)
Edited by P. MICHAEL CONN AND ANTHONY R. MEANS

VOLUME 142. Metabolism of Aromatic Amino Acids and Amines
Edited by SEYMOUR KAUFMAN

VOLUME 143. Sulfur and Sulfur Amino Acids
Edited by WILLIAM B. JAKOBY AND OWEN GRIFFITH

VOLUME 144. Structural and Contractile Proteins (Part D: Extracellular Matrix)
Edited by LEON W. CUNNINGHAM

VOLUME 145. Structural and Contractile Proteins (Part E: Extracellular Matrix)
Edited by LEON W. CUNNINGHAM

VOLUME 146. Peptide Growth Factors (Part A)
Edited by DAVID BARNES AND DAVID A. SIRBASKU

VOLUME 147. Peptide Growth Factors (Part B)
Edited by DAVID BARNES AND DAVID A. SIRBASKU

VOLUME 148. Plant Cell Membranes
Edited by LESTER PACKER AND ROLAND DOUCE

VOLUME 149. Drug and Enzyme Targeting (Part B)
Edited by RALPH GREEN AND KENNETH J. WIDDER

VOLUME 150. Immunochemical Techniques (Part K: *In Vitro* Models of B and T Cell Functions and Lymphoid Cell Receptors)
Edited by GIOVANNI DI SABATO

VOLUME 151. Molecular Genetics of Mammalian Cells
Edited by MICHAEL M. GOTTESMAN

VOLUME 152. Guide to Molecular Cloning Techniques
Edited by SHELBY L. BERGER AND ALAN R. KIMMEL

VOLUME 153. Recombinant DNA (Part D)
Edited by RAY WU AND LAWRENCE GROSSMAN

VOLUME 154. Recombinant DNA (Part E)
Edited by RAY WU AND LAWRENCE GROSSMAN

VOLUME 155. Recombinant DNA (Part F)
Edited by RAY WU

VOLUME 156. Biomembranes (Part P: ATP-Driven Pumps and Related Transport: The Na, K-Pump)
Edited by SIDNEY FLEISCHER AND BECCA FLEISCHER

VOLUME 157. Biomembranes (Part Q: ATP-Driven Pumps and Related Transport: Calcium, Proton, and Potassium Pumps)
Edited by SIDNEY FLEISCHER AND BECCA FLEISCHER

VOLUME 158. Metalloproteins (Part A)
Edited by JAMES F. RIORDAN AND BERT L. VALLEE

VOLUME 159. Initiation and Termination of Cyclic Nucleotide Action
Edited by JACKIE D. CORBIN AND ROGER A. JOHNSON

VOLUME 160. Biomass (Part A: Cellulose and Hemicellulose)
Edited by WILLIS A. WOOD AND SCOTT T. KELLOGG

VOLUME 161. Biomass (Part B: Lignin, Pectin, and Chitin)
Edited by WILLIS A. WOOD AND SCOTT T. KELLOGG

VOLUME 162. Immunochemical Techniques (Part L: Chemotaxis and Inflammation)
Edited by GIOVANNI DI SABATO

VOLUME 163. Immunochemical Techniques (Part M: Chemotaxis and Inflammation)
Edited by GIOVANNI DI SABATO

VOLUME 164. Ribosomes
Edited by HARRY F. NOLLER, JR., AND KIVIE MOLDAVE

VOLUME 165. Microbial Toxins: Tools for Enzymology
Edited by SIDNEY HARSHMAN

VOLUME 166. Branched-Chain Amino Acids
Edited by ROBERT HARRIS AND JOHN R. SOKATCH

VOLUME 167. Cyanobacteria
Edited by LESTER PACKER AND ALEXANDER N. GLAZER

VOLUME 168. Hormone Action (Part K: Neuroendocrine Peptides)
Edited by P. MICHAEL CONN

VOLUME 169. Platelets: Receptors, Adhesion, Secretion (Part A)
Edited by JACEK HAWIGER

VOLUME 170. Nucleosomes
Edited by PAUL M. WASSARMAN AND ROGER D. KORNBERG

VOLUME 171. Biomembranes (Part R: Transport Theory: Cells and Model Membranes)
Edited by SIDNEY FLEISCHER AND BECCA FLEISCHER

VOLUME 172. Biomembranes (Part S: Transport: Membrane Isolation and Characterization)
Edited by SIDNEY FLEISCHER AND BECCA FLEISCHER

VOLUME 173. Biomembranes [Part T: Cellular and Subcellular Transport: Eukaryotic (Nonepithelial) Cells]
Edited by SIDNEY FLEISCHER AND BECCA FLEISCHER

VOLUME 174. Biomembranes [Part U: Cellular and Subcellular Transport: Eukaryotic (Nonepithelial) Cells]
Edited by SIDNEY FLEISCHER AND BECCA FLEISCHER

VOLUME 175. Cumulative Subject Index Volumes 135–139, 141–167

VOLUME 176. Nuclear Magnetic Resonance (Part A: Spectral Techniques and Dynamics)
Edited by NORMAN J. OPPENHEIMER AND THOMAS L. JAMES

VOLUME 177. Nuclear Magnetic Resonance (Part B: Structure and Mechanism)
Edited by NORMAN J. OPPENHEIMER AND THOMAS L. JAMES

VOLUME 178. Antibodies, Antigens, and Molecular Mimicry
Edited by JOHN J. LANGONE

VOLUME 179. Complex Carbohydrates (Part F)
Edited by VICTOR GINSBURG

VOLUME 180. RNA Processing (Part A: General Methods)
Edited by JAMES E. DAHLBERG AND JOHN N. ABELSON

VOLUME 181. RNA Processing (Part B: Specific Methods)
Edited by JAMES E. DAHLBERG AND JOHN N. ABELSON

VOLUME 182. Guide to Protein Purification
Edited by MURRAY P. DEUTSCHER

VOLUME 183. Molecular Evolution: Computer Analysis of Protein and Nucleic Acid Sequences
Edited by RUSSELL F. DOOLITTLE

VOLUME 184. Avidin-Biotin Technology
Edited by MEIR WILCHEK AND EDWARD A. BAYER

VOLUME 185. Gene Expression Technology
Edited by DAVID V. GOEDDEL

VOLUME 186. Oxygen Radicals in Biological Systems (Part B: Oxygen Radicals and Antioxidants)
Edited by LESTER PACKER AND ALEXANDER N. GLAZER

VOLUME 187. Arachidonate Related Lipid Mediators
Edited by ROBERT C. MURPHY AND FRANK A. FITZPATRICK

VOLUME 188. Hydrocarbons and Methylotrophy
Edited by MARY E. LIDSTROM

VOLUME 189. Retinoids (Part A: Molecular and Metabolic Aspects)
Edited by LESTER PACKER

VOLUME 190. Retinoids (Part B: Cell Differentiation and Clinical Applications)
Edited by LESTER PACKER

VOLUME 191. Biomembranes (Part V: Cellular and Subcellular Transport: Epithelial Cells)
Edited by SIDNEY FLEISCHER AND BECCA FLEISCHER

VOLUME 192. Biomembranes (Part W: Cellular and Subcellular Transport: Epithelial Cells)
Edited by SIDNEY FLEISCHER AND BECCA FLEISCHER

VOLUME 193. Mass Spectrometry
Edited by JAMES A. MCCLOSKEY

VOLUME 194. Guide to Yeast Genetics and Molecular Biology
Edited by CHRISTINE GUTHRIE AND GERALD R. FINK

VOLUME 195. Adenylyl Cyclase, G Proteins, and Guanylyl Cyclase
Edited by ROGER A. JOHNSON AND JACKIE D. CORBIN

VOLUME 196. Molecular Motors and the Cytoskeleton
Edited by RICHARD B. VALLEE

VOLUME 197. Phospholipases
Edited by EDWARD A. DENNIS

VOLUME 198. Peptide Growth Factors (Part C)
Edited by DAVID BARNES, J. P. MATHER, AND GORDON H. SATO

VOLUME 199. Cumulative Subject Index Volumes 168–174, 176–194

VOLUME 200. Protein Phosphorylation (Part A: Protein Kinases: Assays, Purification, Antibodies, Functional Analysis, Cloning, and Expression)
Edited by TONY HUNTER AND BARTHOLOMEW M. SEFTON

VOLUME 201. Protein Phosphorylation (Part B: Analysis of Protein Phosphorylation, Protein Kinase Inhibitors, and Protein Phosphatases)
Edited by TONY HUNTER AND BARTHOLOMEW M. SEFTON

VOLUME 202. Molecular Design and Modeling: Concepts and Applications (Part A: Proteins, Peptides, and Enzymes)
Edited by JOHN J. LANGONE

VOLUME 203. Molecular Design and Modeling: Concepts and Applications (Part B: Antibodies and Antigens, Nucleic Acids, Polysaccharides, and Drugs)
Edited by JOHN J. LANGONE

VOLUME 204. Bacterial Genetic Systems
Edited by JEFFREY H. MILLER

VOLUME 205. Metallobiochemistry (Part B: Metallothionein and Related Molecules)
Edited by JAMES F. RIORDAN AND BERT L. VALLEE

VOLUME 206. Cytochrome P450
Edited by MICHAEL R. WATERMAN AND ERIC F. JOHNSON

VOLUME 207. Ion Channels
Edited by BERNARDO RUDY AND LINDA E. IVERSON

VOLUME 208. Protein–DNA Interactions
Edited by ROBERT T. SAUER

VOLUME 209. Phospholipid Biosynthesis
Edited by EDWARD A. DENNIS AND DENNIS E. VANCE

VOLUME 210. Numerical Computer Methods
Edited by LUDWIG BRAND AND MICHAEL L. JOHNSON

VOLUME 211. DNA Structures (Part A: Synthesis and Physical Analysis of DNA)
Edited by DAVID M. J. LILLEY AND JAMES E. DAHLBERG

VOLUME 212. DNA Structures (Part B: Chemical and Electrophoretic Analysis of DNA)
Edited by DAVID M. J. LILLEY AND JAMES E. DAHLBERG

VOLUME 213. Carotenoids (Part A: Chemistry, Separation, Quantitation, and Antioxidation)
Edited by LESTER PACKER

VOLUME 214. Carotenoids (Part B: Metabolism, Genetics, and Biosynthesis)
Edited by LESTER PACKER

VOLUME 215. Platelets: Receptors, Adhesion, Secretion (Part B)
Edited by JACEK J. HAWIGER

VOLUME 216. Recombinant DNA (Part G)
Edited by RAY WU

VOLUME 217. Recombinant DNA (Part H)
Edited by RAY WU

VOLUME 218. Recombinant DNA (Part I)
Edited by RAY WU

VOLUME 219. Reconstitution of Intracellular Transport
Edited by JAMES E. ROTHMAN

VOLUME 220. Membrane Fusion Techniques (Part A)
Edited by NEJAT DÜZGÜNEŞ

VOLUME 221. Membrane Fusion Techniques (Part B)
Edited by NEJAT DÜZGÜNEŞ

VOLUME 222. Proteolytic Enzymes in Coagulation, Fibrinolysis, and Complement Activation (Part A: Mammalian Blood Coagulation Factors and Inhibitors)
Edited by LASZLO LORAND AND KENNETH G. MANN

VOLUME 223. Proteolytic Enzymes in Coagulation, Fibrinolysis, and Complement Activation (Part B: Complement Activation, Fibrinolysis, and Nonmammalian Blood Coagulation Factors)
Edited by LASZLO LORAND AND KENNETH G. MANN

VOLUME 224. Molecular Evolution: Producing the Biochemical Data
Edited by ELIZABETH ANNE ZIMMER, THOMAS J. WHITE, REBECCA L. CANN, AND ALLAN C. WILSON

VOLUME 225. Guide to Techniques in Mouse Development
Edited by PAUL M. WASSARMAN AND MELVIN L. DEPAMPHILIS

VOLUME 226. Metallobiochemistry (Part C: Spectroscopic and Physical Methods for Probing Metal Ion Environments in Metalloenzymes and Metalloproteins)
Edited by JAMES F. RIORDAN AND BERT L. VALLEE

VOLUME 227. Metallobiochemistry (Part D: Physical and Spectroscopic Methods for Probing Metal Ion Environments in Metalloproteins)
Edited by JAMES F. RIORDAN AND BERT L. VALLEE

VOLUME 228. Aqueous Two-Phase Systems
Edited by HARRY WALTER AND GÖTE JOHANSSON

VOLUME 229. Cumulative Subject Index Volumes 195–198, 200–227

VOLUME 230. Guide to Techniques in Glycobiology
Edited by WILLIAM J. LENNARZ AND GERALD W. HART

VOLUME 231. Hemoglobins (Part B: Biochemical and Analytical Methods)
Edited by JOHANNES EVERSE, KIM D. VANDEGRIFF, AND ROBERT M. WINSLOW

VOLUME 232. Hemoglobins (Part C: Biophysical Methods)
Edited by JOHANNES EVERSE, KIM D. VANDEGRIFF, AND ROBERT M. WINSLOW

VOLUME 233. Oxygen Radicals in Biological Systems (Part C)
Edited by LESTER PACKER

VOLUME 234. Oxygen Radicals in Biological Systems (Part D)
Edited by LESTER PACKER

VOLUME 235. Bacterial Pathogenesis (Part A: Identification and Regulation of Virulence Factors)
Edited by VIRGINIA L. CLARK AND PATRIK M. BAVOIL

VOLUME 236. Bacterial Pathogenesis (Part B: Integration of Pathogenic Bacteria with Host Cells)
Edited by VIRGINIA L. CLARK AND PATRIK M. BAVOIL

VOLUME 237. Heterotrimeric G Proteins
Edited by RAVI IYENGAR

VOLUME 238. Heterotrimeric G-Protein Effectors
Edited by RAVI IYENGAR

VOLUME 239. Nuclear Magnetic Resonance (Part C)
Edited by THOMAS L. JAMES AND NORMAN J. OPPENHEIMER

VOLUME 240. Numerical Computer Methods (Part B)
Edited by MICHAEL L. JOHNSON AND LUDWIG BRAND

VOLUME 241. Retroviral Proteases
Edited by LAWRENCE C. KUO AND JULES A. SHAFER

VOLUME 242. Neoglycoconjugates (Part A)
Edited by Y. C. LEE AND REIKO T. LEE

VOLUME 243. Inorganic Microbial Sulfur Metabolism
Edited by HARRY D. PECK, JR., AND JEAN LEGALL

VOLUME 244. Proteolytic Enzymes: Serine and Cysteine Peptidases
Edited by ALAN J. BARRETT

VOLUME 245. Extracellular Matrix Components
Edited by E. RUOSLAHTI AND E. ENGVALL

VOLUME 246. Biochemical Spectroscopy
Edited by KENNETH SAUER

VOLUME 247. Neoglycoconjugates (Part B: Biomedical Applications)
Edited by Y. C. LEE AND REIKO T. LEE

VOLUME 248. Proteolytic Enzymes: Aspartic and Metallo Peptidases
Edited by ALAN J. BARRETT

VOLUME 249. Enzyme Kinetics and Mechanism (Part D: Developments in Enzyme Dynamics)
Edited by DANIEL L. PURICH

VOLUME 250. Lipid Modifications of Proteins
Edited by PATRICK J. CASEY AND JANICE E. BUSS

VOLUME 251. Biothiols (Part A: Monothiols and Dithiols, Protein Thiols, and Thiyl Radicals)
Edited by LESTER PACKER

VOLUME 252. Biothiols (Part B: Glutathione and Thioredoxin; Thiols in Signal Transduction and Gene Regulation)
Edited by LESTER PACKER

VOLUME 253. Adhesion of Microbial Pathogens
Edited by RON J. DOYLE AND ITZHAK OFEK

VOLUME 254. Oncogene Techniques
Edited by PETER K. VOGT AND INDER M. VERMA

VOLUME 255. Small GTPases and Their Regulators (Part A: Ras Family)
Edited by W. E. BALCH, CHANNING J. DER, AND ALAN HALL

VOLUME 256. Small GTPases and Their Regulators (Part B: Rho Family)
Edited by W. E. BALCH, CHANNING J. DER, AND ALAN HALL

VOLUME 257. Small GTPases and Their Regulators (Part C: Proteins Involved in Transport)
Edited by W. E. BALCH, CHANNING J. DER, AND ALAN HALL

VOLUME 258. Redox-Active Amino Acids in Biology
Edited by JUDITH P. KLINMAN

VOLUME 259. Energetics of Biological Macromolecules
Edited by MICHAEL L. JOHNSON AND GARY K. ACKERS

VOLUME 260. Mitochondrial Biogenesis and Genetics (Part A)
Edited by GIUSEPPE M. ATTARDI AND ANNE CHOMYN

VOLUME 261. Nuclear Magnetic Resonance and Nucleic Acids
Edited by THOMAS L. JAMES

VOLUME 262. DNA Replication
Edited by JUDITH L. CAMPBELL

VOLUME 263. Plasma Lipoproteins (Part C: Quantitation)
Edited by WILLIAM A. BRADLEY, SANDRA H. GIANTURCO, AND JERE P. SEGREST

VOLUME 264. Mitochondrial Biogenesis and Genetics (Part B)
Edited by GIUSEPPE M. ATTARDI AND ANNE CHOMYN

VOLUME 265. Cumulative Subject Index Volumes 228, 230–262

VOLUME 266. Computer Methods for Macromolecular Sequence Analysis
Edited by RUSSELL F. DOOLITTLE

VOLUME 267. Combinatorial Chemistry
Edited by JOHN N. ABELSON

VOLUME 268. Nitric Oxide (Part A: Sources and Detection of NO; NO Synthase)
Edited by LESTER PACKER

VOLUME 269. Nitric Oxide (Part B: Physiological and Pathological Processes)
Edited by LESTER PACKER

VOLUME 270. High Resolution Separation and Analysis of Biological Macromolecules (Part A: Fundamentals)
Edited by BARRY L. KARGER AND WILLIAM S. HANCOCK

VOLUME 271. High Resolution Separation and Analysis of Biological Macromolecules (Part B: Applications)
Edited by BARRY L. KARGER AND WILLIAM S. HANCOCK

VOLUME 272. Cytochrome P450 (Part B)
Edited by ERIC F. JOHNSON AND MICHAEL R. WATERMAN

VOLUME 273. RNA Polymerase and Associated Factors (Part A)
Edited by SANKAR ADHYA

VOLUME 274. RNA Polymerase and Associated Factors (Part B)
Edited by SANKAR ADHYA

VOLUME 275. Viral Polymerases and Related Proteins
Edited by LAWRENCE C. KUO, DAVID B. OLSEN, AND STEVEN S. CARROLL

VOLUME 276. Macromolecular Crystallography (Part A)
Edited by CHARLES W. CARTER, JR., AND ROBERT M. SWEET

VOLUME 277. Macromolecular Crystallography (Part B)
Edited by CHARLES W. CARTER, JR., AND ROBERT M. SWEET

VOLUME 278. Fluorescence Spectroscopy
Edited by LUDWIG BRAND AND MICHAEL L. JOHNSON

VOLUME 279. Vitamins and Coenzymes (Part I)
Edited by DONALD B. MCCORMICK, JOHN W. SUTTIE, AND CONRAD WAGNER

VOLUME 280. Vitamins and Coenzymes (Part J)
Edited by DONALD B. MCCORMICK, JOHN W. SUTTIE, AND CONRAD WAGNER

VOLUME 281. Vitamins and Coenzymes (Part K)
Edited by DONALD B. MCCORMICK, JOHN W. SUTTIE, AND CONRAD WAGNER

VOLUME 282. Vitamins and Coenzymes (Part L)
Edited by DONALD B. MCCORMICK, JOHN W. SUTTIE, AND CONRAD WAGNER

VOLUME 283. Cell Cycle Control
Edited by WILLIAM G. DUNPHY

VOLUME 284. Lipases (Part A: Biotechnology)
Edited by BYRON RUBIN AND EDWARD A. DENNIS

VOLUME 285. Cumulative Subject Index Volumes 263, 264, 266–284, 286–289

VOLUME 286. Lipases (Part B: Enzyme Characterization and Utilization)
Edited by BYRON RUBIN AND EDWARD A. DENNIS

VOLUME 287. Chemokines
Edited by RICHARD HORUK

VOLUME 288. Chemokine Receptors
Edited by RICHARD HORUK

VOLUME 289. Solid Phase Peptide Synthesis
Edited by GREGG B. FIELDS

VOLUME 290. Molecular Chaperones
Edited by GEORGE H. LORIMER AND THOMAS BALDWIN

VOLUME 291. Caged Compounds
Edited by GERARD MARRIOTT

VOLUME 292. ABC Transporters: Biochemical, Cellular, and Molecular Aspects
Edited by SURESH V. AMBUDKAR AND MICHAEL M. GOTTESMAN

VOLUME 293. Ion Channels (Part B)
Edited by P. MICHAEL CONN

VOLUME 294. Ion Channels (Part C)
Edited by P. MICHAEL CONN

VOLUME 295. Energetics of Biological Macromolecules (Part B)
Edited by GARY K. ACKERS AND MICHAEL L. JOHNSON

VOLUME 296. Neurotransmitter Transporters
Edited by SUSAN G. AMARA

VOLUME 297. Photosynthesis: Molecular Biology of Energy Capture
Edited by LEE MCINTOSH

VOLUME 298. Molecular Motors and the Cytoskeleton (Part B)
Edited by RICHARD B. VALLEE

VOLUME 299. Oxidants and Antioxidants (Part A)
Edited by LESTER PACKER

VOLUME 300. Oxidants and Antioxidants (Part B)
Edited by LESTER PACKER

VOLUME 301. Nitric Oxide: Biological and Antioxidant Activities (Part C)
Edited by LESTER PACKER

VOLUME 302. Green Fluorescent Protein
Edited by P. MICHAEL CONN

VOLUME 303. cDNA Preparation and Display
Edited by SHERMAN M. WEISSMAN

VOLUME 304. Chromatin
Edited by PAUL M. WASSARMAN AND ALAN P. WOLFFE

VOLUME 305. Bioluminescence and Chemiluminescence (Part C)
Edited by THOMAS O. BALDWIN AND MIRIAM M. ZIEGLER

VOLUME 306. Expression of Recombinant Genes in Eukaryotic Systems
Edited by JOSEPH C. GLORIOSO AND MARTIN C. SCHMIDT

VOLUME 307. Confocal Microscopy
Edited by P. MICHAEL CONN

VOLUME 308. Enzyme Kinetics and Mechanism (Part E: Energetics of Enzyme Catalysis)
Edited by DANIEL L. PURICH AND VERN L. SCHRAMM

VOLUME 309. Amyloid, Prions, and Other Protein Aggregates
Edited by RONALD WETZEL

VOLUME 310. Biofilms
Edited by RON J. DOYLE

VOLUME 311. Sphingolipid Metabolism and Cell Signaling (Part A)
Edited by ALFRED H. MERRILL, JR., AND YUSUF A. HANNUN

VOLUME 312. Sphingolipid Metabolism and Cell Signaling (Part B)
Edited by ALFRED H. MERRILL, JR., AND YUSUF A. HANNUN

VOLUME 313. Antisense Technology
(Part A: General Methods, Methods of Delivery, and RNA Studies)
Edited by M. IAN PHILLIPS

VOLUME 314. Antisense Technology (Part B: Applications)
Edited by M. IAN PHILLIPS

VOLUME 315. Vertebrate Phototransduction and the Visual Cycle (Part A)
Edited by KRZYSZTOF PALCZEWSKI

VOLUME 316. Vertebrate Phototransduction and the Visual Cycle (Part B)
Edited by KRZYSZTOF PALCZEWSKI

VOLUME 317. RNA–Ligand Interactions (Part A: Structural Biology Methods)
Edited by DANIEL W. CELANDER AND JOHN N. ABELSON

VOLUME 318. RNA–Ligand Interactions (Part B: Molecular Biology Methods)
Edited by DANIEL W. CELANDER AND JOHN N. ABELSON

VOLUME 319. Singlet Oxygen, UV-A, and Ozone
Edited by LESTER PACKER AND HELMUT SIES

VOLUME 320. Cumulative Subject Index Volumes 290–319

VOLUME 321. Numerical Computer Methods (Part C)
Edited by MICHAEL L. JOHNSON AND LUDWIG BRAND

VOLUME 322. Apoptosis
Edited by JOHN C. REED

VOLUME 323. Energetics of Biological Macromolecules (Part C)
Edited by MICHAEL L. JOHNSON AND GARY K. ACKERS

VOLUME 324. Branched-Chain Amino Acids (Part B)
Edited by ROBERT A. HARRIS AND JOHN R. SOKATCH

VOLUME 325. Regulators and Effectors of Small GTPases
(Part D: Rho Family)
Edited by W. E. BALCH, CHANNING J. DER, AND ALAN HALL

VOLUME 326. Applications of Chimeric Genes and Hybrid Proteins
(Part A: Gene Expression and Protein Purification)
Edited by JEREMY THORNER, SCOTT D. EMR, AND JOHN N. ABELSON

VOLUME 327. Applications of Chimeric Genes and Hybrid Proteins
(Part B: Cell Biology and Physiology)
Edited by JEREMY THORNER, SCOTT D. EMR, AND JOHN N. ABELSON

VOLUME 328. Applications of Chimeric Genes and Hybrid Proteins (Part C: Protein–Protein Interactions and Genomics)
Edited by JEREMY THORNER, SCOTT D. EMR, AND JOHN N. ABELSON

VOLUME 329. Regulators and Effectors of Small GTPases (Part E: GTPases Involved in Vesicular Traffic)
Edited by W. E. BALCH, CHANNING J. DER, AND ALAN HALL

VOLUME 330. Hyperthermophilic Enzymes (Part A)
Edited by MICHAEL W. W. ADAMS AND ROBERT M. KELLY

VOLUME 331. Hyperthermophilic Enzymes (Part B)
Edited by MICHAEL W. W. ADAMS AND ROBERT M. KELLY

VOLUME 332. Regulators and Effectors of Small GTPases (Part F: Ras Family I)
Edited by W. E. BALCH, CHANNING J. DER, AND ALAN HALL

VOLUME 333. Regulators and Effectors of Small GTPases (Part G: Ras Family II)
Edited by W. E. BALCH, CHANNING J. DER, AND ALAN HALL

VOLUME 334. Hyperthermophilic Enzymes (Part C)
Edited by MICHAEL W. W. ADAMS AND ROBERT M. KELLY

VOLUME 335. Flavonoids and Other Polyphenols
Edited by LESTER PACKER

VOLUME 336. Microbial Growth in Biofilms (Part A: Developmental and Molecular Biological Aspects)
Edited by RON J. DOYLE

VOLUME 337. Microbial Growth in Biofilms (Part B: Special Environments and Physicochemical Aspects)
Edited by RON J. DOYLE

VOLUME 338. Nuclear Magnetic Resonance of Biological Macromolecules (Part A)
Edited by THOMAS L. JAMES, VOLKER DÖTSCH, AND ULI SCHMITZ

VOLUME 339. Nuclear Magnetic Resonance of Biological Macromolecules (Part B)
Edited by THOMAS L. JAMES, VOLKER DÖTSCH, AND ULI SCHMITZ

VOLUME 340. Drug–Nucleic Acid Interactions
Edited by JONATHAN B. CHAIRES AND MICHAEL J. WARING

VOLUME 341. Ribonucleases (Part A)
Edited by ALLEN W. NICHOLSON

VOLUME 342. Ribonucleases (Part B)
Edited by ALLEN W. NICHOLSON

VOLUME 343. G Protein Pathways (Part A: Receptors)
Edited by RAVI IYENGAR AND JOHN D. HILDEBRANDT

VOLUME 344. G Protein Pathways (Part B: G Proteins and Their Regulators)
Edited by RAVI IYENGAR AND JOHN D. HILDEBRANDT

VOLUME 345. G Protein Pathways (Part C: Effector Mechanisms)
Edited by RAVI IYENGAR AND JOHN D. HILDEBRANDT

VOLUME 346. Gene Therapy Methods
Edited by M. IAN PHILLIPS

VOLUME 347. Protein Sensors and Reactive Oxygen Species (Part A: Selenoproteins and Thioredoxin)
Edited by HELMUT SIES AND LESTER PACKER

VOLUME 348. Protein Sensors and Reactive Oxygen Species (Part B: Thiol Enzymes and Proteins)
Edited by HELMUT SIES AND LESTER PACKER

VOLUME 349. Superoxide Dismutase
Edited by LESTER PACKER

VOLUME 350. Guide to Yeast Genetics and Molecular and Cell Biology (Part B)
Edited by CHRISTINE GUTHRIE AND GERALD R. FINK

VOLUME 351. Guide to Yeast Genetics and Molecular and Cell Biology (Part C)
Edited by CHRISTINE GUTHRIE AND GERALD R. FINK

VOLUME 352. Redox Cell Biology and Genetics (Part A)
Edited by CHANDAN K. SEN AND LESTER PACKER

VOLUME 353. Redox Cell Biology and Genetics (Part B)
Edited by CHANDAN K. SEN AND LESTER PACKER

VOLUME 354. Enzyme Kinetics and Mechanisms (Part F: Detection and Characterization of Enzyme Reaction Intermediates)
Edited by DANIEL L. PURICH

VOLUME 355. Cumulative Subject Index Volumes 321–354

VOLUME 356. Laser Capture Microscopy and Microdissection
Edited by P. MICHAEL CONN

VOLUME 357. Cytochrome P450, Part C
Edited by ERIC F. JOHNSON AND MICHAEL R. WATERMAN

VOLUME 358. Bacterial Pathogenesis (Part C: Identification, Regulation, and Function of Virulence Factors)
Edited by VIRGINIA L. CLARK AND PATRIK M. BAVOIL

VOLUME 359. Nitric Oxide (Part D)
Edited by ENRIQUE CADENAS AND LESTER PACKER

VOLUME 360. Biophotonics (Part A)
Edited by GERARD MARRIOTT AND IAN PARKER

VOLUME 361. Biophotonics (Part B)
Edited by GERARD MARRIOTT AND IAN PARKER

VOLUME 362. Recognition of Carbohydrates in Biological Systems (Part A)
Edited by YUAN C. LEE AND REIKO T. LEE

VOLUME 363. Recognition of Carbohydrates in Biological Systems (Part B)
Edited by YUAN C. LEE AND REIKO T. LEE

VOLUME 364. Nuclear Receptors
Edited by DAVID W. RUSSELL AND DAVID J. MANGELSDORF

VOLUME 365. Differentiation of Embryonic Stem Cells
Edited by PAUL M. WASSAUMAN AND GORDON M. KELLER

VOLUME 366. Protein Phosphatases
Edited by SUSANNE KLUMPP AND JOSEF KRIEGLSTEIN

VOLUME 367. Liposomes (Part A)
Edited by NEJAT DÜZGÜNEŞ

VOLUME 368. Macromolecular Crystallography (Part C)
Edited by CHARLES W. CARTER, JR., AND ROBERT M. SWEET

VOLUME 369. Combinational Chemistry (Part B)
Edited by GUILLERMO A. MORALES AND BARRY A. BUNIN

VOLUME 370. RNA Polymerases and Associated Factors (Part C)
Edited by SANKAR L. ADHYA AND SUSAN GARGES

VOLUME 371. RNA Polymerases and Associated Factors (Part D)
Edited by SANKAR L. ADHYA AND SUSAN GARGES

VOLUME 372. Liposomes (Part B)
Edited by NEJAT DÜZGÜNEŞ

VOLUME 373. Liposomes (Part C)
Edited by NEJAT DÜZGÜNEŞ

VOLUME 374. Macromolecular Crystallography (Part D)
Edited by CHARLES W. CARTER, JR., AND ROBERT W. SWEET

VOLUME 375. Chromatin and Chromatin Remodeling Enzymes (Part A)
Edited by C. DAVID ALLIS AND CARL WU

VOLUME 376. Chromatin and Chromatin Remodeling Enzymes (Part B)
Edited by C. DAVID ALLIS AND CARL WU

VOLUME 377. Chromatin and Chromatin Remodeling Enzymes (Part C)
Edited by C. DAVID ALLIS AND CARL WU

VOLUME 378. Quinones and Quinone Enzymes (Part A)
Edited by HELMUT SIES AND LESTER PACKER

VOLUME 379. Energetics of Biological Macromolecules (Part D)
Edited by JO M. HOLT, MICHAEL L. JOHNSON, AND GARY K. ACKERS

VOLUME 380. Energetics of Biological Macromolecules (Part E)
Edited by JO M. HOLT, MICHAEL L. JOHNSON, AND GARY K. ACKERS

VOLUME 381. Oxygen Sensing
Edited by CHANDAN K. SEN AND GREGG L. SEMENZA

VOLUME 382. Quinones and Quinone Enzymes (Part B)
Edited by HELMUT SIES AND LESTER PACKER

VOLUME 383. Numerical Computer Methods (Part D)
Edited by LUDWIG BRAND AND MICHAEL L. JOHNSON

VOLUME 384. Numerical Computer Methods (Part E)
Edited by LUDWIG BRAND AND MICHAEL L. JOHNSON

VOLUME 385. Imaging in Biological Research (Part A)
Edited by P. MICHAEL CONN

VOLUME 386. Imaging in Biological Research (Part B)
Edited by P. MICHAEL CONN

VOLUME 387. Liposomes (Part D)
Edited by NEJAT DÜZGÜNEŞ

VOLUME 388. Protein Engineering
Edited by DAN E. ROBERTSON AND JOSEPH P. NOEL

VOLUME 389. Regulators of G-Protein Signaling (Part A)
Edited by DAVID P. SIDEROVSKI

VOLUME 390. Regulators of G-Protein Signaling (Part B)
Edited by DAVID P. SIDEROVSKI

VOLUME 391. Liposomes (Part E)
Edited by NEJAT DÜZGÜNEŞ

VOLUME 392. RNA Interference
Edited by ENGELKE ROSSI

VOLUME 393. Circadian Rhythms
Edited by MICHAEL W. YOUNG

VOLUME 394. Nuclear Magnetic Resonance of Biological Macromolecules (Part C)
Edited by THOMAS L. JAMES

VOLUME 395. Producing the Biochemical Data (Part B)
Edited by ELIZABETH A. ZIMMER AND ERIC H. ROALSON

VOLUME 396. Nitric Oxide (Part E)
Edited by LESTER PACKER AND ENRIQUE CADENAS

VOLUME 397. Environmental Microbiology
Edited by JARED R. LEADBETTER

VOLUME 398. Ubiquitin and Protein Degradation (Part A)
Edited by RAYMOND J. DESHAIES

VOLUME 399. Ubiquitin and Protein Degradation (Part B)
Edited by RAYMOND J. DESHAIES

VOLUME 400. Phase II Conjugation Enzymes and Transport Systems
Edited by HELMUT SIES AND LESTER PACKER

VOLUME 401. Glutathione Transferases and Gamma Glutamyl Transpeptidases
Edited by HELMUT SIES AND LESTER PACKER

VOLUME 402. Biological Mass Spectrometry
Edited by A. L. BURLINGAME

VOLUME 403. GTPases Regulating Membrane Targeting and Fusion
Edited by WILLIAM E. BALCH, CHANNING J. DER, AND ALAN HALL

VOLUME 404. GTPases Regulating Membrane Dynamics
Edited by WILLIAM E. BALCH, CHANNING J. DER, AND ALAN HALL

VOLUME 405. Mass Spectrometry: Modified Proteins and Glycoconjugates
Edited by A. L. BURLINGAME

VOLUME 406. Regulators and Effectors of Small GTPases: Rho Family
Edited by WILLIAM E. BALCH, CHANNING J. DER, AND ALAN HALL

VOLUME 407. Regulators and Effectors of Small GTPases: Ras Family
Edited by WILLIAM E. BALCH, CHANNING J. DER, AND ALAN HALL

VOLUME 408. DNA Repair (Part A)
Edited by JUDITH L. CAMPBELL AND PAUL MODRICH

VOLUME 409. DNA Repair (Part B)
Edited by JUDITH L. CAMPBELL AND PAUL MODRICH

VOLUME 410. DNA Microarrays (Part A: Array Platforms and Web-Bench Protocols)
Edited by ALAN KIMMEL AND BRIAN OLIVER

VOLUME 411. DNA Microarrays (Part B: Databases and Statistics)
Edited by ALAN KIMMEL AND BRIAN OLIVER

VOLUME 412. Amyloid, Prions, and Other Protein Aggregates (Part B)
Edited by INDU KHETERPAL AND RONALD WETZEL

VOLUME 413. Amyloid, Prions, and Other Protein Aggregates (Part C)
Edited by INDU KHETERPAL AND RONALD WETZEL

VOLUME 414. Measuring Biological Responses with Automated Microscopy
Edited by JAMES INGLESE

VOLUME 415. Glycobiology
Edited by MINORU FUKUDA

VOLUME 416. Glycomics
Edited by MINORU FUKUDA

VOLUME 417. Functional Glycomics
Edited by MINORU FUKUDA

VOLUME 418. Embryonic Stem Cells
Edited by IRINA KLIMANSKAYA AND ROBERT LANZA

VOLUME 419. Adult Stem Cells
Edited by IRINA KLIMANSKAYA AND ROBERT LANZA

VOLUME 420. Stem Cell Tools and Other Experimental Protocols
Edited by IRINA KLIMANSKAYA AND ROBERT LANZA

VOLUME 421. Advanced Bacterial Genetics: Use of Transposons and Phage for Genomic Engineering
Edited by KELLY T. HUGHES

VOLUME 422. Two-Component Signaling Systems, Part A
Edited by MELVIN I. SIMON, BRIAN R. CRANE, AND ALEXANDRINE CRANE

VOLUME 423. Two-Component Signaling Systems, Part B
Edited by MELVIN I. SIMON, BRIAN R. CRANE, AND ALEXANDRINE CRANE

VOLUME 424. RNA Editing
Edited by JONATHA M. GOTT

VOLUME 425. RNA Modification
Edited by JONATHA M. GOTT

VOLUME 426. Integrins
Edited by DAVID CHERESH

VOLUME 427. MicroRNA Methods
Edited by JOHN J. ROSSI

VOLUME 428. Osmosensing and Osmosignaling
Edited by HELMUT SIES AND DIETER HAUSSINGER

VOLUME 429. Translation Initiation: Extract Systems and Molecular Genetics
Edited by JON LORSCH

VOLUME 430. Translation Initiation: Reconstituted Systems and Biophysical Methods
Edited by JON LORSCH

VOLUME 431. Translation Initiation: Cell Biology, High-Throughput and Chemical-Based Approaches
Edited by JON LORSCH

VOLUME 432. Lipidomics and Bioactive Lipids: Mass-Spectrometry–Based Lipid Analysis
Edited by H. ALEX BROWN

VOLUME 433. Lipidomics and Bioactive Lipids: Specialized Analytical Methods and Lipids in Disease
Edited by H. ALEX BROWN

VOLUME 434. Lipidomics and Bioactive Lipids: Lipids and Cell Signaling
Edited by H. ALEX BROWN

VOLUME 435. Oxygen Biology and Hypoxia
Edited by HELMUT SIES AND BERNHARD BRÜNE

VOLUME 436. Globins and Other Nitric Oxide-Reactive Protiens (Part A)
Edited by ROBERT K. POOLE

VOLUME 437. Globins and Other Nitric Oxide-Reactive Protiens (Part B)
Edited by ROBERT K. POOLE

VOLUME 438. Small GTPases in Disease (Part A)
Edited by WILLIAM E. BALCH, CHANNING J. DER, AND ALAN HALL

VOLUME 439. Small GTPases in Disease (Part B)
Edited by WILLIAM E. BALCH, CHANNING J. DER, AND ALAN HALL

VOLUME 440. Nitric Oxide, Part F Oxidative and Nitrosative Stress in Redox Regulation of Cell Signaling
Edited by ENRIQUE CADENAS AND LESTER PACKER

VOLUME 441. Nitric Oxide, Part G Oxidative and Nitrosative Stress in Redox Regulation of Cell Signaling
Edited by ENRIQUE CADENAS AND LESTER PACKER

VOLUME 442. Programmed Cell Death, General Principles for Studying Cell Death (Part A)
Edited by ROYA KHOSRAVI-FAR, ZAHRA ZAKERI, RICHARD A. LOCKSHIN, AND MAURO PIACENTINI

VOLUME 443. Angiogenesis: *In Vitro* Systems
Edited by DAVID A. CHERESH

VOLUME 444. Angiogenesis: *In Vivo* Systems (Part A)
Edited by DAVID A. CHERESH

VOLUME 445. Angiogenesis: *In Vivo* Systems (Part B)
Edited by DAVID A. CHERESH

VOLUME 446. Programmed Cell Death, The Biology and Therapeutic Implications of Cell Death (Part B)
Edited by ROYA KHOSRAVI-FAR, ZAHRA ZAKERI, RICHARD A. LOCKSHIN, AND MAURO PIACENTINI

VOLUME 447. RNA Turnover in Bacteria, Archaea and Organelles
Edited by LYNNE E. MAQUAT AND CECILIA M. ARRAIANO

VOLUME 448. RNA Turnover in Eukaryotes: Nucleases, Pathways and Analysis of mRNA Decay
Edited by LYNNE E. MAQUAT AND MEGERDITCH KILEDJIAN

VOLUME 449. RNA Turnover in Eukaryotes: Analysis of Specialized and Quality Control RNA Decay Pathways
Edited by LYNNE E. MAQUAT AND MEGERDITCH KILEDJIAN

VOLUME 450. Fluorescence Spectroscopy
Edited by LUDWIG BRAND AND MICHAEL L. JOHNSON

VOLUME 451. Autophagy: Lower Eukaryotes and Non-Mammalian Systems (Part A)
Edited by DANIEL J. KLIONSKY

VOLUME 452. Autophagy in Mammalian Systems (Part B)
Edited by DANIEL J. KLIONSKY

VOLUME 453. Autophagy in Disease and Clinical Applications (Part C)
Edited by DANIEL J. KLIONSKY

VOLUME 454. Computer Methods (Part A)
Edited by MICHAEL L. JOHNSON AND LUDWIG BRAND

VOLUME 455. Biothermodynamics (Part A)
Edited by MICHAEL L. JOHNSON, JO M. HOLT, AND GARY K. ACKERS (RETIRED)

VOLUME 456. Mitochondrial Function, Part A: Mitochondrial Electron Transport Complexes and Reactive Oxygen Species
Edited by WILLIAM S. ALLISON AND IMMO E. SCHEFFLER

VOLUME 457. Mitochondrial Function, Part B: Mitochondrial Protein Kinases, Protein Phosphatases and Mitochondrial Diseases
Edited by WILLIAM S. ALLISON AND ANNE N. MURPHY

VOLUME 458. Complex Enzymes in Microbial Natural Product Biosynthesis, Part A: Overview Articles and Peptides
Edited by DAVID A. HOPWOOD

VOLUME 459. Complex Enzymes in Microbial Natural Product Biosynthesis, Part B: Polyketides, Aminocoumarins and Carbohydrates
Edited by DAVID A. HOPWOOD

VOLUME 460. Chemokines, Part A
Edited by TRACY M. HANDEL AND DAMON J. HAMEL

VOLUME 461. Chemokines, Part B
Edited by TRACY M. HANDEL AND DAMON J. HAMEL

VOLUME 462. Non-Natural Amino Acids
Edited by TOM W. MUIR AND JOHN N. ABELSON

VOLUME 463. Guide to Protein Purification, 2nd Edition
Edited by RICHARD R. BURGESS AND MURRAY P. DEUTSCHER

VOLUME 464. Liposomes, Part F
Edited by NEJAT DÜZGÜNEŞ

VOLUME 465. Liposomes, Part G
Edited by NEJAT DÜZGÜNEŞ

VOLUME 466. Biothermodynamics, Part B
Edited by MICHAEL L. JOHNSON, GARY K. ACKERS, AND JO M. HOLT

VOLUME 467. Computer Methods Part B
Edited by MICHAEL L. JOHNSON AND LUDWIG BRAND

VOLUME 468. Biophysical, Chemical, and Functional Probes of RNA Structure, Interactions and Folding: Part A
Edited by DANIEL HERSCHLAG

VOLUME 469. Biophysical, Chemical, and Functional Probes of RNA Structure, Interactions and Folding: Part B
Edited by DANIEL HERSCHLAG

VOLUME 470. Guide to Yeast Genetics: Functional Genomics, Proteomics, and Other Systems Analysis, 2nd Edition
Edited by GERALD FINK, JONATHAN WEISSMAN, AND CHRISTINE GUTHRIE

VOLUME 471. Two-Component Signaling Systems, Part C
Edited by MELVIN I. SIMON, BRIAN R. CRANE, AND ALEXANDRINE CRANE

VOLUME 472. Single Molecule Tools: Fluorescence Based Approaches, Part A
Edited by NILS WALTER

VOLUME 473. Thiol Redox Transitions in Cell Signaling, Part A Chemistry and Biochemistry of Low Molecular Weight and Protein Thiols
Edited by ENRIQUE CADENAS AND LESTER PACKER

VOLUME 474. Thiol Redox Transitions in Cell Signaling, Part B Cellular Localization and Signaling
Edited by ENRIQUE CADENAS AND LESTER PACKER

CHAPTER ONE

ENGINEERING OF FLUORESCENT REPORTERS INTO REDOX DOMAINS TO MONITOR ELECTRON TRANSFERS

Derek Parsonage,* Stacy A. Reeves,* P. Andrew Karplus,[†] and Leslie B. Poole*

Contents

1. Introduction	2
2. The Problem: Low Sensitivity and Improperly Rate-Limited Assays for Redox Functions of Bacterial Peroxiredoxin Systems	3
3. The Solution: Engineering of Fluorescent Redox Reporters into the N-Terminal Domain of AhpF and *E. coli* Grx1	4
4. Engineering of Disulfide-Containing Electron Acceptor Domains to Detect Electron Transfers via Fluorescence Changes; Linkage of Fluorescein to Bacterial AhpC via a Reducible Disulfide Bond	7
5. Materials	8
5.1. Solutions	8
5.2. Chemical modification agents	9
5.3. Proteins	9
6. Methods	10
6.1. Generation of modified AhpC proteins linked to fluorescein via a disulfide bond	10
6.2. Characterization of the fluorescence and activity of the S128W mutant of the NTD *S. typhimurium* AhpF	12
6.3. Fluorescence-based peroxidase activity assays of *S. typhimurium* AhpC with S128W NTD using stopped-flow analysis	13
6.4. Generation and testing of the F6W mutant of *E. coli* Grx1 as an electron donor to *E. coli* BCP	15
7. Summary	19
References	19

* Department of Biochemistry, Wake Forest University School of Medicine, Winston-Salem, North Carolina, USA
[†] Department of Biochemistry and Biophysics, Oregon State University, Corvallis, Oregon, USA

Abstract

The rate of electron transfer through multicomponent redox systems is often monitored by following the absorbance change due to the oxidation of the upstream pyridine nucleotide electron donor (NADPH or NADH) that initiates the process. Such coupled assay systems are powerful, but because of problems regarding the rate-limiting step, they sometimes limit the kinetic information that can be obtained about individual components. For peroxiredoxins, such assays have led to widespread underestimates of their catalytic power. We show here how this problem can be addressed by a protein engineering strategy inspired by some bacterial and eukaryotic thioredoxins for which a significant fluorescence signal is generated during oxidation that provides a highly sensitive tool to directly measure electron transfers into and out of these domains. For the N-terminal domain of AhpF (a flavoprotein disulfide reductase) and *Escherichia coli* glutaredoxin 1, two cases not having such fluorescence signals, we have successfully added "sensor" tryptophan residues using the positions of tryptophan residues in thioredoxins as a guide. In another thioredoxin-fold redox protein, the bacterial peroxiredoxin AhpC, we used chemical modification to introduce a disulfide-bonded fluorophore. This modified AhpC still serves as an excellent substrate for the upstream AhpF electron donor but now generates a strong fluorescence signal during electron transfer. These tools have fundamentally changed our understanding of the catalytic power of peroxiredoxin systems and should also be widely applicable for improving quantitative assay capabilities in other electron transfer systems.

1. Introduction

Reactive cysteine-containing redox centers within proteins are important to cellular metabolic and signaling processes and are often found within the CXXC-related motif of the thioredoxin (Trx)-like fold common to a broad range of redox proteins (e.g., Trx, glutaredoxin (Grx), glutathione-S-transferase (GST), peroxiredoxin (Prx), and glutathione peroxidase (Gpx) families) (Atkinson and Babbitt, 2009). In some members of this group, including *Escherichia coli* Trx, tryptophan (Trp) residues are sufficiently close and appropriately oriented to act as sensitive fluorescent reporters of the active-site redox state (Holmgren, 1972). In these cases, Trp fluorescence provides a powerful way to conduct kinetic studies focused on one or a few steps of reaction, without having to rely on linked spectral assays (e.g., by including NADPH and Trx reductase with Trx-dependent assays). However, in many proteins in the group, the conversion between disulfide and dithiol forms is spectrally silent or minimally detectable (e.g., through observation of a thiolate anion absorbing at 240 nm), so such assays cannot be used.

As described herein, sensitive fluorescent reporters of redox activity can be engineered into redox domains, either by utilizing mutation to introduce strategically located Trp residues, or by chemically converting a cysteinyl residue of interest into a disulfide center which includes a highly fluorescent leaving group. While these approaches may not be successful in all enzymes, they have been quite successful in the bacterial Prx systems that are the focus of this chapter. In these cases, the fluorescence-based assays not only allowed insight into mechanistic details but also transformed perceptions in the field by showing that the enzymes in question had an intrinsic activity 100 times greater than had been previously appreciated.

2. THE PROBLEM: LOW SENSITIVITY AND IMPROPERLY RATE-LIMITED ASSAYS FOR REDOX FUNCTIONS OF BACTERIAL PEROXIREDOXIN SYSTEMS

Peroxiredoxins (EC 1.11.1.15) are a widespread class of thiol-dependent peroxidases with roles in the control of damaging and signaling-relevant reactive oxygen species such as hydrogen peroxide, organic hydroperoxides, and peroxynitrite (Hall *et al.*, 2009a). Structurally, they are Trx-fold proteins with insertions and modifications that support their peroxidase functions, and include conserved Arg, Pro, and Thr residues in addition to the peroxidatic cysteine (Karplus and Hall, 2007). The catalytic cycle of Prx enzymes is initiated with the nucleophilic attack of the thiolate from the peroxidatic Cys on the peroxide substrate, forming the alcohol (or water), which is released, and a sulfenic acid at the active-site Cys (R-SOH) (Poole, 2005, 2007). The resolution and reductive recycling of the Cys-sulfenic acid back to the thiol form for further catalysis differs in various Prx enzymes. Formation of a disulfide bond with a second, resolving Cys residue occurs in 2-Cys Prx, whereas 1-Cys enzymes form an intermolecular disulfide with a redox donor such as Trx, Grx, or glutathione.

The most widespread and well-characterized form of Prx are the typical 2-Cys Prx, where the resolving Cys is positioned near the C-terminus of a second monomer, hence a dimeric (or higher order) form of the enzyme is required for activity. Trx is the typical cellular reductant of this class of Prx in eukaryotes, whereas in prokaryotes a gene encoding a specialized flavoprotein disulfide reductase, AhpF, is typically found downstream of the gene for AhpC-like Prx proteins (Wood *et al.*, 2003). Plant Prx proteins may in some cases use Grxs as electron donors (Rouhier *et al.*, 2002). With each of these reductive recycling systems, reduction of the Prx is linked to the cellular reductants NADPH or NADH through a flavoprotein (i.e., Trx reductase, AhpF, or glutathione reductase); the coupled oxidation of NAD(P)H at 340 nm is followed to measure the catalytic activity of these Prx enzymes.

Using such a linked assay (e.g., including NADH, AhpF, AhpC, and peroxide), measurements of enzymatic activity of AhpC from *Salmonella typhimurium* revealed no variations in rate for a wide range of concentrations of different hydroperoxide substrates (Parsonage *et al.*, 2005, 2008; Poole and Ellis, 1996), showing that the rates obtained were limited by the ability of AhpF to reduce the disulfide form of AhpC (Poole *et al.*, 2000b). This provided accurate V_{max} and K_m values for the AhpF–AhpC electron transfer interaction (Poole *et al.*, 2000b), but underestimated the reactivity of AhpC with peroxide substrates, promoting the view that Prx proteins are poor enzymes in comparison with heme-dependent peroxidases and catalases (Dietz *et al.*, 2006; Hofmann *et al.*, 2002). Usually, where the coupling enzyme is rate limiting, the solution is to increase the concentration of the coupling enzyme in the assay. In this case, however, the assay is complicated by the high NADH oxidase activity of AhpF, whereby the FAD noncovalently bound to the enzyme is reduced by NADH and subsequently reacts directly with molecular oxygen to produce hydrogen peroxide and oxidized NAD^+ (Niimura *et al.*, 1995; Poole and Ellis, 1996). It is possible to carry out these activity measurements under anaerobic conditions to obviate this problem, but only at the expense of making these assays very time-consuming. For these and issues of better sensitivity, a more direct assay to assess catalytic turnover of bacterial AhpC was sought.

3. The Solution: Engineering of Fluorescent Redox Reporters into the N-Terminal Domain of AhpF and *E. coli* Grx1

Amino acid sequence analysis of AhpF revealed that this protein contains a C-terminal domain homologous to prokaryotic Trx reductase and an N-terminal domain (NTD) homologous to Trx, fused into one polypeptide (Poole *et al.*, 2000a). These two fragments of AhpF could be separated by limited proteolysis using trypsin (Poole, 1996), or separately expressed and purified as distinct Trx reductase- and Trx-like fragments (Poole *et al.*, 2000a). The fold of the NTD of AhpF is, in fact, a tandem duplication of Trx folds in which only the second repeat retains the characteristic CXXC redox-active dithiol motif (Hall *et al.*, 2009b; Wood *et al.*, 2001). The overall AhpC reductase activity of AhpF could be reconstituted, albeit much less efficiently, by mixing the two fragments (Poole *et al.*, 2000a). This structural homology of AhpF to Trx and Trx reductase was later confirmed by solution of the three-dimensional structure of AhpF by X-ray crystallography (Wood *et al.*, 2001).

Unlike Trx, which displays an increase in intrinsic Trp fluorescence when the enzyme is reduced, the NTD of AhpF shows no corresponding redox-dependent fluorescence changes. The reason for this is apparent in Fig. 1.1, which shows an alignment of the amino acid sequences surrounding the CXXC motif of the NTD, Trx, and several other redox-active Trx-like proteins (Tlps). A Trp residue is found preceding the CXXC active site in both prokaryotic and eukaryotic Trxs, as well as protein disulfide isomerase (Pdi) and several extracytoplasmic Tlps. A second Trp residue is also located three residues upstream of the first in prokaryotic Trxs and Tlps; in *E. coli* Trx, these residues are Trp28 and Trp31. Mutagenesis of these residues in *E. coli* Trx identified Trp31 as the major contributor to fluorescence of oxidized Trx (Krause and Holmgren, 1991), whereas the fluorescence of reduced Trx is dominated by the emission from Trp28 (Slaby *et al.*, 1996). Prokaryotic and eukaryotic Trxs differ in the extent to which their

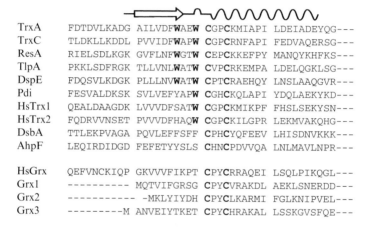

Figure 1.1 Sequence alignment of thioredoxin (Trx)-fold proteins. Partial sequences surrounding the active CXXC motif of Trx and related redox-active proteins were aligned using CLUSTALW (Larkin *et al.*, 2007), showing the position of Trp residues upstream of the CXXC motif (in bold font). The representative structure above the sequences indicates the β-strand (arrow), turn (inverted U), and α-helix (zig-zag) of the *E. coli* Trx 1 structure. Sequences shown are for: TrxA, *E. coli* Trx 1 (THIO_ECOLI); TrxC, *E. coli* Trx 2 (THIO2_ECOLI); ResA, a Trx-like protein involved in cytochrome *c* maturation in *Bacillus subtilis* (RESA_BACSU); TlpA, a membrane-anchored protein thiol:disulfide oxidoreductase essential for cytochrome aa3 maturation in *Bradyrhizobium japonicum* (TLPA_BRAJA); DspE, an *E. coli* protein required for cytochrome *c* maturation (DSBE_ECOLI); Pdi, the protein disulfide isomerase from *Drosophila melanogaster* (PDI_DROME); HsTrx1 and HsTrx2, two human thioredoxins (THIO_HUMAN and THIOM_HUMAN); and AhpF from *Salmonella typhimurium*, the flavoprotein reductase for AhpC (AHPF_SALTY). Also shown are the sequences of one human and three *E. coli* dithiol-containing glutaredoxins of (GLRX1_HUMAN; GLRX1_ECOLI; GLRX2_ECOLI; and GLRX3_ECOLI).

Trp fluorescence increases when they are reduced; *E. coli* Trx undergoes a greater than threefold increase in fluorescence, whereas mammalian and yeast enzymes display only a 50–70% increase (Merola *et al.*, 1989).

Another redox-active enzyme with a Trx-fold structure, DsbA (the key prokaryotic enzyme responsible for the formation of disulfide bonds in periplasmic proteins) (Kadokura *et al.*, 2003), also exhibits redox-dependent changes in fluorescence. Although DsbA lacks Trp residues adjacent to the CXXC motif or within the Trx-like domain, two Trp residues are present within a second α-helical domain inserted into the Trx fold. Communication between the disulfide of the oxidized active site and Trp76 is mediated by Phe26, resulting in quenching of the Trp fluorescence by the disulfide through an intramolecular, dynamic-quenching process (Hennecke *et al.*, 1997). This is responsible for the threefold increase in Trp fluorescence when the active-site disulfide of DsbA is reduced (Wunderlich and Glockshuber, 1993).

In order to use the AhpF NTD in fluorescence-monitored assays of AhpC-dependent peroxidase activities, we mutated the NTD (comprised of residues 1–202 of AhpF) to contain the equivalent of *E. coli* Trx1 Trp31, generating S128W NTD (Fig. 1.2A) (Parsonage *et al.*, 2005). Described in this chapter are the characterization and use of this mutant to very sensitively and accurately measure AhpC activity. As another potential electron donor to Prx proteins, *E. coli* Grx1 was also of interest and was under investigation as a reductant of another bacterial Prx, the *E. coli* "bacterioferritin comigratory protein" (BCP). Wild-type Grx1 exhibits only a low level of fluorescence, a characteristic that was not useful for monitoring its redox state changes during assays with BCP. As shown in Fig. 1.1, placing a Trp residue in *E. coli* Grx1 corresponding to Trp31 of *E. coli* Trx would involve replacing the smallest possible amino acid (Gly) with the largest (Trp). Because this position is adjacent to the redox center, it is possible that

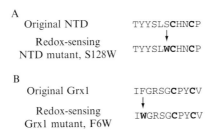

Figure 1.2 Strategies for introducing Trp residues as fluorophores into electron transferring proteins. Shown are the original sequences surrounding the active-site cysteine residues for the N-terminal domain (NTD) of *S. typhimurium* AhpF (A) and *E. coli* Grx1 (B) as well as their respective mutated forms used to provide redox-sensitive fluorophores near the active site, S128W and F6W, respectively.

this Gly or Trp contributes to the specificity of each of these redox domains for given substrates. As an alternative, we chose to mutate a nearby site in Grx1, adjacent to the location of Trp28 in *E. coli* Trx, which already includes a bulky, aromatic residue, creating the F6W mutant of Grx1 (Fig. 1.2B). In fact, both the AhpF S128W and the Grx1 F6W constructs were successful in yielding highly active, CXXC-containing redox domains that could be used to directly monitor electron transfer reactions in simple, three component (donor–Prx–peroxide) assays conducted using a stopped-flow spectrofluorometer, as described in detail below.

4. Engineering of Disulfide-Containing Electron Acceptor Domains to Detect Electron Transfers via Fluorescence Changes; Linkage of Fluorescein to Bacterial AhpC via a Reducible Disulfide Bond

During the course of our studies of the AhpF–AhpC system, we also sought a way to sensitively detect electron transfers from the NTD to AhpC by modifying the electron acceptor instead of the donor. Incorporation of a disulfide-bonded fluorescent reporter into each of the Cys residues in turn generated two potential AhpC-based substrates for AhpF that produced a strong fluorescent signal when reduced. This approach allowed us to compare the activity of both the native and S128W mutated forms of the NTD (Parsonage *et al.*, 2005), and to determine which cysteine within the disulfide bond of AhpC was the site of attack by the electron donor (Jönsson *et al.*, 2007; Poole, 1999).

Standard chemical modification procedures were employed to introduce the chosen fluorescein reporter into each single Cys mutant of AhpC [lacking either the peroxidatic (C46S) or resolving (C165S) Cys residues]. Equilibration of single Cys-containing proteins with N,N'-didansyl-L-cystine can also be used to incorporate disulfide-linked fluorescent groups, but the low efficiency and long times needed for labeling protein with this probe can make this a suboptimal approach. Instead, we made a fluorescein-linked methane thiosulfinate (MTS) reagent that rapidly and efficiently modifies the target protein with disulfide-bonded fluorescein (Fig. 1.3). Then when another protein or small molecule reductant attacks that disulfide bond, the release of fluorescein is accompanied by a fluorescence change, providing a sensitive way to measure reduction rates. Because this engineered AhpC was used to assay the S128W NTD mutant described above, our description of the methods used in this research begins with the engineering of AhpC to generate the modified, disulfide-containing test proteins.

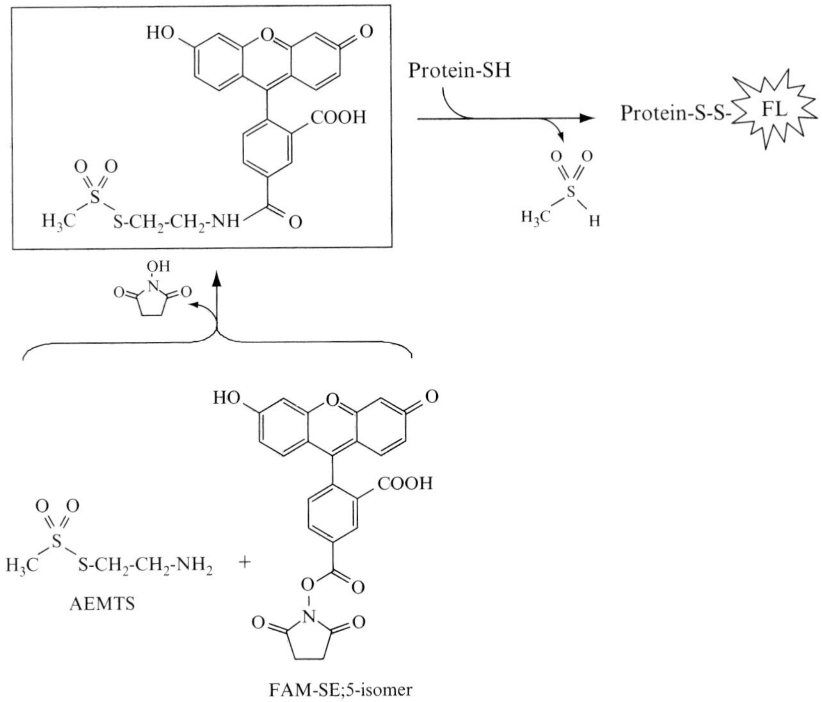

Figure 1.3 Labeling of AhpC mutants with fluorescein via a disulfide bond. Succinimidyl ester-linked carboxyfluorescein (FAM-SE; 5-isomer from Invitrogen is shown, but a mixture of 5- and 6-isomers can be used) was incubated with 2-aminoethyl methane thiosulfonate (AEMTS), then with free protein thiols, resulting in fluorescence quenching.

5. MATERIALS

5.1. Solutions

25 mM potassium phosphate, pH 7.0, 1 mM EDTA (standard buffer)
50 mM potassium phosphate, pH 7.0, 0.5 mM EDTA, 100 mM ammonium sulfate (AhpC reaction buffer)
50 mM H$_2$O$_2$ (~456 μl 30% solution in 100 ml H$_2$O)
100 mM cumene hydroperoxide (reagent diluted approximately 60-fold into dimethyl sulfoxide)
100 mM t-butyl hydroperoxide
~30 mM NADH or NADPH (~2.5 mg per 100 μl 50 mM Tris–HCl, pH 8.0, stored at 4 °C in the dark for ≤ 1 day)

100 mM 1,4-dithio-DL-threitol (DTT), 154.2 g/mol (aliquots stored at −80 °C)
Glutathione, reduced
Hydroxyethyldisulfide (HED) (disulfide-bonded form of 2- mercaptoethanol).

5.2. Chemical modification agents

Carboxyfluorescein succinimidyl ester (FAM-SE; 5- and 6-isomers) from Molecular Probes (now Invitrogen, Carlsbad, CA), 473.39 g/mol, dissolved in DMSO
2-Aminoethyl methane thiosulfonate hydrobromide (AEMTS), from Anatrace, Inc. (Maumee, OH), 236.15 g/mol
5,5′-Dithiobis(2-nitrobenzoic acid) (DTNB), 396.4 g/mol
2-Nitro-5-thiobenzoic acid (TNB) solution, equimolar DTNB and DTT mixed) (Poole and Ellis, 2002).

5.3. Proteins

- *S. typhimurium* AhpC purified essentially as described previously (Parsonage *et al.*, 2008). Aliquots were stored at −80 °C at a concentration of 10 mg/ml.
- *S. typhimurium* C165S and C46S mutants of AhpC, expressed and purified as described previously (Ellis and Poole, 1997; Nelson *et al.*, 2008) and stored at −20 °C in 5 mM DTT. Prior to conducting experiments, DTT is removed using a Sephadex G-50 gel filtration column (monitor A_{280}, pool and if necessary concentrate protein eluting in first peak).
- *S. typhimurium* AhpF purified essentially as described previously (Poole and Ellis, 1996). Aliquots were stored at −80 °C at a concentration of 10 mg/ml.
- NTD of *S. typhimurium* AhpF and S128W mutant of the NTD were purified as described previously (Poole *et al.*, 2000a). The S128W mutant was created using a QuikChange mutagenesis kit (Parsonage *et al.*, 2005). Aliquots of purified protein were stored at 10 mg/ml at −80 °C.
- *E. coli* Grx 1 and the F6W mutant of Grx1 were expressed as His-tagged versions, purified first using a His-trap chelating column (GE Healthcare), cleaved with biotinylated-thrombin (Novagen), then further purified by gel filtration chromatography as described previously (Yamamoto *et al.*, 2008). The F6W mutant was generated and expressed essentially as described previously for S128W NTD, and purified by the same method as wild-type Grx1.
- *E. coli* BCP was expressed in an *ahpC E. coli* strain (lacking AhpC expression) and purified by a combination of Q-Sepharose, Superose 12 prep grade, and ceramic hydroxyapatite columns.
- Glutathione reductase from Baker's yeast.

6. Methods

6.1. Generation of modified AhpC proteins linked to fluorescein via a disulfide bond

A two-step procedure—first linking the fluorescein to a thiol-reactive methane thiosulfonate (MTS) reagent, then incubating this reagent with the protein—is used to introduce the fluorescein reporter into the target protein via a disulfide bond (Fig. 1.3). First, amine-reactive FAM-SE (5- and/or 6-isomers, 56 μmol) and AEMTS (from Anatrace, 37.5 μmol) are predissolved in small volumes of DMSO and then diluted together into a volume of 2.5 ml of 60 mM sodium bicarbonate (a final DMSO concentration of up to 70% DMSO can be used). The reaction is incubated for 2–3 h at room temperature and then quenched with excess glycine (120 μmol in a small volume of water) and incubated overnight at 4 °C in order to block the unreacted fluorescein reagent. Because only the MTS species, which is completely modified with fluorescein, is reactive, the components of the mixture do not need to be separated before adding to protein. Due to the highly efficient reaction of MTS reagents with free thiol groups, any protein with an accessible or partially accessible thiol group can be quickly and efficiently labeled by this reagent. If desirable, thiol reactivity of the carboxyfluorescein-linked AEMTS can be assessed by titrating TNB (generated through addition of DTT to a slight excess of DTNB) with this reagent and observing changes in A_{412} ($\varepsilon_{412} = 14{,}150\ M^{-1}\ cm^{-1}$ for TNB) (Riddles et al., 1979).

To modify the target proteins (in this case the C46S and C165S mutants of AhpC which each contain a single reactive thiol group), a 1.2- to 2-fold molar excess of the reagent is incubated with 6 μmol of AhpC protein in the standard buffer for 10 min at room temperature followed by gel filtration chromatography on a Sephadex G-50 column to remove small molecules. Concentrations of DMSO as high as 20% gave good results with AhpC mutants, probably due to the very short exposure time, but lower concentrations (at 2% or less) may help prevent protein precipitation. Both mutants of AhpC were very efficiently labeled with the reagent; this created the carboxyfluorescein-modified constructs designated AhpC-FAM(165) (with fluorescein attached to Cys165 of the C46S mutant) and AhpC-FAM(46) (with fluorescein attached to Cys46 of the C165S mutant).

The fluorescence of the fluorescein moiety ($\lambda_{max,ex} = 498$ nm, $\lambda_{max,em} = 522$ nm; $\varepsilon_{495} \sim 40{,}000\ M^{-1}\ cm^{-1}$ in standard buffer) was highly quenched upon incorporation into both AhpC mutants. In the presence of excess oxidized C46S AhpC, fluorescein fluorescence decreased 70% upon linkage to the protein. Such quenching due to the surrounding protein environment allows for extremely sensitive detection by fluorescence of the reduction of the engineered disulfide bond.

By measuring the initial and final fluorescence values when the fluorescein-labeled protein is incubated with excess reductant, the arbitrary units of fluorescence intensity can be converted to fluorophore concentration. An independent measure of enzymatic activity may also be conducted to validate the fluorescence results. For AhpC-FAM(165) (ranging from 0 to 72 μM), assays with 0.06 μM intact AhpF and 150 μM NADH were carried out aerobically in an Applied Photophysics SX.18MV stopped-flow spectrophotometer in the standard AhpC reaction buffer without peroxide. Rates were determined based on both absorbance changes at 340 nm and fluorescence changes using excitation at 493 nm and a 510 nm filter for emission. Rates of fluorescence changes were directly proportional to rates of absorbance changes up to (at least) 29 μM AhpC-FAM(165); plotting the observed rate against the concentration of AhpC-FAM(165) gave a straight line passing through the origin, consistent with an irreversible reaction.

For the two AhpC mutants labeled by this procedure, reactivity toward AhpF is considerably different due to the differential protein environments and roles for each of these residues in the wild-type protein (Ellis and Poole, 1997; Jönsson et al., 2007). AhpF reduced AhpC-FAM(165) with very similar kinetic parameters as those for wild-type AhpC (only a sevenfold difference in catalytic efficiency); the k_{cat} was about ninefold lower (25.5 vs. 237 s^{-1}, respectively) and K_m for the AhpC substrate approximately the same (14.3 vs. 19.0 μM) as for wild-type AhpC (Poole et al., 2000a,b). In contrast, AhpF reduced AhpC-FAM(46) at a rate more than two orders of magnitude lower than the rate with wild-type AhpC (k_{cat}/K_m of 5×10^4 $M^{-1} s^{-1}$ vs. 1.2×10^7 $M^{-1} s^{-1}$ for AhpC-FAM(46) and wild-type AhpC, respectively) (Poole, 1999). Very similar results were obtained using a TNB conjugate with C46S AhpC (analogous to AhpC-FAM(165)), and we have used these systems to investigate the roles of the different Cys in AhpF (Jönsson et al., 2007). These results demonstrated that AhpF can attack a disulfide-bonded Cys165 (but not C46) with high efficiency even with large differences in the moiety to which this residue is disulfide bonded.

To test for activity of the NTD of AhpF with this substrate, the NTD protein was prereduced overnight at 4 °C with a 100-fold excess of DTT and separated from the excess reagent by gel filtration chromatography on a Sephadex G-50 column.

On mixing reduced NTD with excess AhpC-FAM(165) in the stopped-flow spectrophotometer at concentrations up to 29 μM, fluorescence changes were readily fit to a single exponential at each concentration of AhpC-FAM(165) (Fig. 1.4). Over these concentrations, the fluorescein label was released with a bimolecular rate constant of 1.2×10^6 $M^{-1} s^{-1}$ (Fig. 1.4, inset), nearly identical with the k_{cat}/K_m of 1.8×10^6 $M^{-1} s^{-1}$ of AhpC-FAM(165) with intact AhpF (Poole et al., 2000a). The separately expressed C-terminal fragment of AhpF (residues 208–521) exhibited less than 0.1% of the reactivity of intact AhpF with AhpC-FAM(165), clearly

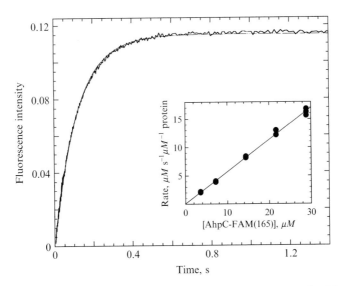

Figure 1.4 Reduction of AhpC-FAM(165) by prereduced NTD. The N-terminal domain (residues 1–202) of AhpF (0.35 μM of NTD) prereduced with dithiothreitol was mixed with AhpC-FAM(165) at varying concentrations in peroxidase assay buffer (50 mM potassium phosphate, pH 7.0, with 0.5 mM EDTA and 100 mM ammonium sulfate) at 25 °C in the stopped-flow spectrophotometer. Shown are the fluorescence changes observed on mixing reduced NTD with 14.4 μM AhpC-FAM(165) and the fitted single-exponential curve. The inset shows the pseudo-first-order rates of fluorescence change observed over varying concentrations of AhpC-FAM(165). Reprinted with permission from Poole *et al.* (2000a).

illustrating the ability of the N-terminus, but not the C-terminus, of AhpF to rapidly reduce AhpC (Poole *et al.*, 2000a).

6.2. Characterization of the fluorescence and activity of the S128W mutant of the NTD *S. typhimurium* AhpF

The S128W mutant of the independently-expressed NTD of *S. typhimurium* AhpF was highly overexpressed from a T7 promoter using several *E. coli* strains, and was readily purified in large amounts (Parsonage *et al.*, 2005). Introduction of a third Trp into the NTD is expected to increase the 280 nm extinction coefficient, so the 280 nm absorbance of a solution of S128W NTD was compared to the protein content measured by the microbiuret assay. This gave an experimentally determined extinction coefficient of 21,250 M^{-1} cm^{-1} at 280 nm for S128W NTD, higher than the value of 15,100 M^{-1} cm^{-1} for wild-type NTD.

One immediate question is whether the S128W mutant NTD is capable of reducing AhpC, and how this activity compares to wild-type NTD.

In order to answer this question, we examined the ability of prereduced S128W NTD to release the fluorescein label from the artificial substrate AhpC-FAM(165) described above. Wild-type or S128W NTD (0.35 μM) was mixed with a range from 5 to 30 μM AhpC-FAM(165) in a stopped-flow spectrofluorometer. All concentrations are the final values after mixing. As above, fluorescence intensity was measured by excitation at 493 nm and emission at wavelengths greater than 510 nm. When reactions of S128W NTD with AhpC-FAM(C165) were compared with those of wild-type NTD, identical rates were observed, indicating that the NTD's ability to reduce AhpC was not perturbed.

Comparison of the fluorescence of equal concentrations of reduced and oxidized S128W NTD revealed that the reduced form was 12% more fluorescent than the oxidized form (excitation at 280 nm, emission peak at 343 nm). This is a smaller difference between redox states than is seen with *E. coli* Trx, but was sufficient for the measurement of AhpC reacting with 0.5 μM H_2O_2 in the presence of 40 μM S128W (conditions exhibiting the smallest change in fluorescence measured).

Reduced S128W NTD was prepared by adding an approximately 40-fold excess of DTT; typically 0.2 ml of 100 mM DTT was added to 1 ml of 10 mg/ml S128W NTD and incubated at room temperature for a minimum of 1 h. Excess DTT was removed by passing the S128W NTD over a PD10 desalting column (GE Healthcare) equilibrated with standard AhpC reaction buffer. The peak protein fractions were determined by 280 nm absorbance, pooled and filtered through a 0.2 or 0.45 μm syringe-tip filter. The protein concentration of the pool was found by measuring the 280 nm absorbance of a 15-fold dilution.

6.3. Fluorescence-based peroxidase activity assays of *S. typhimurium* AhpC with S128W NTD using stopped-flow analysis

For a bisubstrate kinetic characterization of AhpC, reaction rates have to be determined over varying concentrations of both reductant (S128W NTD) and hydroperoxide. As there is no regeneration of the reductant, it is necessary to use a stopped-flow spectrofluorometer to measure this reaction as the initial rates are only linear for the first few seconds of the reaction. Solutions containing reduced S128W NTD and AhpC in one syringe are mixed with solutions of different concentrations of hydroperoxide (including hydrogen peroxide, ethyl hydroperoxide, cumene hydroperoxide, and *t*-butyl hydroperoxide).

For the stopped-flow experiments, solutions contained 5–100 μM concentrations of the prereduced S128W NTD along with 0.1–1 μM AhpC (wild type or mutant). The solutions used in the stopped flow are all prepared in AhpC reaction buffer. Typically, using an Applied Photophysics stopped-flow instrument, 5–7 ml is sufficient to determine the rate of reaction at one concentration of S128W NTD and a range of peroxide substrate concentrations (7–9 solutions) that bracket the apparent K_m. This allows for at least four replicates for each rate determination. A 15-fold dilution of the stock AhpC with AhpC reaction buffer is made directly in a cuvette to measure the 280-nm absorbance. This primary dilution is removed from the cuvette and used to prepare the solutions with S128W NTD. If the AhpC has been stored in a buffer containing a component that will interfere with the reaction assay, DTT for example, a small desalting column should be used to transfer the enzyme into AhpC reaction buffer, or a compatible buffer if necessary. Care should be taken to minimize changes in buffer composition during the measurements. Four to five sets of experiments with each set at a different concentration of S128W NTD can be carried out in a normal working day.

The solutions containing a range of peroxide concentrations are prepared from the 50 mM stock or a dilution of the stock in AhpC reaction buffer. The initial 50 mM dilution is freshly prepared and made in water or DMSO in the case of cumene hydroperoxide; further dilutions are made using AhpC reaction buffer. Sufficient volumes of each dilution are made for all the sets of experiments to be carried out on that day. Aliquots of each concentration are removed to load into the drive syringes for each rate determination, then discarded. This minimizes any possible contamination of the peroxide solutions with enzyme. It is important to remember that stopped-flow instruments usually mix equal volumes of reactants, so that all solutions need to be made at twice the final concentration.

For best performance, the stopped-flow instrument should be allowed to warm up in order to stabilize the lamp output, photomultiplier detectors, and the temperature of the mixing chamber. The stopped-flow spectrophotometer is set up to measure fluorescence with excitation at 280 nm and emission monitored at 90° and at wavelengths >320 nm using an emission filter. The solution of S128W NTD and AhpC is first mixed in the stopped flow with AhpC reaction buffer and the photomultiplier voltage is set to give 80% of maximum signal. The signal should be monitored to ensure that the fluorescence signal is stable in the absence of any reaction. The reaction buffer is then replaced by a peroxide solution of concentration higher than the S128W NTD, and after temperature equilibration ($\sim$5 min) mixed with the solution of enzymes. The reaction is followed until no further change occurs. After extrapolating the signal back to time $= 0$ s, the total change in fluorescence is calculated; this is defined as the change in fluorescence (signal voltage) caused by the oxidation of the known amount of S128W NTD. Hence, the rates measured in V/s can now be converted to rates in terms of μM NTD-oxidized s^{-1} μM^{-1} AhpC. This determination should be carried out at least in duplicate. The initial rate of fluorescence change for each assay is observed for the first 5 s. This reaction is repeated until at least three consistent traces are obtained. The next concentration of peroxide substrate is loaded into the drive syringe and allowed to thermally equilibrate while the previously obtained traces are averaged and the initial linear rate calculated by linear regression. This is repeated until rates have been measured with all the different peroxide solutions or fail to further increase with higher peroxide concentrations. At this point, the next concentration of S128W NTD with AhpC is loaded into the drive syringe, and the photomultiplier voltage adjusted and the calibration of the fluorescence signal carried out as before.

At high concentrations of peroxide substrate, there may be a significant background rate of reaction directly between the S128W NTD (or other) reductant and the peroxide. This rate was only apparent with millimolar concentrations of hydrogen peroxide and S128W NTD, and the rate of reaction between hydrogen peroxide and S128W NTD was not linearly dependent on the peroxide concentration. Therefore, to correct for this background rate, identical series of experiments were carried out in the absence of AhpC, and the observed background rate was subtracted from the rate measured in the presence of AhpC.

The first step in analyzing the data is to determine the appropriate kinetic mechanism to apply. This is done by plotting the data using a linearized form of the Michaelis–Menten equation, for example, the Hanes plot (Cornish-Bowden, 2004). An example of a Hanes plot of data obtained with wild-type AhpC and t-butyl hydroperoxide in shown in Fig. 1.5A. The intersection of the lines at the y-axis indicates a substituted enzyme (ping-pong) mechanism for AhpC. Consequently, all the rate data obtained for one peroxide substrate can be fit to Eq. (1.1),

$$\text{Rate} = \frac{k_{\text{cat}} \times [\text{S128W}] \times [\text{ROOH}]}{K_m^{\text{S128W}} \times [\text{ROOH}] + K_m^{\text{ROOH}} \times [\text{S128W}] + [\text{ROOH}] \times [\text{S128W}]} \quad (1.1)$$

using the multiple-function nonlinear regression capability of Sigmaplot (Systat Software, Inc. San Jose, CA) to calculate a global fit for k_{cat} and K_m of the two substrates (Fig. 1.5B). Kinetic analyses of a range of hydroperoxide substrates and AhpC mutants have been conducted using these approaches (Parsonage et al., 2005, 2008).

6.4. Generation and testing of the F6W mutant of *E. coli* Grx1 as an electron donor to *E. coli* BCP

As with the S128W NTD described above, generation of the pure, mutated Grx1 protein was straightforward. The F6W mutant Grx1 was created using the QuikChange method; mutant and wild-type Grx1 were expressed and

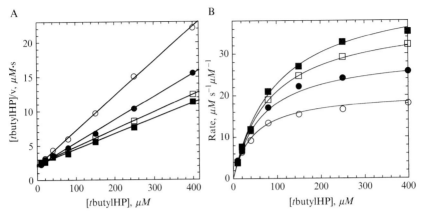

Figure 1.5 Kinetics of *t*-butyl hydroperoxide reduction by AhpC. S128W NTD (prereduced by dithiothreitol) and 100 nM AhpC were mixed with 10–400 μM *t*-butyl hydroperoxide in a stopped-flow spectrophotometer at 25 °C. All concentrations given are after mixing. The buffer was 50 mM potassium phosphate, pH 7.0, 0.5 mM EDTA, and 100 mM ammonium sulfate. The reaction rate was measured by monitoring the fluorescence change of S128W NTD with excitation at 280 nm, and emission at >320 nm. The fixed concentrations of S128W NTD in assays over a range of *t*-butyl hydroperoxide were 2.5 μM (○), 5 μM (●), 10 μM (□), and 20 μM (■). A Hanes plot of the data (panel A) shows that lines drawn through the data points intersect on the *y*-axis, indicating that AhpC is following a substituted enzyme mechanism. The rate data with global nonlinear regression fits to the data are plotted in panel B with calculated values for $K_m^{t-\text{ButylHP}}$ of 119 ± 4 μM, K_m^{S128W} of 4.1 ± 0.2 μM, and k_{cat} of 55 ± 1 s^{-1}.

purified as described previously (Yamamoto *et al.*, 2008). Briefly, the protein was expressed from a T7 expression vector with an N-terminal His-tag. After purification on a His-Trap chelating column, the His-tag was removed by cleaving with thrombin, followed by gel filtration. As had been intended, the F6W mutant displays higher overall fluorescence than wild-type Grx1; in addition, F6W shows a nearly twofold increase in fluorescence upon reduction (Fig. 1.6). This allows a dramatic increase in the sensitivity in assays where the redox state of Grx1 is monitored.

To evaluate functional features of the new mutant protein, the catalytic activities of wild-type and F6W Grx1 were compared using an assay based upon the ability of Grx to catalyze the reduction of a mixed disulfide formed between glutathione and the small molecule HED (Holmgren and Åslund, 1995). The resulting oxidized glutathione is reduced by glutathione reductase, allowing the reaction to be monitored by oxidation of NADPH measured spectrophotometrically at 340 nm. The activities of wild-type and F6W mutant Grx1 using this assay were found to be essentially identical (Fig. 1.7).

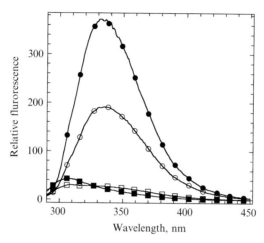

Figure 1.6 Fluorescence emission spectra of oxidized and reduced forms of *E. coli* Grx1 and the F6W mutant of Grx1. Fluorescence intensity was measured with excitation at 280 nm using a Varian Cary Eclipse fluorescence spectrophotometer. Proteins in 50 mM potassium phosphate buffer (pH 7) with 1 mM EDTA included 10 μM each oxidized (□) or reduced (■) Grx1, and oxidized (○) or reduced (●) F6W Grx1.

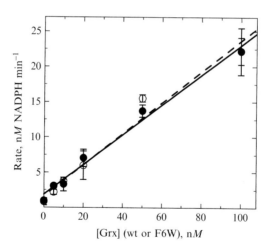

Figure 1.7 A comparison of activity of Grx1 and Grx1 F6W with GSH in the GSH-dependent disulfide reductase assay with hydroxyethyldisulfide (HED) indicates that F6W is as active as wild type. The HED assay (Holmgren and Åslund, 1995) couples the reduction of a glutathione-linked mixed disulfide to oxidation of NADPH, monitored by the decrease in 340 nm absorbance. The reaction mixture (0.5 ml) contained 100 mM Tris–Cl, pH 8.0 buffer with 2 mM EDTA, 0.7 mM HED, 1 mM GSH, 0.2 mM NADPH, 0.1 mg/ml BSA, 0.12 μM glutathione reductase, and 5–100 nM wild-type (●, solid line) or F6W(○, dashed line) Grx1.

Earlier attempts to demonstrate that Grx1 could function as a reducing substrate for BCP in the presence of hydrogen peroxide using a coupled assay with glutathione, glutathione reductase, and NADPH were unsuccessful because of a high background rate. Assays carried out in the absence of BCP revealed that the background rate arises from the reaction between reduced glutathione and hydrogen peroxide. By including BCP in high amounts and leaving out the peroxide, this background reaction was avoided; using this assay, F6W Grx1 was found to reduce BCP at 65% of the rate of wild-type Grx1.

The sensitive, redox-dependent fluorescence of F6W Grx1 can be utilized in stopped-flow spectrophotometric assays similar to those conducted with S128W NTD and AhpC. To conduct these assays, F6W Grx1 was reduced with excess DTT, which was subsequently removed using a small Sephadex G-25 column, added to a solution with BCP, and mixed with different concentrations of hydrogen peroxide. Examples of the resulting fluorescence changes are shown in Fig. 1.8 with varying amounts of hydrogen peroxide, demonstrating the utility of this approach.

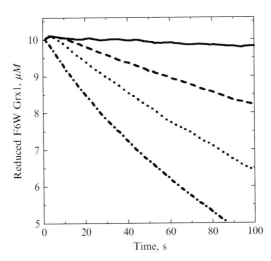

Figure 1.8 Reaction of BCP with hydrogen peroxide, monitored by F6W Grx1 fluorescence. BCP (0.5 mM) and 10 mM F6W Grx1 were combined with 0, 5, 20, or 200 μM hydrogen peroxide (in order of increasing downward slope) in a stopped-flow spectrofluorometer. The buffer in both syringes was 50 mM potassium phosphate, pH 7, 0.5 mM EDTA at 25 °C. Fluorescence excitation was at 280 nm, with emission measured at >320 nm.

7. Summary

We have described three systems in which we have used an introduced fluorophore in order to measure particular enzymatic rates. These include the release of a fluorophore leaving group (fluorescein) from a chemically modified residue, and two systems in which an amino acid near the redox-active CXXC motif has been mutated to Trp, mimicking the Trp(s) at the active site of Trx. The resultant redox-dependent changes in fluorescence of the introduced Trp allow the oxidation of the protein to be monitored directly. We have used these three systems to examine the kinetics of two different Prx enzymes. These principles could be extended to other Prxs with non-Trx reductants, and also to other redox systems where there is little or no detectable spectral signal.

REFERENCES

Atkinson, H. J., and Babbitt, P. C. (2009). An atlas of the thioredoxin fold class reveals the complexity of function-enabling adaptations. *PLoS Comput. Biol.* **5,** e1000541.

Cornish-Bowden, A. (2004). Fundamentals of Enzyme Kinetics. Portland Press, London.

Dietz, K. J., Jacob, S., Oelze, M. L., Laxa, M., Tognetti, V., de Miranda, S. M., Baier, M., and Finkemeier, I. (2006). The function of peroxiredoxins in plant organelle redox metabolism. *J. Exp. Bot.* **57,** 1697–1709.

Ellis, H. R., and Poole, L. B. (1997). Roles for the two cysteine residues of AhpC in catalysis of peroxide reduction by alkyl hydroperoxide reductase from *Salmonella typhimurium*. *Biochemistry* **36,** 13349–13356.

Hall, A., Karplus, P. A., and Poole, L. B. (2009a). Typical 2-Cys peroxiredoxins—structures, mechanisms and functions. *FEBS J.* **276,** 2469–2477.

Hall, A., Parsonage, D., Horita, D., Karplus, P. A., Poole, L. B., and Barbar, E. (2009b). Redox-dependent dynamics of a dual thioredoxin fold protein: Evolution of specialized folds. *Biochemistry* **48,** 5984–5993.

Hennecke, J., Sillen, A., Huber-Wunderlich, M., Engelborghs, Y., and Glockshuber, R. (1997). Quenching of tryptophan fluorescence by the active-site disulfide bridge in the DsbA protein from *Escherichia coli*. *Biochemistry* **36,** 6391–6400.

Hofmann, B., Hecht, H.-J., and Flohé, L. (2002). Peroxiredoxins. *Biol. Chem.* **383,** 347–364.

Holmgren, A. (1972). Tryptophan fluorescence study of conformational transitions of the oxidized and reduced form of thioredoxin. *J. Biol. Chem.* **247,** 1992–1998.

Holmgren, A., and Åslund, F. (1995). Glutaredoxin. *Methods Enzymol.* **252,** 283–292.

Jönsson, T. J., Ellis, H. R., and Poole, L. B. (2007). Cysteine reactivity and thiol-disulfide interchange pathways in AhpF and AhpC of the bacterial alkyl hydroperoxide reductase system. *Biochemistry* **46,** 5709–5721.

Kadokura, H., Katzen, F., and Beckwith, J. (2003). Protein disulfide bond formation in prokaryotes. *Annu. Rev. Biochem.* **72,** 111–135.

Karplus, P. A., and Hall, A. (2007). Structural survey of the peroxiredoxins. *In* "Peroxiredoxin Systems," (L. Flohé and J. R. Harris, eds.), pp. 41–60. Springer, New York.

Krause, G., and Holmgren, A. (1991). Substitution of the conserved tryptophan 31 in *Escherichia coli* thioredoxin by site-directed mutagenesis and structure–function analysis. *J. Biol. Chem.* **266**, 4056–4066.

Larkin, M. A., Blackshields, G., Brown, N. P., Chenna, R., McGettigan, P. A., McWilliam, H., Valentin, F., Wallace, I. M., Wilm, A., Lopez, R., Thompson, J. D., Gibson, T. J., and Higgins, D. G. (2007). Clustal W and Clustal X version 2.0. *Bioinformatics* **23**, 2947–2948.

Merola, F., Rigler, R., Holmgren, A., and Brochon, J. C. (1989). Picosecond tryptophan fluorescence of thioredoxin: Evidence for discrete species in slow exchange. *Biochemistry* **28**, 3383–3398.

Nelson, K. J., Parsonage, D., Hall, A., Karplus, P. A., and Poole, L. B. (2008). Cysteine pK_a values for the bacterial peroxiredoxin AhpC. *Biochemistry* **47**, 12860–12868.

Niimura, Y., Poole, L. B., and Massey, V. (1995). *Amphibacillus xylanus* NADH oxidase and *Salmonella typhimurium* alkyl-hydroperoxide reductase flavoprotein components show extremely high scavenging activity for both alkyl hydroperoxide and hydrogen peroxide in the presence of *S. typhimurium* alkyl-hydroperoxide reductase 22-kDa protein component. *J. Biol. Chem.* **270**, 25645–25650.

Parsonage, D., Youngblood, D. S., Sarma, G. N., Wood, Z. A., Karplus, P. A., and Poole, L. B. (2005). Analysis of the link between enzymatic activity and oligomeric state in AhpC, a bacterial peroxiredoxin. *Biochemistry* **44**, 10583–10592.

Parsonage, D., Karplus, P. A., and Poole, L. B. (2008). Substrate specificity and redox potential of AhpC, a bacterial peroxiredoxin. *Proc. Natl. Acad. Sci. USA* **105**, 8209–8214.

Poole, L. B. (1996). Flavin-dependent alkyl hydroperoxide reductase from *Salmonella typhimurium*. 2. Cystine disulfides involved in catalysis of peroxide reduction. *Biochemistry* **35**, 65–75.

Poole, L. B. (1999). Flavin-linked redox components required for AhpC reduction in alkyl hydroperoxide reductase systems. *In* "Flavins and Flavoproteins," (S. Ghisla, *et al.*, eds.), pp. 195–202. Agency for Scientific Publications, Berlin.

Poole, L. B. (2005). Bacterial defenses against oxidants: Mechanistic features of cysteine-based peroxidases and their flavoprotein reductases. *Arch. Biochem. Biophys.* **433**, 240–254.

Poole, L. B. (2007). The catalytic mechanism of peroxiredoxins. *In* "Peroxiredoxin Systems," (L. Flohé and J. R. Harris, eds.), pp. 61–81. Springer, New York.

Poole, L. B., and Ellis, H. R. (1996). Flavin-dependent alkyl hydroperoxide reductase from *Salmonella typhimurium*. 1. Purification and enzymatic activities of overexpressed AhpF and AhpC proteins. *Biochemistry* **35**, 56–64.

Poole, L. B., and Ellis, H. R. (2002). Identification of cysteine sulfenic acid in AhpC of alkyl hydroperoxide reductase. *Methods Enzymol.* **348**, 122–136.

Poole, L. B., Godzik, A., Nayeem, A., and Schmitt, J. D. (2000a). AhpF can be dissected into two functional units: Tandem repeats of two thioredoxin-like folds in the N-terminus mediate electron transfer from the thioredoxin reductase-like C-terminus to AhpC. *Biochemistry* **39**, 6602–6615.

Poole, L. B., Higuchi, M., Shimada, M., Calzi, M. L., and Kamio, Y. (2000b). *Streptococcus mutans* H_2O_2-forming NADH oxidase is an alkyl hydroperoxide reductase protein. *Free Radic. Biol. Med.* **28**, 108–120.

Riddles, P. W., Blakeley, R. L., and Zerner, B. (1979). Ellman's reagent: 5,5'-dithiobis(2-nitrobenzoic acid)—a reexamination. *Anal. Biochem.* **94**, 75–81.

Rouhier, N., Gelhaye, E., and Jacquot, J. P. (2002). Glutaredoxin-dependent peroxiredoxin from poplar: Protein–protein interaction and catalytic mechanism. *J. Biol. Chem.* **277**, 13609–13614.

Slaby, I., Cerna, V., Jeng, M. F., Dyson, H. J., and Holmgren, A. (1996). Replacement of Trp28 in *Escherichia coli* thioredoxin by site-directed mutagenesis affects thermodynamic stability but not function. *J. Biol. Chem.* **271**, 3091–3096.

Wood, Z. A., Poole, L. B., and Karplus, P. A. (2001). Structure of intact AhpF reveals a mirrored thioredoxin-like active site and implies large domain rotations during catalysis. *Biochemistry* **40**, 3900–3911.

Wood, Z. A., Schröder, E., Harris, J. R., and Poole, L. B. (2003). Structure, mechanism and regulation of peroxiredoxins. *Trends Biochem. Sci.* **28**, 32–40.

Wunderlich, M., and Glockshuber, R. (1993). Redox properties of protein disulfide isomerase (DsbA) from *Escherichia coli*. *Protein Sci.* **2**, 717–726.

Yamamoto, Y., Ritz, D., Planson, A. G., Jonsson, T. J., Faulkner, M. J., Boyd, D., Beckwith, J., and Poole, L. B. (2008). Mutant AhpC peroxiredoxins suppress thiol-disulfide redox deficiencies and acquire deglutathionylating activity. *Mol. Cell* **29**, 36–45.

CHAPTER TWO

Blot-Based Detection of Dehydroalanine-Containing Glutathione Peroxidase with the Use of Biotin-Conjugated Cysteamine

Sue Goo Rhee *and* Chun-Seok Cho

Contents

1. Introduction 24
2. Oxidative Inactivation of Glutathione Peroxidase and the Conversion of Its Active Site Sec to DHA 25
3. Preparation of Biotin-Conjugated Cysteamine 27
4. Blot-Based Detection of DHA–GPx1 in RBCs 28
5. Effects of Oxidative Stress on the Formation of DHA–GPx1 in RBCs 29
6. Concluding Remarks 30
Acknowledgment 32
References 32

Abstract

Dehydroalanine (DHA), α,β-unsaturated amino acid, is found in the position corresponding to the serine, cysteine, and selenocysteine (Sec) residues of various proteins. Proteinaceous Sec is readily oxidized and subsequently undergoes β-elimination to produce DHA. Glutathione peroxidase (GPx), which contains a Sec at the active site, is irreversibly inactivated by its own substrate as the result of the oxidation of selenium atom followed by the conversion of oxidized Sec to DHA.

We developed a convenient method for estimation of the amount of DHA–GPx1 in cell homogenates. This blot-based method depends on specific addition of biotin-conjugated cysteamine to the DHA residue followed by detection of biotinylated protein based on its interaction with streptavidin. The method required an immunoprecipitation of GPx1 before labeling with the cysteamine derivative because many other proteins contain DHA. With the use of this method, we found that conversion of the Sec residue at the active site of GPx1 to DHA occurred during aging of red blood cells (RBCs) *in vivo* as well

Division of Life and Pharmaceutical Sciences, Ewha Womans University, Seodaemun-gu, Seoul, Korea

Methods in Enzymology, Volume 474 © 2010 Elsevier Inc.
ISSN 0076-6879, DOI: 10.1016/S0076-6879(10)74002-7 All rights reserved.

as in RBCs exposed to H_2O_2 generated either externally by glucose oxidase or internally as a result of aniline-induced Hb autoxidation. Accordingly, the content of DHA–GPx1 in each RBC likely reflects total oxidative stress experienced by the cell during its lifetime of 120 days. Previous studies suggested that the activity of GPx1 in RBCs is most influenced by lifestyle and environmental factors such as the use of dietary supplements and smoking habit. Therefore, DHA–GPx1 in RBCs might be a suitable surrogate marker for evaluation of oxidative stress in the body. Our blot-based method for the detection of DHA–GPx1 will be very useful for evaluation of such stress. In addition, similar blot detection method can be devised for other proteins for which immunoprecipitating antibodies are available.

1. INTRODUCTION

Dehydroalanine (DHA) is an α,β-unsaturated amino acid which is found in a variety of naturally occurring antibiotic and phytotoxic peptides (Gross and Morell, 1967). The presence of DHA residue confers the unique peptide conformation that is required for exhibition of biological activities of those peptides. DHA is also found in the catalytic sites of several amino acid ammonia-lyases (Schuster and Retey, 1995; Wickner, 1969). For example, phenylalanine ammonia-lyase from parsley is posttranslationally modified by dehydrating its Ser202 to the catalytically essential DHA prosthetic group (Schuster and Retey, 1995). DHA is also produced during thyroid hormone synthesis in thyroglobulin (Gavaret et al., 1980; Ohmiya et al., 1990).

Certain cysteine and selenocysteine (Sec) residues of proteins are slowly converted to DHA as the result of β-elimination in nonphysiological processes (Bernardes et al., 2008; Jones et al., 1983; Ma et al., 2003). The β-elimination is facilitated when those residues are alkylated or oxidized. Sec residue is more favorable for the β-elimination reaction compared with cysteine residue because selenol is more sensitive to oxidation and the C–Se bond (234 kJ/mol) is weaker than the C–S bond (272 kJ/mol) (Krief, 1987). Indeed, synthesis of peptides containing phenylSec [Ph–Se–CH$_2$–CH (–COOH)–NH$_2$] and exposure of them to a mild oxidative condition to remove phenylselenenic acid (Ph–SeOH) via β-elimination underlie a standard method for the preparation of DHA-containing peptides (Levengood and van der Donk, 2006). In addition to Sec and cysteine residues, site-specific dehydration of serine generates DHA in various peptides and proteins (Strumeyer et al., 1963).

A number of methods are available for detection of DHA in proteins (Bartone et al., 1991). Under the conditions of conventional acid hydrolysis, DHA residues are converted to pyruvic acid and ammonia, and a spectrophotometric assay of pyruvate, based on an NADH-coupled reaction catalyzed by lactic dehydrogenase, has been used to measure DHA contents (Bartone *et al.*, 1991). Treatment of DHA-containing proteins with [^{3}H] sodium borohydride, [^{14}C] sodium cyanide, and [^{35}S] sodium sulfite, followed by acid hydrolysis, results in radiolabeled alanine, aspartic acid, and cysteic acid, respectively (Consevage and Phillips, 1985; Gavaret *et al.*, 1980; Mega *et al.*, 1990; Wickner, 1969). Owing to its electrophilicity, DHA reacts with nucleophiles via Michael-type addition and addition of thiols such as methane thiol, benzyl mercaptane, 4-aminothiophenol, 4-pyridoethanethiol has been successfully used to detect DHA in proteins (Bartone *et al.*, 1991; Gross and Kiltz, 1973; Masri and Friedman, 1982; Mega *et al.*, 1990; Ohmiya *et al.*, 1990). The Michael addition reaction has also provided a very effective means to map the sites of serine or threonine phosphorylation (Knight *et al.*, 2003). In this method, phosphoserine (and phosphothreonine) residues were converted to DHA (and β-methylDHA) in the presence of Ba(OH)$_2$ and the resulting DHA (and β-methylDHA) was selectively transformed into aminoethylcysteine (and β-methylaminoethylcysteine) by reacting with cysteamine. Because aminoethylcysteine (and β-methylaminoethylcysteine) is isosteric with lysine, proteases that recognize lysine (e.g., trypsin and Lys-C) cleave proteins at this residue.

2. Oxidative Inactivation of Glutathione Peroxidase and the Conversion of Its Active Site Sec to DHA

Glutathione peroxidase (GPx) catalyzes the reduction of hydrogen peroxide (H$_2$O$_2$) and lipid peroxides by glutathione (GSH). Selenium is present at the active site of GPx as Sec (Flohe *et al.*, 1973). During catalysis, the selenolate (GPx–Se$^-$) reacts with hydroperoxides to yield selenenic acid (GPx–SeOH), which, in the presence of GSH, is rapidly converted to a glutathionylated intermediate (GPx–Se–S–G). This intermediate then reacts with another GSH molecule to produce GPx–Se$^-$ plus oxidized glutathione (GSSG).

GPx is susceptible to inactivation by its own substrates. Exposure of purified GPx1 to various hydroperoxides gradually results in its irreversible inactivation (Blum and Fridovich, 1985; Pigeolet *et al.*, 1990). To elucidate the mechanism of inactivation, GPx1 purified from human red blood cells (RBCs) was incubated with 1 mM H$_2$O$_2$ for 1 h at 37 °C. Such treatment resulted in an $\sim$40% loss of peroxidase activity and mass spectral analysis of tryptic peptides derived from inactivated GPx1 indicated that Sec at the active site was converted to DHA. The conversion is believed to be achieved via the oxidation of Sec by H$_2$O$_2$ followed by the loss of selenium oxide. This conversion reaction is similar to the reaction in which the synthesis of DHA-containing peptides is achieved by incorporating phenylselenocysteine into growing peptide chains via standard peptide synthesis procedures, followed by oxidative β-elimination of phenylselenol to yield a DHA at the desired position (Levengood and van der Donk, 2006). In the conversion of GPx Sec to DHA, the catalytic intermediate GPx1–SeOH itself can be the source of DHA. However, further oxidation to –SeO$_2$H will provide better opportunity for efficient β-elimination because –SeO$_2$H is a better leaving group than –SeOH. Treatment of GPx with H$_2$O$_2$ has been shown to generate GPx–SeO$_2$H (Wendel et al., 1975).

About 3% of the hemoglobin undergoes autoxidation every day and this constant flux of superoxide anion from hemoglobin autoxidation produces hydroperoxides (Winterbourn, 1985). Oxygen transport by RBCs is thus a substantial contributor to oxidative stress. To cope with the oxidative stress, RBCs are equipped both with metHb reductase, which converts metHb back to Hb to allow continued O$_2$ transport, and with various antioxidant enzymes that eliminate reactive oxygen species (ROS). RBCs contain a high concentration of CuZn–superoxide dismutase, which converts O$_2^{\bullet-}$ to H$_2$O$_2$. Enzymes responsible for the elimination of H$_2$O$_2$ in RBCs include catalase, GPx1, and several peroxiredoxins (Cho et al., 2010). Mammalian cell expresses five different GPx enzymes, but GPx1 is the only type present in RBCs. The average amount of GPx1 measured from 16 healthy adults was 6.7 μg/mg of RBC lysate protein, which corresponds to a concentration of 110 μM (Cho et al., 2010).

In addition to H$_2$O$_2$ molecules produced from hemoglobin autoxidation, H$_2$O$_2$ molecules produced by other tissues are also taken up by RBCs, thereby providing protection against oxidative damage (Winterbourn and Stern, 1987). The average life span of human RBCs is 120 days and each RBC travels $\sim$400 km during its life span. Given the limited capacity of RBCs to replace damaged proteins by *de novo* synthesis, inactivation of GPx1 would be expected to perturb the balance between oxidant production and elimination and thereby to accelerate the accumulation of ROS. RBC membranes are rich in polyunsaturated fatty acids, which are highly susceptible to oxidation by ROS. Lipid peroxidation and damage to membrane proteins result in deformation of RBCs and consequent impairment of their passage through small capillaries. The irreversible inactivation of GPx1 in RBCs appears to be associated with conversion of the Sec residue at the active site to DHA. The content of DHA–GPx1 in each RBC, which likely reflects total oxidative stress experienced by the cell during its lifetime, may provide a new type of health risk information.

It was possible to detect the DHA-containing peptide of GPx1 by mass spectrometry because it was obtained from the purified protein. Similar detection of DHA-containing GPx1 in cell extracts is not feasible, however. Moreover, quantification of the amount of the DHA-containing peptide by MS is a cumbersome process. To address this problem, we developed a method for specific labeling of DHA-containing proteins with biotin-

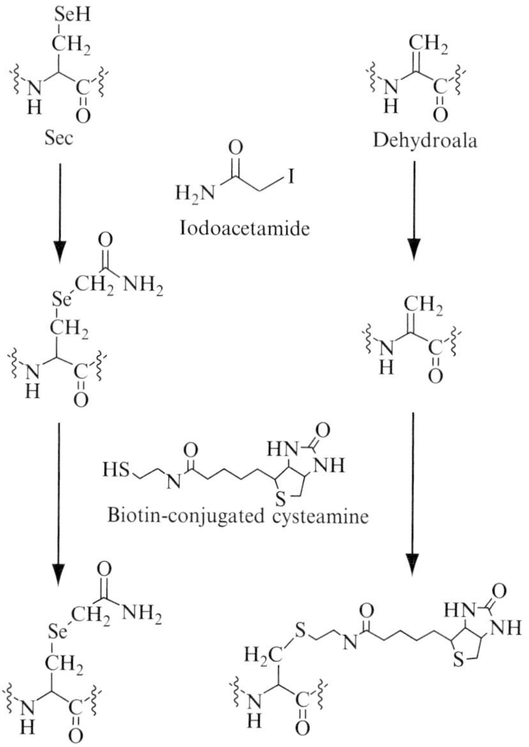

Figure 2.1 Chemical reactions underlying the biotinylation of GPx1-containing DHA. After alkylation of free SH and SeH groups by iodoacetamide, DHA residues are biotinylated with biotin-conjugated cysteamine.

conjugated cysteamine. This method, which relies on the well-established Michael-type addition of cysteamine to the DHA moiety (Fig. 2.1), involves the immunoprecipitation of GPx1 from cell extracts and the alkylation of Cys–SH and intact Sec–SeH in the precipitated proteins with iodoacetamide. The precipitated proteins are then incubated with biotin-conjugated cysteamine to biotinylate DHA-containing GPx1, which is detected by SDS–PAGE followed by blot analysis with HRP-conjugated streptavidin.

3. PREPARATION OF BIOTIN-CONJUGATED CYSTEAMINE

To synthesize biotin-conjugated cystamine, 150 μl of 50 mM cystamine·2HCl (Sigma-Aldrich; 1.14 mg dissolved in 100 μl of 0.1 M NaHCO$_3$, pH 9.0) was incubated for 1 h at 25 °C with 50 μl of 50 mM EZ-linked N-hydroxysuccinimide biotin (Pierce; 1.71 mg dissolved in 100 μl of DMF) with shaking. Biotin-conjugated cystamine was then reduced to a cysteamine derivative by the addition of 50 µl of 0.5 M dithiothreitol into the 200 µl reaction solution and incubating the reaction mixture for an additional 1 h. After drying the reaction mixture in a SpeedVac, the dried mixture was dissolved in 20 µl of a 3:1 mixture of 0.1% TFA and DMSO and subjected to a fractionation by HPLC on a C_{18} column (4.6 × 25 cm; Vydac). The column was eluted with 0.1% trifluoroacetic acid in distilled water for 10 min then with a 40 min linear gradient of 0–100% acetonitrile at a flow rate of 1 ml/min. Elution was monitored at 215 nm with a UV detector. A 500 µl of fractions corresponding to the peak of biotin-conjugated cysteamine eluted at 20.5–21 min was pooled, dried in a SpeedVac, and stored at −80 °C. The mass of biotin-conjugated cysteamine was verified by MALDI-TOF mass spectrometry.

4. Blot-Based Detection of DHA–GPx1 in RBCs

Whole-cell lysates of human RBCs (1 mg of proteins in 800 µl of phosphate-buffered saline (PBS)) were precleared by incubating with 50 µl of Protein G-Sepharose beads (Amersham Bioscience, Sweden, 16 µl of 50% bead slurry) for 2 h at 4 °C with gentle rotation to remove proteins that may bind nonspecifically. The precleared cell lysates were then incubated with mouse monoclonal anti-GPx1 antibody (Young-In Frontier, Seoul, Korea, 3 µg in 3 µl PBS) for overnight at 4 °C with gentle rotation. Protein G-Sepharose beads (16 µl of 50% bead slurry) were added and the mixture was incubated for additional 4 h at 25 °C with gentle rotation. Immunoprecipitated GPx1complex was pelleted by microcentrifugation for 30 s at 25 °C. The pellets were washed three times with 1 ml of ice-cold PBS, and the bead-bound proteins were eluted from the beads by incubating in 100 µl of 10 mM Tris–Cl (pH 6.8) containing 1% SDS and 4 M urea for 5 min at 25 °C. This elution step was repeated four times and the resulting eluants were combined. To the 400 µl of combined eluants, 40 µl of 0.4 M iodoacetamide was added and incubated for 30 min at room temperature to alkylate sulfhydryl and selenol groups. Proteins were then precipitated by adding 50 µl of 100% trichloroacetic acid, and the precipitated proteins were washed twice with 1 ml of ice-cold acetone, and dried in a SpeedVac. Dried proteins were suspended by a brief sonication in 90 µl of 0.1 M $NaHCO_3$ (pH 10.0) containing 1% SDS followed by addition of 10 µl of 5 mM biotin-conjugated cysteamine. The biotinylation reaction mixture was kept at 37 °C for 18 h. After the incubation, 20 µl of the biotinylation reaction mixture was mixed with 5 µl of 5× SDS sample buffer, boiled at 100 °C for 5 min, and fractionated on a 14% SDS–polyacrylamide gel. The separated proteins were transferred to a nitrocellulose membrane for blot analysis with

HRP-conjugated streptavidin (Pierce) or for immunoblot analysis with rabbit polyclonal antibodies to human GPx1 (Young-In Frontier).

5. EFFECTS OF OXIDATIVE STRESS ON THE FORMATION OF DHA–GPx1 IN RBCs

We applied the DHA detection method to determine whether the conversion of Sec to DHA in GPx1 occurs with aging of RBCs. The density of RBCs increases with aging. We, therefore, fractionated human RBCs from healthy adult donors by centrifugation on a discontinuous density gradient of Percoll to obtain cells of four different mean ages (Cho et al., 2010). The activity of G6PDH, a marker of aging in RBCs, decreased gradually with aging, whereas the abundance of G6PDH as determined by immunoblot analysis remained constant (Fig. 2.2A). The activity of GPx1 also decreased with aging, with the amount of GPx1 as determined by immunoblot analysis remaining constant or increasing slightly (Fig. 2.2B). To determine whether

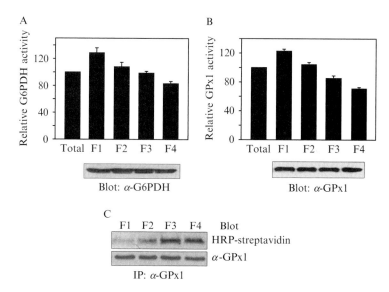

Figure 2.2 Effects of *in vivo* RBC aging on the formation of DHA–GPx1 in RBCs. Fresh RBCs obtained from a healthy human adult were separated into four fractions on the basis of their age (F1–F4 for youngest to oldest, respectively) by centrifugation on a discontinuous density gradient of Percoll. The activities of GPx1 (right panels) in each fraction were measured and normalized by the corresponding value for RBCs before fractionation. GPx1 was immunoprecipitated from the lysate of each of the F1–F4 fractions and analyzed for DHA content by sequential reaction with iodoacetamide and biotin-conjugated cysteamine. Equal loading of proteins was confirmed by immunoblot analysis with antibodies to GPx1.

the loss of GPx1 activity during aging is accompanied by formation of DHA, GPx1 was immunoprecipitated from the lysates of different mean ages and subjected to blot analysis using biotin-conjugated cysteamine. Indeed, the blot intensity of the band recognized by HRP-conjugated streptavidin increased gradually with aging of RBCs (Fig. 2.2C), indicating that the Sec residue of GPx1 is converted to DHA in a time-dependent manner during exposure to the mild oxidative stress resulting from heme autoxidation.

The combined concentration of reduced and oxidized forms of glutathione (GSH + GSSG) is ~ 2 mM in RBCs (Mueller et al., 1997). A decrease in the GSH:GSSG ratio would be expected to increase the production of DHA–GPx1 because GPx1–SeOH would have a longer time to lose H_2SeO before it reacts with GSH or to react with H_2O_2. The GSH:GSSG ratio in RBCs has been shown to decrease with age (Imanishi et al., 1985). Given that maintenance of a high GSH:GSSG ratio requires NADPH, inactivation of G6PDH, a key enzyme in the pentose phosphate pathway, may be a cause of the decrease in the GSH:GSSG ratio associated with aging.

A variety of drugs including dapsone, sulfonamides, phenacetin, and primaquine as well as industrial chemicals such as aniline induce hemolytic anemia (Harrison and Jollow, 1986). These arylamine compounds are metabolized in the liver, and the resulting N-hydroxyarylamines react with oxyHb by reducing oxygen to the superoxide anion, thus generating additional ROS in RBCs. To examine the effects of such extra oxidative stress produced internally by environmental chemicals, we incubated a 50% hematocrit of RBCs with 0.2 mM N-phenylhydroxylamine. The DHA content of GPx1 increased on exposure of RBCs to N-phenylhydroxylamine (Fig. 2.3A).

H_2O_2 passes through the plasma membrane of RBCs, and antioxidant enzymes in RBCs eliminate ROS that originate from the external environment and thereby protect other cells from oxidative injury induced by phagocytic cells or toxins (Winterbourn and Stern, 1987). To examine the effects of extracellular H_2O_2 on RBCs, we added various amounts of glucose oxidase to these cells (50% hematocrit) suspended in Dulbecco's modified Eagle's medium (DMEM) containing a high concentration of glucose. Glucose oxidase catalyzes the oxidation β-D-glucose to D-glucono-δ-lactone with concomitant reduction of O_2 to H_2O_2. Incubation of RBCs with glucose oxidase at 37 °C for 3 h resulted in concentration-dependent increase in the DHA content of GPx1 (Fig. 2.3C).

6. Concluding Remarks

DHA has been detected in the position corresponding to the serine, cysteine, and Sec residues of various proteins. The sulfur and selenium atoms of cysteine and Sec residues of many proteins are oxidized to various

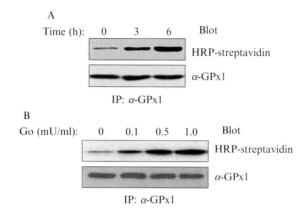

Figure 2.3 Effects on the formation of DHA–GPx1 in RBCs of internally elevated ROS (A) and externally produced ROS (B). (A) N-Phenylhydroxylamine (200 μM) was added to a 50% hematocrit of RBCs in DMEM containing a high glucose concentration (4500 mg/l). After incubation for various times (0, 3, or 6 h) with gentle shaking at 37 °C, the RBCs were lysed and subjected to the determination of DHA–GPx1. (B) Various amounts (0, 0.1, 0.5, or 1 mU) of glucose oxidase (GO) were added to 1 ml of RBCs at a 50% hematocrit in DMEM containing a high concentration (4500 mg/l) of glucose. After incubation for 3 h with gentle shaking at 37 °C, the RBCs were lysed and subjected to the determination of DHA–GPx1. Equal loading of proteins in panels (A) and (B) was confirmed by immunoblot analysis with antibodies to GPx1.

state of oxidation under oxidative stress, and the oxidized Secs appear to readily undergo β-elimination to produce DHA. Oxidized cysteines also undergo β-elimination, though not as readily as oxidized Secs. In order to visualize DHA-containing protein after SDS gel electrophoresis, we devised a blot-based method that depends on specific addition of biotin-conjugated cystamine to DHA moiety followed by detection of biotinylated proteins with the use of HRP-conjugated streptavidin. It was necessary to immunoprecipitate GPx1 before labeling with biotin-conjugated cysteamine in order to measure DHA specifically in GPx1, because many other proteins also contain DHA.

We applied this method to determine whether the conversion of Sec to DHA in GPx1 occurs with aging of RBCs. The intensity of DHA–GPx1 increased gradually with aging of RBCs, indicating that the Sec residue of GPx1 is converted to DHA in a time-dependent manner during exposure to the mild oxidative stress resulting from heme autoxidation. We found that the Sec to DHA conversion also occurred in RBCs exposed to H_2O_2 generated either externally by glucose oxidase or internally as a result of aniline-induced Hb autoxidation.

During 120 days of life span, RBCs protect other tissues against oxidative damage by taking up and metabolizing peroxides (Winterbourn and

Stern, 1987). Given that the rate of DHA–GPx1 accumulation in RBCs depends on peroxide flux, the content of DHA–GPx1 in each RBC likely reflects total oxidative stress experienced by the cell while traveling ~400 km during its lifetime. In addition to genetic polymorphisms, exposure to toxic chemicals such as aniline and sulfonamides, pathological conditions such as diabetes and local inflammation, and an insufficient intake of antioxidants are all expected to affect the rate of GPx1 inactivation. In this regard, among the several antioxidant enzymes in RBCs, the activity of GPx1 was shown to be most influenced by lifestyle and environmental factors such as the use of dietary supplements and smoking habit: GPx1 activity was higher among users of dietary supplements and lower among smokers (Andersen et al., 1997). GPx1 activity in RBCs has also been proposed as a strong predictor of cardiovascular risk, which is associated with oxidative stress (Blankenberg et al., 2003). Therefore, DHA–GPx1 in RBCs might be a suitable surrogate marker for evaluation of oxidative stress in the body. Our blot-based method for the detection of DHA–GPx1 would be very useful for evaluation of such stress. In addition, similar blot detection method can be devised for other proteins for which immunoprecipitating antibodies are available.

ACKNOWLEDGMENT

This study was supported by Korean Science and Engineering Foundation grants (National Honor Scientist Program grant 2006-05106 and grant FPR0502-470 of the 21C Frontier Functional Proteomics Projects) to S. G. R.

REFERENCES

Andersen, H. R., Nielsen, J. B., Nielsen, F., and Grandjean, P. (1997). Antioxidative enzyme activities in human erythrocytes. *Clin. Chem.* **43,** 562–568.

Bartone, N. A., Bentley, J. D., and Maclaren, J. A. (1991). Determination of dehydroalanine residues in proteins and peptides: An improved method. *J. Protein Chem.* **10,** 603–607.

Bernardes, G. J., Chalker, J. M., Errey, J. C., and Davis, B. G. (2008). Facile conversion of cysteine and alkyl cysteines to dehydroalanine on protein surfaces: Versatile and switchable access to functionalized proteins. *J. Am. Chem. Soc.* **130,** 5052–5053.

Blankenberg, S., Rupprecht, H. J., Bickel, C., Torzewski, M., Hafner, G., Tiret, L., Smieja, M., Cambien, F., Meyer, J., and Lackner, K. J. (2003). Glutathione peroxidase 1 activity and cardiovascular events in patients with coronary artery disease. *N. Engl. J. Med.* **349,** 1605–1613.

Blum, J., and Fridovich, I. (1985). Inactivation of glutathione peroxidase by superoxide radical. *Arch. Biochem. Biophys.* **240,** 500–508.

Cho, C.-S., Lee, S., Lee, G., Woo, H., Choi, E.-J., and Rhee, S. (2010). Irreversible inactivation of glutathione peroxidase 1 and reversible inactivation of peroxiredoxin II by H_2O_2 in red blood cells. *Antioxid. Redox Signal.* **12,** 1235–1246.

Consevage, M. W., and Phillips, A. T. (1985). Presence and quantity of dehydroalanine in histidine ammonia-lyase from *Pseudomonas putida*. *Biochemistry* **24**, 301–308.
Flohe, L., Gunzler, W. A., and Schock, H. H. (1973). Glutathione peroxidase: A selenoenzyme. *FEBS Lett.* **32**, 132–134.
Gavaret, J. M., Nunez, J., and Cahnmann, H. J. (1980). Formation of dehydroalanine residues during thyroid hormone synthesis in thyroglobulin. *J. Biol. Chem.* **255**, 5281–5285.
Gross, E., and Kiltz, H. H. (1973). The number and nature of α,β-unsaturated amino acids in subtilin. *Biochem. Biophys. Res. Commun.* **50**, 559–565.
Gross, E., and Morell, J. L. (1967). The presence of dehydroalanine in the antibiotic nisin and its relationship to activity. *J. Am. Chem. Soc.* **89**, 2791–2792.
Harrison, J. H. Jr., and Jollow, D. J. (1986). Role of aniline metabolites in aniline-induced hemolytic anemia. *J. Pharmacol. Exp. Ther.* **238**, 1045–1054.
Imanishi, H., Nakai, T., Abe, T., and Takino, T. (1985). Glutathione metabolism in red cell aging. *Mech. Ageing Dev.* **32**, 57–62.
Jones, A. J., Helmerhorst, E., and Stokes, G. B. (1983). The formation of dehydroalanine residues in alkali-treated insulin and oxidized glutathione. A nuclear-magnetic-resonance study. *Biochem. J.* **211**, 499–502.
Knight, Z. A., Schilling, B., Row, R. H., Kenski, D. M., Gibson, B. W., and Shokat, K. M. (2003). Phosphospecific proteolysis for mapping sites of protein phosphorylation. *Nat. Biotechnol.* **21**, 1047–1054.
Krief, A. (1987). *In* "The Chemistry of Organic Selenium and Tellurium Compounds," (S. Patai and Z. Rappoport, eds.), p. 675. John Wiley and Sons, Chichester.
Levengood, M. R., and van der Donk, W. A. (2006). Dehydroalanine-containing peptides: Preparation from phenylselenocysteine and utility in convergent ligation strategies. *Nat. Protoc.* **1**, 3001–3010.
Ma, S., Caprioli, R. M., Hill, K. E., and Burk, R. F. (2003). Loss of selenium from selenoproteins: Conversion of selenocysteine to dehydroalanine *in vitro*. *J. Am. Soc. Mass Spectrom.* **14**, 593–600.
Masri, M. S., and Friedman, M. (1982). Transformation of dehydroalanine residues in casein to S-beta-(2-pyridylethyl)-L-cysteine side chains. *Biochem. Biophys. Res. Commun.* **104**, 321–325.
Mega, T., Nakamura, N., and Ikenaka, T. (1990). Modifications of substituted seryl and threonyl residues in phosphopeptides and a polysialoglycoprotein by beta-elimination and nucleophile additions. *J. Biochem.* **107**, 68–72.
Mueller, S., Riedel, H. D., and Stremmel, W. (1997). Direct evidence for catalase as the predominant H2O2-removing enzyme in human erythrocytes. *Blood* **90**, 4973–4978.
Ohmiya, Y., Hayashi, H., Kondo, T., and Kondo, Y. (1990). Location of dehydroalanine residues in the amino acid sequence of bovine thyroglobulin. Identification of "donor" tyrosine sites for hormonogenesis in thyroglobulin. *J. Biol. Chem.* **265**, 9066–9071.
Pigeolet, E., Corbisier, P., Houbion, A., Lambert, D., Michiels, C., Raes, M., Zachary, M. D., and Remacle, J. (1990). Glutathione peroxidase, superoxide dismutase, and catalase inactivation by peroxides and oxygen derived free radicals. *Mech. Ageing Dev.* **51**, 283–297.
Schuster, B., and Retey, J. (1995). The mechanism of action of phenylalanine ammonia-lyase: The role of prosthetic dehydroalanine. *Proc. Natl. Acad. Sci. USA* **92**, 8433–8437.
Strumeyer, D. H., White, W. N., and Koshland, D. E. Jr. (1963). Role of serine in chymotrypsin action. Conversion of the active serine to dehydroalanine. *Proc. Natl. Acad. Sci. USA* **50**, 931–935.
Wendel, A., Pilz, W., Ladenstein, R., Sawatzki, G., and Weser, U. (1975). Substrate-induced redox change of selenium in glutathione peroxidase studied by X-ray photoelectron spectroscopy. *Biochim. Biophys. Acta* **377**, 211–215.

Wickner, R. B. (1969). Dehydroalanine in histidine ammonia lyase. *J. Biol. Chem.* **244**, 6550–6552.

Winterbourn, C. C. (1985). Free-radical production and oxidative reactions of hemoglobin. *Environ. Health Perspect.* **64**, 321–330.

Winterbourn, C. C., and Stern, A. (1987). Human red cells scavenge extracellular hydrogen peroxide and inhibit formation of hypochlorous acid and hydroxyl radical. *J. Clin. Invest.* **80**, 1486–1491.

CHAPTER THREE

ANALYSIS OF THE REDOX REGULATION OF PROTEIN TYROSINE PHOSPHATASE SUPERFAMILY MEMBERS UTILIZING A CYSTEINYL-LABELING ASSAY

Benoit Boivin *and* Nicholas K. Tonks

Contents

1. Introduction — 36
2. Active-Site Structure, Catalysis, and Oxidation — 37
3. Detection Methods — 39
4. General Principle of the Assay — 40
5. Solutions — 42
6. Preparation of the Lysis Buffer — 42
7. Preparation of the Hypoxic Glove Box — 43
8. Preparation of Cell Lysates — 43
9. Cysteinyl-Labeling Assay — 44
10. Acute Stimulus-Induced Reversible Oxidation of PTPs — 44
11. Perspectives — 45
12. Conclusion — 48
References — 48

Abstract

The catalytic activity of protein tyrosine phosphatase (PTP) superfamily members is regulated by the reversible oxidation of their invariant catalytic Cys residue *in vivo*. Transient and specific regulation of PTP activity by reactive oxygen species (ROS) attenuates dephosphorylation and, thereby, promotes phosphorylation, hence facilitating signal transduction. We have recently developed a modified cysteinyl-labeling assay [Boivin, B., Zhang, S., Arbiser, J. L., Zhang, Z. Y., and Tonks, N. K. (2008). *Proc. Natl. Acad. Sci. USA* 105, 9959–9964.] that showed broad selectivity in detecting reversible oxidation of members from different PTP subclasses in platelet-derived growth factor (PDGF)-BB overexpressing cells. Herein, we applied this assay, which utilizes the unique chemistry of the invariant catalytic Cys residue to enrich and identify PTPs that are reversibly oxidized upon

Cold Spring Harbor Laboratory, Cold Spring Harbor, New York, USA acute growth factor stimulation. Performing the cysteinyl-labeling assay with Rat-1 fibroblasts enabled us to capture both PTEN and SHP-2 as a consequence to acute PDGF-BB stimulation. Given the ability of this assay to detect reversible oxidation of a broad array of members of the PTP family, we anticipate that it should permit profiling of the entire ROS-regulated PTPome in a wide array of signaling paradigms.

1. INTRODUCTION

Nearly 20 years separates the two discoveries leading to a redox-regulated view of phosphorylation-dependent signal transduction. A first hint that dynamic sulfhydryl oxidation could function as a mechanism controlling biological functions came from observations by Czech et al. in a 1974 study reporting that a component involved in glucose transport was regulated by oxidation following insulin receptor activation (Czech et al., 1974). Two decades later, a seminal study from Sundaresan et al. revealed that the dynamic production of oxidants, such as hydrogen peroxide, exerted a new tier of regulation over tyrosine phosphorylation in growth factor signaling (Sundaresan et al., 1995). The ensuing search for mechanisms underlying the dynamic redox regulation of phosphorylation-dependent signaling led to enzymes known to regulate phosphoryl hydrolysis and to be sensitive to oxidation, i.e. the protein tyrosine phosphatases (PTPs) (Sullivan et al., 1994; Tonks et al., 1988).

The sequencing of the human genome revealed a comparable number of PTP and protein tyrosine kinase (PTK) genes, illustrating a similar level of complexity in the two families of enzymes. It is now clear that PTPs and PTKs act in a synchronized, complementary manner to regulate tyrosine phosphorylation-dependent signaling. This view is supported by mathematical models suggesting that PTKs establish the amplitude of the signal whereas PTPs control the rate and the duration of the response (Tonks, 2006). Members of the PTP superfamily display structural diversity, but are all characterized by a common signature motif, $HC(X)_5R$, in which the Cys residue is the nucleophile at the core of the phosphoryl hydrolysis mechanism. Part of the structural diversity imparted on PTPs is reflected in their phosphoamino acid specificity, dividing the family in two broad classes: the 37 classical PTPs dephosphorylate Tyr residues, whereas the 65 dual-specificity phosphatases (DSPs) may also dephosphorylate Ser/Thr residues, inositol phospholipids, and mRNA (Deshpande et al., 1999; Tonks, 2006). PTPs are membrane spanning or cytosolic proteins possessing noncatalytic flanking motifs that control activity either by a direct interaction with the active site or by controlling subcellular localization and substrate availability (e.g. kinase-interaction motif (KIM), Src-homology-2 (SH2), PDZ-binding sequence, and pleckstrin homology (PH)) (Tonks, 2006). In addition to the various non-catalytic motifs, PTP activity and phosphoryl hydrolysis are also directly perturbed by oxidation of the sulfhydryl side chain of the catalytic Cys residue. Indeed, the architecture of the active-site cleft confers an unusually low pK_a to the catalytic Cys residue, with the result that it is a thiolate ion at physiological pH (Denu and Dixon, 1998). Ionization of the sulfur atom of the cysteinyl side chain is the keystone of its cellular reactivity as a nucleophile toward phosphoamino acid residues, toward reactive oxygen species (ROS) or, as we will discuss later, toward certain alkylating agents (Fig. 3.1A).

2. Active-Site Structure, Catalysis, and Oxidation

The general architecture of the active site is conserved among PTPs and DSPs. The PTP active site forms a crevice on the molecular surface of the enzyme, with the signature motif, $HC(X)_5R$, located on the PTP loop at the base (Barford et al., 1994). The walls of the active-site cleft are formed by a "pTyr loop" that determines depth of the cleft and thus the nature of the phosphoamino acid residue being dephosphorylated; a "Q loop" containing a conserved Gln residue involved in coordinating a water molecule in the hydrolysis of the cysteinyl–phosphate intermediate; and a "WPD loop" containing an essential invariant Asp residue involved in the protonation of the Tyr leaving groups (Barford et al., 1995). The architecture of the conserved PTP signature motif creates an environment that lowers the pK_a of the conserved catalytic Cys residue. The positive dipole from the adjacent α-helix, the peptide dipole induced by the H bond from the conserved His residue toward the carbonyl oxygen of the active-site Cys, the amides of the five residues following the conserved Cys residue, and the conserved Arg residue itself all contribute to lowering its pK_a. In addition to lowering the pK_a of the conserved Cys residue, the network of H bonds from the polar groups on neighboring conserved residues orients the PTP loop, and the charged sulfur atom to act as a nucleophile on a bound phosphorylated substrate (Barford et al., 1994; Denu and Dixon, 1998). After enzyme–substrate complex formation, PTP-mediated catalysis proceeds via a two-step mechanism. In the initial step, the sulfur atom of the thiolate ion of the conserved Cys residue performs a nucleophilic attack on the phosphorus center of the substrate. This step is coupled to the expulsion of the Tyr leaving group upon protonation by the conserved aspartic acid residue from the "WPD loop." Finally, hydrolysis of the phosphoenzyme intermediate occurs by attack of a water molecule and the deprotonated Asp acid residue now acting as a general base.

ROS are now established second messenger molecules and have been shown to act directly on several aspects of signal transduction, including

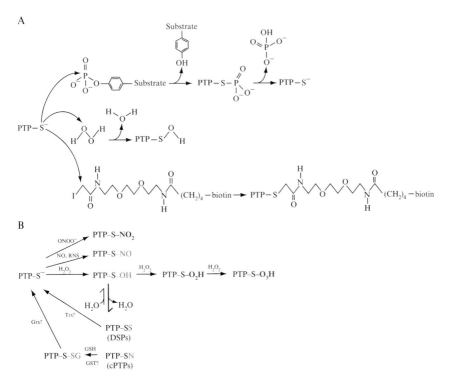

Figure 3.1 Redox regulation of PTP. (A) Schematic mechanism of PTP reactivity toward phosphoamino acid residues, hydrogen peroxide, and a biotinylated iodoacetyl-polyethylene glycol (IAP) probe. (B) Schematic outline of PTP regulation by oxidation and nitration. In resting cells, the architecture of the active site confers a low pK_a to PTP catalytic Cys residues indicating that the Cys residue is a thiolate ion at physiological pH. Ionization of the sulfur atom of the cysteinyl side chain makes it a good nucleophile toward ROS and RNS. Hence, the transient production of ROS and RNS in the vicinity of the PTP following growth factor stimulation, favors these nucleophilic reactions with concomitant inactivation of PTP function. Briefly, oxidation of the thiolate ion by hydrogen peroxide (H_2O_2), and nitration by nitric oxide (NO), leads to the formation of transitory states (reversible cysteinyl modifications are represented in light grey). However, transient oxidation only occurs at low H_2O_2 levels. This leads to the formation of sulfenic acid (SOH), as a primary transitory state whereas higher H_2O_2 levels lead to the formation of sulfonic acid (SO_2H), and sulfinic acid (SO_3H), which are essentially irreversible modifications (irreversible cysteinyl modifications are represented in bold). Interestingly, it has also been shown that sulfenic acids undergo a rapid condensation reaction to form secondary transitory states such as cyclic sulfenamides (SN) for classical PTPs and intramolecular disulfides (SS) for DSPs. These secondary transitory states, as well as the direct nitrosylation of PTP active-site Cys residues (SNO), protect PTPs from undergoing further irreversible oxidation reactions to SO_2H and SO_3H forms. Irreversible PTP inactivation also occurs upon nitration of the catalytic Cys residues with peroxynitrite (SNO_2). Ultimately, the formation of these secondary transitory states allows the redox cycle to be completed by cellular reductants, such as glutathione, and PTPs to be reduced back to their active form by thioredoxins or glutaredoxins.

PTPs, protein kinases, and transcription factors (Rhee *et al.*, 2003; Valko *et al.*, 2007). Regarding PTPs, the thiolate ion of the catalytic Cys residue readily reacts with oxidants *in vitro* and *in vivo* resulting in the transient inactivation of the enzyme and *de facto* facilitation of phosphotransfer events. PTP inactivation occurs *in vitro* following incubation with superoxide (Barrett *et al.*, 1999), hydrogen peroxide (Denu and Tanner, 1998; Lee *et al.*, 1998), peroxynitrite (Takakura *et al.*, 1999), and nitric oxide (Chen *et al.*, 2008). In addition, PTP oxidation is observed in cells in response to exogenous hydrogen peroxide (Sullivan *et al.*, 1994), upon NADPH oxidase activation by several growth factor receptors (Boivin *et al.*, 2008; Lee *et al.*, 1998; Mahadev *et al.*, 2001; Meng *et al.*, 2002), and in tumor lines (Lou *et al.*, 2007) (Fig. 3.1B). Interestingly, PTP oxidation has been shown to occur exclusively on the catalytic Cys residue (Lou *et al.*, 2007). The reversibility of oxidation is facilitated by the rapid conversion of sulfenic acid to a cyclic sulfenamide in the classical PTPs (Salmeen *et al.*, 2003; Yang *et al.*, 2007a) or to an intramolecular disulfide with a vicinal Cys residue in DSPs (Buhrman *et al.*, 2005; Salmeen and Barford, 2005). These transitory states, as well as PTP cysteinyl nitrosylation (Chen *et al.*, 2008; Li and Whorton, 2003), protect the enzyme from irreversible oxidation and allow the redox cycle regulating PTP activity to be completed by cellular reductants such as glutathione, thioredoxin, or glutaredoxin (Barrett *et al.*, 1999; Lee *et al.*, 1998; Li and Whorton, 2003).

Although several oxidants have been shown to inactivate PTPs *in vitro*, it appears that hydrogen peroxide is the relevant second messenger ROS molecule *in vivo*. NADPH oxidases generate a highly reactive superoxide anion in the extracellular milieu or inside endosomes, via a one electron reduction of molecular oxygen. Although superoxide reacts toward protein thiols faster than hydrogen peroxide, the latter possesses greater stability and diffusibility to act on its targets *in vivo* (Brown and Griendling, 2009). Supporting this concept, studies have shown that decomposition of hydrogen peroxide to water and oxygen using catalase greatly attenuated growth factor signaling (Sundaresan *et al.*, 1995). Furthermore, inhibition of superoxide dismutase (SOD) 1 (i.e., inhibiting production of hydrogen peroxide) impaired the reversible oxidation of PTP1B upon EGFR activation in A431 epithelial carcinoma cells (Juarez *et al.*, 2008). Thus, these biophysical and biochemical observations suggest that hydrogen peroxide is a second messenger molecule that diffuses into the intracellular milieu, inhibits PTPs, and enhances phosphorylation-dependent signaling *in vivo* (Rhee *et al.*, 2003; Xu *et al.*, 2002).

3. Detection Methods

Several methods have been utilized to measure cysteine oxidation. Our interest being PTPs, reversible oxidation of this superfamily has been studied using broad screening approaches that address the family as a whole, as well as methods that are specific for individual PTPs. PTP-specific methods, such as the electrophoretic mobility shift assay in nonreducing conditions (Kamata *et al.*, 2005; Kwon *et al.*, 2004; Leslie *et al.*, 2003) or a spectrophotometric assay using sulfenic acid reacting compounds (Denu and Tanner, 1998) and PTP-specific analysis by mass spectrometry (Lou *et al.*, 2007), allow an investigator to focus on PTPs that are known to be involved in a given pathway. However, other approaches, such as the activity-based in-gel phosphatase assay (Meng *et al.*, 2002, 2004), antibodies directed against terminally oxidized PTP active-site Cys residues (Persson *et al.*, 2004), reactivity toward sulfhydryl-directed compounds (Boivin *et al.*, 2008; Kwon *et al.*, 2005; Lee *et al.*, 1998), or suicide substrates (Boivin *et al.*, 2008; Kumar *et al.*, 2006), allow one to screen for novel targets of ROS signaling.

4. General Principle of the Assay

The aim of this assay was to develop a chemical approach that allows detection of all reversibly oxidized PTP superfamily members. To achieve this goal, we devised a three-step strategy in which all, including less abundant, PTPs can be enriched and detected in a gain of signal readout that expands the dynamic range of detection (i.e., by measuring the reversibly oxidized PTPs rather than the reduction in overall PTP reactivity that accompanies oxidation and inhibition Wu and Terada, 2006). In addition, this approach should be adaptable for proteomic screening or *in situ* visualization of oxidized PTPs. Contrary to existing techniques, the conditions set up for our cysteinyl-labeling assay take advantage of the unique characteristics of the PTP active site. We utilized a mildly acidic labeling buffer (pH 5.5) and opted for a nondenaturing strategy, to exploit fully the active-site architecture, and low pK_a of the invariant catalytic Cys residue of PTPs as the basis for a labeling strategy. Under these conditions, the thiolate ion of the catalytic Cys residue acts as a potent nucleophile and nonspecific labeling of other Cys (pK_a 8.5), His, or Met residues is limited (Chung and Lewis, 1986; Jullien and Garel, 1981; Lou *et al.*, 2007). In addition, keeping the PTPs in their folded, active conformation minimizes potential labeling of buried residues.

The three-step strategy devised for our cysteinyl-labeling assay specifically biotinylates reversibly oxidized catalytic cysteines, allowing those PTPs to be enriched by streptavidin pull-down (Fig. 3.2). Following a physiological stimulus, the first step consists of cellular lysis performed with a pH 5.5 buffer containing iodoacetic acid (IAA). This leads to alkylation of the thiol of the active-site cysteine in those PTPs that were

Analysis of the Redox-Regulated PTPs

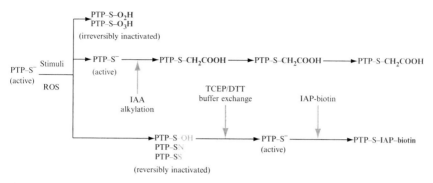

Figure 3.2 Schematic outline of the cysteinyl-labeling assay. In resting cells, active PTPs possess an ionized sulfur atom at the catalytic Cys residue. This results from the enzyme's unique active-site architecture which lowers the pK_a of this Cys residue, and is essential for its reactivity toward phosphoamino acid residues, ROS, or certain sulfhydryl-reactive compounds. Following a ROS-promoting stimulus, specific PTPs are inactivated by oxidation. Reversible oxidation [sulfenic acids (SOH), cyclic sulfenamides (SN), and intramolecular disulfides (SS)] are depicted in light grey, whereas irreversible oxidation [sulfenic acid (SO$_2$H) and sulfinic acid (SO$_3$H)] are depicted in bold. In the first step of the cysteinyl-labeling assay, cellular lysis is performed with a degassed buffer at pH 5.5 containing iodoacetic acid (IAA). PTPs that remained in their active state following cell stimulation react and are terminally inactivated by IAA, whereas oxidized PTPs are protected from alkylation. The second step consists of a buffer exchange to remove excess IAA from the lysate and to reduce the oxidized Cys residues back to their thiolate form using TCEP [tris(2-carboxyethyl)phosphine] or dithiothreitol (DTT). In the third step of the assay, a biotinylated sulphydryl-reactive probe [iodoacetyl-polyethylene oxide (IAP)] is added to the reduced lysate to react with reactivated PTPs. Purification by streptavidin pull-down and immunoblotting permits identification of ROS-targeted PTPs.

untouched by ROS, leaving the oxidized PTPs protected from alkylation. As a second step, the cell extract is then cleared of the alkylating agent by rapid desalting on size-exclusion chromatography and supplemented with a reducing agent. This is crucial in reducing the reversibly oxidized and inactivated PTPs, and in reactivating the nucleophilic potential of their active-site Cys residues. Finally, the third step consists of a nucleophilic attack of the active-site cysteinyl side chain of PTPs toward biotinylated sulfhydryl-reactive compounds or biotinylated suicide substrates. We have successfully employed iodoacetyl-polyethylene glycol (IAP) and α-bromobenzylphosphonate (BBP) (Kumar et al., 2004) as sulfhydryl-reactive compound and biotinylated suicide substrate, respectively, for this purpose. Once this second labeling step is completed, the biotinylated proteins are enriched by streptavidin pull-down, subjected to SDS gel electrophoresis, and identified by immunoblotting with PTP-specific or anti-biotin antibodies.

5. Solutions

Lysis buffer: 50 mM sodium acetate, 150 mM sodium chloride, 10% glycerol (v/v), and degassed ddH$_2$O. *Refer to the instruction section to prepare the lysis buffer.*
IAA stock: 2 M IAA in ethanol. *Prepare immediately prior to use. Protect from light.*
SOD stock: 250 U/µl in dH$_2$O. *Aliquot in 1.5 ml amber tubes, store at $-20\,°C$.*
Catalase stock: 125 U/µl in dH$_2$O. *Aliquot in 1.5 ml amber tubes, store at $-20\,°C$.*
IAP-biotin stock: 25 mM in dH$_2$O

6. Preparation of the Lysis Buffer

Air oxygen dissolved in the lysis buffer is sufficient to cause spontaneous oxidation and inactivation of PTPs in the absence of reducing agents (Tonks *et al.*, 1988; Zhu *et al.*, 2001). Hence, it is crucial for the success of this assay to degas the lysis buffer thoroughly prior to undertaking the assay. There are several ways to degas buffers. We have opted for a boiling and vacuuming step. Boil double distilled water (150–200 ml) for a one-hour period in a glass bottle containing glass beads (3-mm glass beads, cat. no. 11-312A, Fisher Scientific). Once dissolved oxygen is partly released, close hermetically and allow to cool down. Carefully transfer 80 ml of this partially degassed ddH$_2$O into a Büchner flask, add sodium acetate (final, 50 mM), sodium chloride (final, 150 mM), and 10 ml glycerol (i.e., calculated for a final volume of 100 ml). Allow to mix under mild vacuuming conditions (i.e., 5–10 mmHg) until the components are dissolved. The vacuuming at this step is aimed to prevent regassing and should be mild in order to avoid evaporation of your buffer: we usually allow the solution to dissolve with gentle stirring overnight at 4 °C. Add 10 ml of a 10% nonoxidized NP-40 solution (Surfact-Amps Nonidet P-40, cat. no. 28324, Thermo Scientific) and rapidly adjust the pH to 5.5 using 2 N HCl. We use nitrogen packaged NP-40 ampoules to avoid detergent-induced PTP oxidation by peroxides contaminants. Then, transfer 20–25 ml of lysis buffer to a precooled 50-ml Büchner flask and degas with mild stirring for 20 min on ice and 30–60 min at room temperature at 35–40 mmHg. Once the degassing is over (i.e., when bubbles stop surfacing, and when bubbles at the interface have vanished) and the buffer presents a flat surface for a 10 min period, close the flask hermetically using a Hoffman open-side tubing clamp (Fisher Scientific) and transfer it into an argon-filled anaerobic glove box (Fig. 3.3, hypoxic glove box, COY Laboratory products Inc.). Immediately before use, transfer 10 ml of the lysis buffer in a 15 ml foil-covered tube and supplement it with 2500 U catalase (final, 250 U/mL)

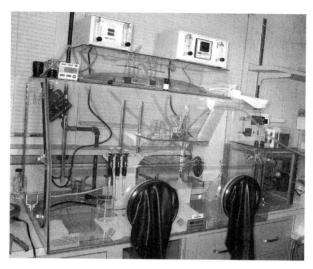

Figure 3.3 The hypoxic glove box used to generate cell lysate in an anaerobic environment.

(Cat. no. 219261, Calbiochem), 1250 U SOD (final, 125 U/mL) (Cat. no. 574593, Calbiochem), leupeptin (final, 5 μg/mL), aprotinin (final, 5 μg/mL), and of freshly prepared IAA (final, 10 mM) (Cat. no. 35603, Thermo Scientific).

7. PREPARATION OF THE HYPOXIC GLOVE BOX

As a first step, place all the necessary equipment in the box (e.g., microcentrifuge tube rack, labeled 1.5-ml microcentrifuge tubes, pipets, pipet tips, cell scrapers, and beakers). Equilibrate the anaerobic glove box station with ultrapure argon gas while the lysis buffer is being prepared. Once the 20–25 ml of lysis buffer is completely degassed, place the sealed Büchner flask in the hypoxic glove box and complete it as detailed above. All steps, including the completion of the lysis buffer and the cellular lysis, from the moment the cells are scraped off their culture dish until the lysates are being transferred to the shaker to complete the lysis and alkylation, are performed under strict anaerobic conditions in the argon-equilibrated glove box.

8. PREPARATION OF CELL LYSATES

Prior to performing this assay, low-passage, healthy cells grown in 10% serum, low-glucose DMEM are serum starved in low-glucose DMEM without phenol red (Cat. no. D5921, Sigma Aldrich Inc.). Once the lysis buffer is assembled, stimulate the cells with a ROS-promoting stimulus of choice and transfer the culture dish into the anaerobic chamber through an airlock. Discard rapidly the excess DMEM from the dish. Remove the remaining DMEM using a pipetman, and add 600 μl of IAA-supplemented degassed lysis buffer to the 10-cm culture dish. Scrape and transfer the cells into an amber 1.5-ml centrifuge tube (Cat. no. 05-408-134, Fisher Scientific). Some cells are more resistant to lysis and can be disrupted through a 25-gauge needle, three times. Transfer the cells to a shaker for an hour at 25 °C to allow lysis and complete alkylation of the active PTPs. Then, centrifuge the lysates at 14,000 × g for 5 min in a table top microcentrifuge and determine the protein concentration of the supernatant by the method of Bradford.

9. Cysteinyl-Labeling Assay

On your regular working area, apply slowly the cell lysate (0.6–1.0 mg determined empirically depending on the cell line being used) to a size-exclusion column (Zeba Desalt Spin Columns, 2 ml, cat. no. 89889, Thermo Scientific) previously equilibrated with three bed volumes of IAA-free lysis buffer. Allow your lysate to enter into the resin bed. Proceed to the buffer exchange by centrifuging at 1000 × g 2 min, and transfer the IAA-cleared lysate to clear microfuge tubes. Reduce your cell extract by adding DTT (dithiothreitol, final 1 M) or TCEP [tris(2-carboxyethyl)phosphine, final 1 M] for a 30 min period: keep the tubes in the dark when shaking. Preclear the reduced lysates with 15 μl of streptavidin-Sepharose (Cat. no. 17-5113-01, GE Healthcare) for 15 min and transfer the precleared extract to amber-colored microfuge tubes. Add IAP probe (iodoacetyl-polyethylene glycol, final 5 mM) to the reduced lysate and place the tubes on a shaker at 25 °C for an additional hour. In the final step of the assay, transfer the samples to clear-colored microfuge tubes, and add streptavidin-Sepharose beads to enrich the biotinylated PTPs by pull-down (25 μl of a 50% slurry, overnight at 4 °C on a rotating wheel). Briefly centrifuge the beads (1 min, 4 °C), wash three times with 1 ml PBS, resuspend in 20 μl 4× Laemmli sample buffer, and heat at 95 °C for 90–120 s.

10. Acute Stimulus-Induced Reversible Oxidation of PTPs

We have previously performed the cysteinyl-labeling assay using angio-myolipoma cells overexpressing platelet-derived growth factor (PDGF)-BB. This sustained PDGF-BB expression system allowed us to identify SHP-2,

PTEN, MKP-1, and PTP-LAR as downstream targets of ROS signaling in angiomyolipomas (Boivin *et al.*, 2008). However, we sought to verify whether reversible oxidation of PTPs could be detected using this approach upon acute growth factor stimulation in cell lines that are used commonly. ROS production and SHP-2 reversible oxidation have previously been observed in Rat-1 fibroblasts following PDGF-BB stimulation (Meng and Tonks, 2003; Meng *et al.*, 2002). Hence, Rat-1 fibroblasts were cultured to confluency in low-glucose DMEM supplemented with 10% fetal bovine serum and serum deprived for a 16 h period in low-glucose DMEM without phenol red. Cells were then incubated with 100 ng/ml PDGF-BB for 2 min and transferred in the argon-equilibrated hypoxic glove box to be lysed with the degassed lysis buffer. The cysteinyl-labeling assay was performed as described above and proteins were separated by electrophoresis on a 10% SDS–polyacrylamide gel and transferred electrophoretically onto nitrocellulose at 60 V and 4 °C for 2 h. Following transfer, nitrocellulose membranes were blocked for 2 h in TBST containing 5% nonfat milk powder and incubated for 16 h at 4 °C with PTEN or SHP-2 antisera diluted in TBST containing 1% BSA. Following extensive washing with TBST, membranes probed for PTEN and SHP-2 were briefly reblocked for 5–10 min in TBST containing 5% nonfat milk powder (to reduce the background from the secondary antibodies) and incubated for 2 h with horseradish peroxidase (HRP)-conjugated secondary antibodies, washed extensively with TBST and the immunoreactive bands were visualized by ECL (Fig. 3.4A). Reversible oxidation of SHP-2 and PTEN was observed upon PDGFR activation. Interestingly, IAP labeling of SHP-2 was not observed in absence of PDGF-BB when IAA was left out of the lysis buffer. This suggests that the active-site Cys residue of the autoinhibited closed form of SHP-2 (Barford and Neel, 1998) is not available to react with the sulfhydryl-reactive compound. Profiling of reversibly oxidized PTPs was also visualized by reprobing the membrane with streptavidin–HRP (Fig. 3.4B). Interestingly, probing for biotinylated PTPs using streptavidin–HRP revealed a significant increase staining of several bands, consistent with reversible oxidation of several PTPs.

11. Perspectives

Our approach, which involves a blocking step followed by reactivation and labeling steps, is similar to the biotin-switch technique for identifying *S*-nitrosylated proteins (Jaffrey and Snyder, 2001) and the acyl–biotinyl exchange chemistry for identifying protein palmitoylation (Drisdel *et al.*, 2006). A similar protocol for identifying reversibly oxidized PTPs has been established by Kwon *et al.* (2004, 2005), which utilizes neutral pH and sodium dodecyl sulfate (i.e., an anionic surfactant) in order to denature proteins from the cellular extract and for the alkylating agents to gain access

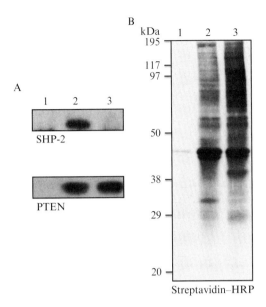

Figure 3.4 Use of biotinylated IAP probes to detect reversible oxidation of PTEN and SHP-2 following acute stimulation of Rat-1 fibroblasts. Serum-deprived (16 h) Rat-1 fibroblasts were lysed in presence of 10 mM IAA without stimulation (lane 1) or following a 2 min incubation with PDGF-BB at 37 °C (lane 2, 100 ng/ml), or lysed in absence of IAA (lane 3) and subjected to the cysteinyl-labeling assay using a biotinylated IAP probe. Biotinylated proteins were purified on streptavidin-Sepharose beads, resolved by SDS–PAGE and probed using antibodies against PTEN or SHP-2 (A) or streptavidin-HRP (B). Reversible PTP oxidation was significantly increased in PDGF-BB stimulated cells, as measured with both probes.

to hidden sulfhydryl-reactive Cys residues. However, these denaturing conditions disrupt the active-site architecture (and low pK_a), thus making the method less specific for PTPs. Another affinity-based approach employs antibodies raised against the terminally oxidized VHCSAG peptide from the PTP signature motif. This strategy also has potential to be applied for a proteomic screening of reversibly oxidized PTPs (Persson *et al.*, 2004). However, the terminal oxidation of all reversibly oxidized PTPs, which is essential for antibody recognition, does not allow distinction between reversibly oxidized PTPs and the terminally oxidized PTPs that are found at high levels in cancer cells (Lou *et al.*, 2007). Direct sulfenic acid labeling methods using dimedone or close derivatives have also been used to detect reversible oxidation of proteins *in vivo* (Benitez and Allison, 1974; Charles *et al.*, 2007; Conway *et al.*, 2004; Ellis and Poole, 1997; Leonard *et al.*, 2009). However, no reversibly oxidized PTPs have been identified yet using this approach in complex protein extracts.

As we move beyond the sequence of the human genome, it is essential to develop methods to define and characterize the proteome so as to decipher those enzymatic activities that underlie fundamental biological processes. Hence, the conception of assays using protein modifying compounds that allow specific enrichment and profiling of enzyme families within a complex proteome is a cornerstone to achieve this objective. In addition, such probes could also potentially be used *in situ* to identify the subcellular localization of these activated or inactivated enzymes. Different PTP-specific activity-based probes have been described for this purpose. Among them, phosphotyrosine mimetics, such as 4-difluoromethylphenyl phosphate, have been reported for the study of protein phosphatases (Betley *et al.*, 2002; Wang *et al.*, 1994). The nucleophilic attack of 4-difluoromethylphenyl phosphate by the PTP reactive thiolate ion leads to its dephosphorylation and the production of a reactive quinone methide that alkylates nearby residues. In contrast, the BBP that we have used in our previous study (Boivin *et al.*, 2008) is a suicide substrate probe for PTPs (Kumar *et al.*, 2004). Hence, the cysteinyl-phosphate bond formed between BBP and the PTP catalytic cysteine provides increased labeling specificity. Both probes have yet to be tested in approaches coupling the cysteinyl-labeling assay with proteomics-based detection; however, our initial studies using biotinylated BBP as a suicide substrate shows great promise and should allow us to obtain a cell-specific molecular snapshot of the redox status of all PTP superfamily members.

By combining cysteinyl labeling, to identify those PTPs that are oxidized in response to a defined stimulus, with specific knockdown strategies and the use of substrate trapping mutants (Flint *et al.*, 1997), to define physiological PTP substrate specificity, we have produced a strategy with which to delineate and refine novel regulatory roles for PTPs in the control of cell signaling. For signaling purposes, disrupting the expression of a PTP gene by RNAi reproduces oxidation-mediated PTP inactivation and elevates the phosphorylation of the sites targeted by this specific PTP, as well as downstream events in the signaling cascade. These changes can be studied by immunoblotting of known targets or by measuring the phosphoproteome using mass spectrometry (Hilger *et al.*, 2009). In addition, this cysteinyl-labeling assay may foreseeably be upgraded for *in situ* visualization of PTP oxidation. Interestingly, a probe in which a BBP moiety linked to a rhodamine B fluorophore has been used for detecting PTP activity (Kumar *et al.*, 2006). Thus, using a three-step approach in fixed cells in which reactive PTPs are blocked, followed by reactivation of the oxidized PTPs, and a labeling step involving a rhodamine-BBP probe could allow us to visualize discrete sites of PTP reversible oxidation *in vivo*. Two similar three-step methods have recently been used to monitor general thiol oxidation (disulfide formation) *in situ* (Mastroberardino *et al.*, 2008; Yang *et al.*, 2007b). Although those methods are aimed at general disulfide formation, some technical modifications and the use of PTP-activity-based probes should allow to detect the dynamic reversible oxidation of

PTPs *in situ*. The growth factor-mediated dynamic ROS production, the redox status of subcellular organelles, and the redox variations during the cell cycle itself presumably all contribute to regulating PTP oxidation and inactivation, overall protein phosphorylation, and signal transduction. Hence, the imaging of discrete sites of PTP reversible oxidation, combined with organelle-specific mass spectrometry should prove to be greatly informative and provide insight into mechanisms of the redox biology.

12. Conclusion

Overall, there are over 100 PTP genes in the human genome; however, only a few of these enzymes have been characterized extensively. This cysteinyl-labeling assay shows great promise in enabling us to perform profiling of the redox regulation of the PTP superfamily. This direct insight on regulation of all PTP members jointly with the strategies to delineate and refine cell signaling pathways will prove to be useful in studying their role in redox biology.

REFERENCES

Barford, D., and Neel, B. G. (1998). *Structure* **6**, 249–254.
Barford, D., Flint, A. J., and Tonks, N. K. (1994). *Science* **263**, 1397–1404.
Barford, D., Jia, Z., and Tonks, N. K. (1995). *Nat. Struct. Biol.* **2**, 1043–1053.
Barrett, W. C., DeGnore, J. P., Keng, Y. F., Zhang, Z. Y., Yim, M. B., and Chock, P. B. (1999). *J. Biol. Chem.* **274**, 34543–34546.
Benitez, L. V., and Allison, W. S. (1974). *J. Biol. Chem.* **249**, 6234–6243.
Betley, J. R., Cesaro-Tadic, S., Mekhalfia, A., Rickard, J. H., Denham, H., Partridge, L. J., Plückthun, A., and Blackburn, G. M. (2002). *Angew. Chem. Int. Ed. Engl.* **41**, 775–777.
Boivin, B., Zhang, S., Arbiser, J. L., Zhang, Z. Y., and Tonks, N. K. (2008). *Proc. Natl. Acad. Sci. USA* **105**, 9959–9964.
Brown, D. I., and Griendling, K. K. (2009). *Free Radic. Biol. Med.* **47**, 1239–1253.
Buhrman, G., Parker, B., Sohn, J., Rudolph, J., and Mattos, C. (2005). *Biochemistry* **44**, 5307–5316.
Charles, R. L., Schröder, E., May, G., Free, P., Gaffney, P. R., Wait, R., Begum, S., Heads, R. J., and Eaton, P. (2007). *Mol. Cell. Proteomics* **6**, 1473–1484.
Chen, Y. Y., Chu, H. M., Pan, K. T., Teng, C. H., Wang, D. L., Wang, A. H., Khoo, K. H., and Meng, T. C. (2008). *J. Biol. Chem.* **283**, 35265–35272.
Chung, D. G., and Lewis, P. N. (1986). *Biochemistry* **25**, 5036–5042.
Conway, M. E., Poole, L. B., and Hutson, S. M. (2004). *Biochemistry* **43**, 7356–7364.
Czech, M. P., Lawrence, J. C. J., and Lynn, W. S. (1974). *Proc. Natl. Acad. Sci. USA* **71**, 4173–4177.
Denu, J. M., and Dixon, J. E. (1998). *Curr. Opin. Chem. Biol.* **2**, 633–641.
Denu, J. M., and Tanner, K. G. (1998). *Biochemistry* **37**, 5633–5642.
Deshpande, T., Takagi, T., Hao, L., Buratowski, S., and Charbonneau, H. (1999). *J. Biol. Chem.* **274**, 16590–16594.

Drisdel, R. C., Alexander, J. K., Sayeed, A., and Green, W. N. (2006). *Methods* **40,** 127–134.
Ellis, H. R., and Poole, L. B. (1997). *Biochemistry* **36,** 15013–15018.
Flint, A. J., Tiganis, T., Barford, D., and Tonks, N. K. (1997). *Proc. Natl. Acad. Sci. USA* **94,** 1680–1685.
Hilger, M., Bonaldi, T., Gnad, F., and Mann, M. (2009). *Mol. Cell. Proteomics* **8,** 1908–1920.
Jaffrey, S. R., and Snyder, S. H. (2001). *Sci. STKE* **86,** 1–9.
Juarez, J. C., Manuia, M., Burnett, M. E., Betancourt, O., Boivin, B., Shaw, D. E., Tonks, N. K., Mazar, A. P., and Doñate, F. (2008). *Proc. Natl. Acad. Sci. USA* **105,** 7147–7152.
Jullien, M., and Garel, J. R. (1981). *Biochemistry* **20,** 7021–7026.
Kamata, H., Honda, S., Maeda, S., Chang, L., Hirata, H., and Karin, M. (2005). *Cell* **120,** 649–661.
Kumar, S., Zhou, B., Liang, F., Wang, W. Q., Huang, Z., and Zhang, Z. Y. (2004). *Proc. Natl. Acad. Sci. USA* **101,** 7943–7948.
Kumar, S., Zhou, B., Liang, F., Yang, H., Wang, W. Q., and Zhang, Z. Y. (2006). *J. Proteome Res.* **5,** 1898–1905.
Kwon, J., Lee, S. R., Yang, K. S., Ahn, Y., Kim, Y. J., Stadtman, E. R., and Rhee, S. G. (2004). *Proc. Natl. Acad. Sci. USA* **101,** 16419–16424.
Kwon, J., Qu, C. K., Maeng, J. S., Falahati, R., Lee, C., and Williams, M. S. (2005). *EMBO J.* **24,** 2331–2341.
Lee, S. R., Kwon, K. S., Kim, S. R., and Rhee, S. G. (1998). *J. Biol. Chem.* **273,** 15366–15372.
Leonard, S. E., Reddie, K. G., and Carroll, K. S. (2009). *ACS Chem. Biol.* **4,** 783–799.
Leslie, N. R., Bennett, D., Lindsay, Y. E., Stewart, H., Gray, A., and Downes, C. P. (2003). *EMBO J.* **22,** 5501–5510.
Li, S., and Whorton, A. R. (2003). *Arch. Biochem. Biophys.* **410,** 269–279.
Lou, Y. W., Chen, Y. Y., Hsu, S. F., Chen, R. K., Lee, C. L., Khoo, K. H., Tonks, N. K., and Meng, T. C. (2007). *FEBS J.* **275,** 69–88.
Mahadev, K., Zilbering, A., Zhu, L., and Goldstein, B. J. (2001). *J. Biol. Chem.* **276,** 21938–21942.
Mastroberardino, P. G., Orr, A. L., Hu, X., Na, H. M., and Greenamyre, J. T. (2008). *Free Radic. Biol. Med.* **45,** 971–981.
Meng, T. C., and Tonks, N. K. (2003). *Methods Enzymol.* **366,** 304–318.
Meng, T. C., Fukada, T., and Tonks, N. K. (2002). *Mol. Cell* **9,** 387–399.
Meng, T. C., Buckley, D. A., Galic, S., Tiganis, T., and Tonks, N. K. (2004). *J. Biol. Chem.* **279,** 37716–37725.
Persson, C., Sjöblom, T., Groen, A., Kappert, K., Engström, U., Hellman, U., Heldin, C. H., den Hertog, J., and Ostman, A. (2004). *Proc. Natl. Acad. Sci. USA* **1001,** 1886–1891.
Rhee, S. G., Chang, T. S., Bae, Y. S., Lee, S. R., and Kang, S. W. (2003). *J. Am. Soc. Nephrol.* **14,** S211–S215.
Salmeen, A., and Barford, D. (2005). *Antioxid. Redox Signal.* **7,** 560–577.
Salmeen, A., Andersen, J. N., Myers, M. P., Meng, T. C., Hinks, J. A., Tonks, N. K., and Barford, D. (2003). *Nature* **423,** 769–773.
Sullivan, S. G., Chiu, D. T., Errasfa, M., Wang, J. M., Qi, J. S., and Stern, A. (1994). *Free Radic. Biol. Med.* **16,** 399–403.
Sundaresan, M., Yu, Z. X., Ferrans, V. J., Irani, K., and Finkel, T. (1995). *Science* **270,** 296–299.
Takakura, K., Beckman, J. S., MacMillan-Crow, L. A., and Crow, J. P. (1999). *Arch. Biochem. Biophys.* **369,** 197–207.
Tonks, N. K. (2006). *Nat. Rev. Mol. Cell Biol.* **7,** 833–846.

Tonks, N. K., Diltz, C. D., and Fischer, E. H. (1988). *J. Biol. Chem.* **263,** 6731–6737.
Valko, M., Leibfritz, D., Moncol, J., Cronin, M. T., Mazur, M., and Telser, J. (2007). *Int. J. Biochem. Cell Biol.* **39,** 44–84.
Wang, Q., Dechert, U., Jirik, F., and Withers, S. G. (1994). *Biochem. Biophys. Res. Commun.* **200,** 577–583.
Wu, R. F., and Terada, L. S. (2006). *Sci. STKE* pl2.
Xu, D., Rovira, I. I., and Finkel, T. (2002). *Dev. Cell* **2,** 251–252.
Yang, J., Groen, A., Lemeer, S., Jans, A., Slijper, M., Roe, S. M., den Hertog, J., and Barford, D. (2007a). *Biochemistry* **46,** 709–719.
Yang, Y., Song, Y., and Loscalzo, J. (2007b). *Proc. Natl. Acad. Sci. USA* **104,** 10813–10817.
Zhu, L., Zilbering, A., Wu, X., Mahadev, K., Joseph, J. I., Jabbour, S., Deeb, W., and Goldstein, B. J. (2001). *FASEB J.* **15,** 1637–1639.

CHAPTER FOUR

MEASURING THE REDOX STATE OF CELLULAR PEROXIREDOXINS BY IMMUNOBLOTTING

Andrew G. Cox, Christine C. Winterbourn, *and* Mark B. Hampton

Contents

1. Introduction	52
2. Measurement of Prx Dimerization	54
2.1. Principle of the method	54
2.2. General method	55
2.3. Examples	57
3. Measurement of Prx Hyperoxidation	59
3.1. Principle of method	59
3.2. General method	60
3.3. Examples	61
4. Discussion	61
Acknowledgments	64
References	64

Abstract

The peroxiredoxins (Prxs) are a family of thiol peroxidases that scavenge hydroperoxides and peroxynitrite. The abundance and reactivity of these proteins makes them primary targets for cellular H_2O_2. The catalytic cycle of typical 2-Cys Prxs involves formation of an intermolecular disulfide bond between peroxidatic and resolving cysteines on opposing subunits. Rapid alterations in the ratio of reduced monomer and oxidized dimer have been detected in the cytoplasm and mitochondria of cultured cells exposed to various exogenous and endogenous sources of oxidative stress. Here we describe immunoblot methods to monitor the interconversion of individual 2-Cys Prxs in cultured cells. We also outline an adaptation of this method to measure the extent to which individual 2-Cys Prxs become hyper oxidized in treated cells. Together, these methods enable the redox status of cellular Prxs to be assessed and quantified in a rapid and robust manner.

Free Radical Research Group, Department of Pathology, and National Research Centre for Growth and Development, University of Otago, Christchurch, New Zealand

Methods in Enzymology, Volume 474
ISSN 0076-6879, DOI: 10.1016/S0076-6879(10)74004-0
© 2010 Elsevier Inc.
All rights reserved.

1. Introduction

The peroxiredoxins (Prxs) have received considerable attention in recent years as regulators of cellular H_2O_2 and modulators of H_2O_2-dependent signaling pathways (Hall *et al.*, 2009; Rhee *et al.*, 2005). Prxs are highly abundant proteins constituting up to 1% of the total soluble protein in most cells. Unlike other peroxidases that require a prosthetic heme group or selenocysteine moiety, the Prxs use a conserved cysteine residue present as a thiolate anion to decompose hydroperoxides. Recent kinetic studies have revealed that many Prxs react with H_2O_2 at rates comparable to catalase or glutathione peroxidase ($k \sim 10^7$–$10^8\ M^{-1}\ s^{-1}$) (Cox *et al.*, 2009b; Ogusucu *et al.*, 2007; Parsonage *et al.*, 2008; Peskin *et al.*, 2007). As such, the Prxs are increasingly being seen as primary targets of cellular H_2O_2 (Winterbourn and Hampton, 2008).

The initial reaction of the peroxidatic cysteine with H_2O_2 forms a sulfenic acid (Cys_P-SOH). The fate of this sulfenic acid differs depending on whether the Prx is a typical 2-Cys, atypical 2-Cys or 1-Cys enzyme. The most common class is the typical 2-Cys Prxs, which exist as obligate homodimers orientated in a head-to-tail manner. The sulfenic acid condenses with the resolving cysteine of the second subunit to form an intermolecular disulfide bond (Fig. 4.1A). The disulfide bonds in the oxidized dimer are reduced by thioredoxin (Trx) to complete the catalytic cycle. In contrast, atypical 2-Cys Prxs form an intramolecular disulfide bond upon oxidation (Seo *et al.*, 2000), whereas the 1-Cys Prxs form a mixed disulfide with glutathione (Manevich *et al.*, 2004).

One interesting aspect of these proteins is their sensitivity to hyperoxidation, with H_2O_2 reacting with the sulfenic acid intermediate to form a sulfinic acid (Fig. 4.1B). Intriguingly, eukaryotic typical 2-Cys Prxs have evolved structural features that dramatically enhance susceptibility to hyperoxidation compared to their prokaryotic counterparts (Wood *et al.*, 2003). The hyperoxidized Prxs are enzymatically inactive, although they can be regenerated slowly by the sulfinyl reductase sulfiredoxin (Biteau *et al.*, 2003; Jonsson and Lowther, 2007). It has been proposed that Prx hyperoxidation may enable H_2O_2 levels to accumulate and function as second messengers in cell signaling (Wood *et al.*, 2003). Alternatively, studies suggest that the hyperoxidation of Prxs may represent a molecular switch that abolishes peroxidase activity, while enhancing chaperone activity (Jang *et al.*, 2004).

Cellular Prxs are maintained in their reduced form by the Trx system. However, several studies have reported increases in the ratio of oxidized to reduced Prxs through increased exposure to H_2O_2 (Baty *et al.*, 2005; Cox and Hampton, 2007; Cox *et al.*, 2009a; Low *et al.*, 2007), chloramines (Stacey *et al.*, 2009), auranofin (Cox *et al.*, 2008a), isothiocyanates (Brown

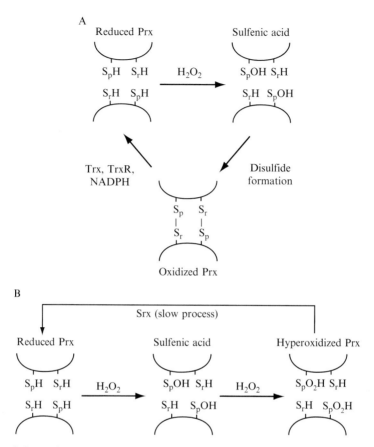

Figure 4.1 Redox transformations of typical 2-Cys Prxs. (A) Catalytic cycle of the typical 2-Cys Prxs. The peroxidatic cysteine (S_PH) reacts with H_2O_2 to form a sulfenic acid intermediate, which can condense with the resolving cysteine (S_RH) on another Prx to form an intermolecular disulfide. The oxidized Prx dimer is reduced by thioredoxin (Trx) to complete the catalytic cycle. Oxidized Trx is reduced by thioredoxin reductase (TrxR) using NADPH as an electron donor. (B) Hyperoxidation of typical 2-Cys Prxs and retroreduction by Srx. In the presence of excess substrate, H_2O_2 reacts with the sulfenic acid intermediate to form the hyperoxidized sulfinic acid. Hyperoxidized Prxs are slowly repaired by the sulfinyl reductase enzyme Srx.

et al., 2008), hexavalent chromium (Myers and Myers, 2009), and induction of receptor-mediated apoptosis (Cox *et al.*, 2008b). Prx oxidation has also been observed in hearts during ischemia (Kumar *et al.*, 2009) or upon perfusion with H_2O_2 (Schroder *et al.*, 2008). Mammals have six Prxs that are widely distributed in different cellular compartments, including cytosol (Prx 1, 2, and 6), mitochondria (Prx 3 and 5), endoplasmic reticulum

(Prx 4), and peroxisomes (Prx 5). In some studies where Prx oxidation was observed, it was restricted to a subset of Prxs, suggesting organelle specificity. Also, Prx oxidation occurred in the absence of widespread oxidative damage, for example, protein carbonylation or lipid peroxidation. Analysis of the redox transformations of endogenously expressed Prxs may provide valuable insight as a marker of disruptions in redox homeostasis. The methods described in this chapter provide a means to monitor the dimerization or hyperoxidation of individual typical 2-Cys Prxs in cells or tissue by immunoblotting under nonreducing conditions. These simple methods are well suited to the testing of multiple samples in parallel and overcome some of the technical challenges associated with previous methodologies.

2. Measurement of Prx Dimerization

2.1. Principle of the method

During their catalytic cycle, typical 2-Cys Prxs are oxidized to form an intermolecular disulfide bond, with the oxidized protein running as a dimer under nonreducing conditions (Fig. 4.2). The principle of the method is to trap the oxidized and reduced forms during cell or tissue extraction and monitor the proportions of monomer and dimer for each Prx with selective antibodies.

A critical step in the process is rapid alkylation of reduced cysteines to prevent artefactual Prx oxidation during sample preparation. Alkylation will also prevent Trx from reducing oxidized Prxs. This is unlikely to be a problem after cell lysis, but removal of cells from a source of oxidative stress can lead to rapid reduction of oxidized Prxs (Low et al., 2007), and therefore an underestimation of oxidized Prx levels.

While thiol-trapping by alkylation is common in redox proteomic techniques, for Prxs this step is challenging. Investigations with purified mammalian Prxs have shown that they react extremely rapidly with hydroperoxides, particularly H_2O_2 ($k \sim 10^7 \, M^{-1} \, s^{-1}$), but the rate of alkylation is five to eight orders of magnitude slower (Cox et al., 2009b; Peskin et al., 2007). N-Ethylmaleimide (NEM) is the most effective alkylating agent so far tested, but even so adventitious peroxides present at low micromolar concentrations in buffers will cause oxidation despite the presence of millimolar NEM.

There are a variety of approaches that can assist in limiting artefactual oxidation. Protein denaturation or acidification dramatically reduces the reactivity of the peroxidatic cysteine, thereby making alkylation more competitive. While we have exploited this methodology with pure protein, it is less applicable in complex cell extracts. We currently utilize three other approaches to limit artefactual oxidation. First, the impact of

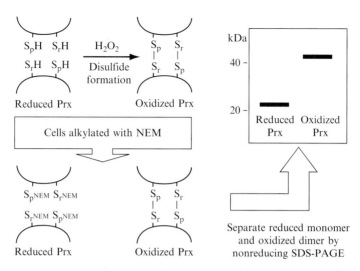

Figure 4.2 Redox immunoblot method to detect Prx oxidation. The method involves alkylating reduced thiols on Prxs with NEM, then resolving the alkylated monomer from the oxidized dimer by SDS–PAGE in nonreducing conditions.

adventitious peroxides is limited by preparing lysates in small volumes so that Prx concentrations are higher than that of the peroxides. Second, catalase is included in extract buffers to limit H_2O_2 levels. It should be emphasized that the catalase is not effective at competing with the Prxs for H_2O_2 present at low concentration, indeed endogenous peroxisomal catalase will be released during cell lysis; rather it is used to remove H_2O_2 from buffers before use. The final step for optimizing Prx alkylation is to add NEM several minutes before cell lysis, thereby promoting alkylation before the Prxs are released and exposed to exogenous peroxides. The length of incubation time should be determined empirically for each cell type, and should be long enough to enable effective alkylation but not so long as to disrupt cell integrity and therefore prevent the ability to harvest and pellet cells where necessary.

2.2. General method

The exact conditions for trapping Prxs have to be optimized for each cell type, but here we offer specific protocols for cells grown in suspension and adherent cells.

2.2.1. Alkylation of suspension cells

1. Culture cells under the experimental conditions of choice.
2. At the end of the incubation period, pellet cells (1×10^6) rapidly by centrifugation at $10,000 \times g$ for 30 s.

3. Remove the supernatant and resuspend in 100 μl of alkylation buffer (40 mM HEPES, 50 mM NaCl, 1 mM EGTA, Complete™ protease inhibitors, pH 7.4, 10 μg/ml catalase, 100 mM NEM). Add the catalase 30 min prior to use, and the NEM immediately prior to use. Incubate at room temperature for 10 min.
4. Add 4 μl of 25% CHAPS (1% final) to lyse the cells. The type of detergent and most effective concentration may differ between cell types. Incubate for 10 min with periodic vortex mixing.
5. Remove insoluble material by centrifugation at $16,000 \times g$ for 5 min. Measure the protein concentration using the detergent-compatible BioRad™ protein assay. Store at $-80\ °C$ before electrophoresis.

2.2.2. Alkylation of adherent cells

(A) For short-term treatments in which cells remain attached (i.e., no "floating cells"):
 1. Culture cells under the experimental conditions of choice.
 2. Remove culture media and add sufficient alkylation buffer to cover all of the adherent cells on the plate (e.g., 100 μl/well of a 24-well plate). Incubate for 10 min at room temperature.
 3. Add CHAPS directly to the well at a final concentration of 1%. The type of detergent and most effective concentration may differ between cell types. Incubate for 10 min then follow step 5 in 2.2.1

(B) For treatments that cause cells to detach from plasticware:
 1. Culture cells under the experimental conditions of choice.
 2. Collect "floating cells" into a centrifuge tube.
 3. Wash adherent cells with PBS and cover bottom of plate with 0.05% trypsin solution containing EDTA (e.g., 100 μl/well of a 24-well plate). Incubate for 2 min.
 4. Collect trypsinized cells and combine with "floating cells." Pellet cells rapidly by centrifugation at $10,000 \times g$ for 30 s. Follow steps 3–5 in 2.2.1 above.

2.2.3. Nonreducing electrophoresis and Prx Western blotting

1. Dilute protein extracts (5 μg final) in nonreducing sample buffer (62.5 mM Tris–HCl, pH 6.8, 10% glycerol, 2% SDS, and 0.025% bromophenol blue) at a 2:1 ratio.
2. Resolve samples by SDS–PAGE in a 15% gel in running buffer (25 mM Tris, 192 mM glycine, 0.1% SDS, pH 8.3).
3. Transfer proteins from SDS–PAGE gels to Hybond PVDF membranes by Western blotting in blot buffer (25 mM Tris, 192 mM glycine, 10% methanol) for 50 min at 100 V.

4. Block PVDF membranes in 5% skim milk TBST$_{20}$ (150 mM NaCl, 50 mM Tris–HCl, pH 8.0, and 0.05% Tween 20) for 1 h.
5. Wash and probe with the appropriate Prx primary antibody in 2% skim milk TBST$_{20}$ overnight at 4 °C. Rabbit polyclonal antibodies to Prx 1–3 can be obtained from Abfrontier (Seoul, Korea) or Abcam (Cambridge, UK), and can be used at a concentration of 1 in 10,000.
6. Wash the blots thoroughly in TBST$_{20}$ and probe with HRP-conjugated goat antirabbit (1:20,000 dilution) antibodies in 2% skim milk TBST$_{20}$ for 1 h.
7. Wash the membranes thoroughly in TBST$_{20}$, and detect the bound secondary antibody by a chemiluminescence system of choice.
8. Quantify the extent of oxidation by measuring the relative band densities of monomer and dimer. The extent of oxidation is presented as a percentage of the Prx in the oxidized state. As a control, gels can include lanes with samples that have 5% β-mercaptoethanol added, to ensure that the Prx is indeed reversibly oxidized.

2.3. Examples

The technique is illustrated with the Jurkat human T-lymphoma cell line acquired from the American Type Culture Collection, Rockville, MD, USA, and SV40-immortalized mouse embryonic fibroblasts (MEFs) generously provided by Dr. David Huang, WEHI, Melbourne, Australia.

Jurkat cells are grown in RPMI 1640 supplemented with 10% FBS, 100 U/ml penicillin, and 100 μg/ml and are maintained in a humidified incubator at 37 °C and 5% CO$_2$/air and passaged regularly to maintain optimal growth (between 10^5 and 10^6 cells/ml). Jurkat cells are harvested and resuspended in fresh media at a concentration of 1×10^6 cells/ml and left to equilibrate for 30 min prior to treatment. MEFs are grown in DMEM containing 10% FBS, 100 μM asparagine, 50 μM β-mercaptoethanol, 100 U/ml penicillin, and 100 μg/ml streptomycin. MEFs are maintained at optimal growth by passaging every 3 days at a 1:6 ratio. Subculturing involves aspiration of used medium, washing with PBS, aspiration and then incubation of the cell monolayer with 0.05% trypsin–EDTA. After 2 min, dissociated cells are counted, resuspended in fresh medium, and plated out. MEFs are treated at 80% confluency in fresh DMEM containing 10% FBS. MEFs are seeded the day before experimentation at 10^5 cells/well of a 24-well plate.

In this example, Jurkat cells (1×10^6 per sample) were rapidly pelleted by centrifugation and alkylated as described for suspension cells above. MEFs were cultured in a 24-well plate and alkylated as described for adherent cells (A) above. In our experience with various cell types, Prxs are predominantly in the reduced form under resting conditions.

The efficiency by which cellular Prxs are alkylated was determined by altering the concentration of NEM present in the buffer (Fig. 4.3). Under nonreducing conditions, monomeric Prx migrates at ∼21 kDa, whereas the dimeric oxidized Prxs migrate at ∼42 kDa. It is apparent in both Jurkats and MEFs that 100 mM NEM should be used for alkylation step, as lower concentrations lead to impaired trapping of the reduced monomer and an

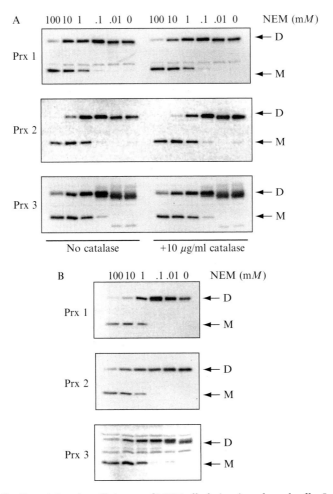

Figure 4.3 Examining the efficiency of NEM alkylation in cultured cells. Jurkat cells (A) and MEFs (B) were harvested in alkylation buffer containing different concentrations of NEM. Jurkat samples were alkylated with NEM buffer in the presence or absence of 10 μg/ml catalase. MEFs were harvested as described in 2.2.2A. The redox state of Prx 1–3 was determined by immunoblotting in nonreducing conditions. "D" indicates dimeric Prx, whereas "M" indicates monomeric Prx. Immunoblots shown are representative of three independent experiments.

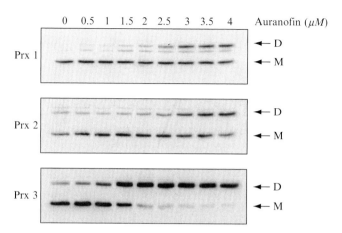

Figure 4.4 Exposure to the TrxR inhibitor auranofin leads to the accumulation of oxidized Prxs. Jurkat cells were exposed to auranofin for 40 min before being harvested in NEM alkylation buffer. The redox state of Prx 1–3 was determined by immunoblotting in nonreducing conditions. "D" indicates dimeric Prx, whereas "M" indicates monomeric Prx. Data adapted from Cox *et al.* (2008a), with permission.

increase in the amount of dimer. The absence of catalase in the NEM alkylation buffer was tested in Jurkat cells. There was a subtle increase in oxidation, especially when alkylation is inefficient (e.g., Prx 2 with 1 and 10 mM NEM).

Figure 4.4 demonstrates the ability of this method to measure Prx dimerization following exposure of Jurkat cells to the TrxR inhibitor auranofin. Inhibition of TrxR prevents the reduction of Prxs and leads to accumulation of the oxidized dimer. Although auranofin exposure leads to the oxidation of all Prxs, it is evident that the individual Prxs behave differently, with mitochondrial Prx 3 being the most sensitive to oxidation.

3. Measurement of Prx Hyperoxidation

3.1. Principle of method

The hyperoxidation of typical 2-Cys Prxs can be monitored through an adaptation of the method described above. It takes advantage of the fact that in the absence of alkylation, reduced Prxs immediately dimerize following cell lysis. However, hyperoxidized Prxs are unable to dimerize, and will remain as monomers (Fig. 4.5).

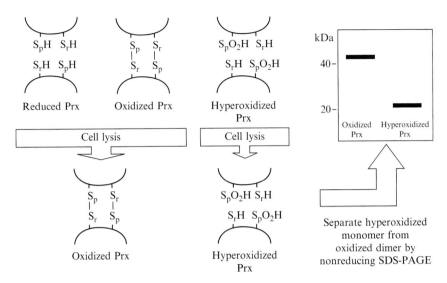

Figure 4.5 Redox immunoblot method to detect Prx hyperoxidation. The method involves resolving cell extracts in nonreducing conditions and differentiating the hyperoxidized monomer from the oxidized dimer.

We described above the difficulty in alkylating reduced Prxs before they are oxidized by adventitious peroxides. That complication is the basis of this method, and is what makes it relatively simple. It is possible to check the efficacy of this uncontrolled oxidation step by adding the reductant dithiothreitol to a sample to reduce the Prxs, and then passing this material through a Micro Bio-Spin® 6 chromatography spin column (Bio-Rad, Hercules, CA, USA) into the extract buffer of choice to determine if complete oxidation occurs. The greater the volume of lysate buffer the more effective oxidation will be, but the sample should not be so dilute as to limit Prx detection. In general we use a volume that yields a lysate with a protein concentration of 1–2 mg/ml. We have also noted that spontaneous Prx dimerization upon cell lysis is not as efficient with trypsinized cells.

3.2. General method

1. Culture cells under the experimental conditions of choice.
2. At the end of the incubation period, pellet cells (1×10^6) rapidly by centrifugation at $10,000 \times g$ for 30 s.
3. Remove the supernatant and resuspend in 100 μl of lysis buffer (40 mM HEPES, 50 mM NaCl, 1 mM EGTA, Complete^{TM} protease inhibitors,

pH 7.4, no NEM) and add 4 μl of 25% CHAPS (1% final) to lyse the cells. The type of detergent and most effective concentration may differ between cell types.
4. Incubate at room temperature for 10 min with periodic vortex mixing.
5. Remove insoluble material by centrifugation at 16,000×g for 5 min. Measure the protein concentration using the detergent-compatible BioRad^{TM} protein assay. Store at −80 °C before electrophoresis.
6. Perform nonreducing electrophoresis and Prx Western blotting as described in section 2.2.3 above.

3.3. Examples

Figure 4.6 illustrates Prx hyperoxidation in Jurkat cells exposed to increasing doses of H_2O_2. In this example, we compared the nonreducing immunoblot method with an alternate approach that uses an antibody raised against the hyperoxidized cysteine-containing sequence (Woo et al., 2003b). Although this antibody is useful, it cannot determine which Prx is being hyperoxidized, nor can it determine the extent of hyperoxidation. In contrast, the nonreducing methodology exhibits a similar sensitivity (both methods can detect Prx hyperoxidation at 20 μM H_2O_2), while providing information on the degree of hyperoxidation. Using this method, it is also possible to determine which Prx family members are becoming hyperoxidized.

4. Discussion

We have described a method for quantifying the interconversion of typical 2-Cys Prxs between their three main redox forms: reduced, oxidized, and hyperoxidized. The method was illustrated for cultured mammalian cells, but the principles have been adapted for use with purified protein (Cox et al., 2009a,b; Peskin et al., 2007), isolated mitochondria (Requejo et al., 2010), and whole tissue (Kumar et al., 2009; Schroder et al., 2008). It should also be possible to explore this approach in model organisms such as drosophila, nematodes, and zebrafish, depending on the availability of selective antibodies against the Prxs of these species.

Traditional measures of oxidative stress rely on measurement of end-stage damage to lipid, proteins, and nucleic acids. The Prxs appear to have antioxidant and signaling properties and may be key cellular redox sensors. Measurement of their functional interconversions may enable small and transient disruptions of redox homeostasis to be assessed. A variety of new

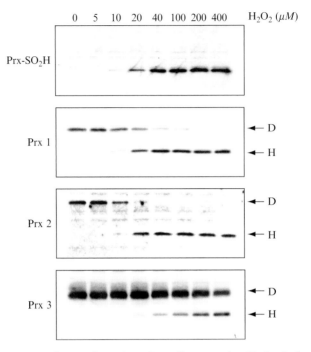

Figure 4.6 Hyperoxidation of Prxs in Jurkat cells exposed to H_2O_2. Jurkat cells were exposed to H_2O_2 (0–400 μM) for 10 min before being harvested in extract buffer and analyzed by nonreducing immunoblotting. "D" indicates dimeric Prx, whereas "H" indicates hyperoxidized Prx. Immunoblots shown are representative of three independent experiments. Data adapted from Cox et al. (2009a), with permission.

generation redox-sensitive probes are coming into use. Older probes such as dichlorofluorescein have major problems with regards specificity and susceptibility to artefactual oxidation (Bonini et al., 2006; Burkitt and Wardman, 2001; Zielonka et al., 2008), that hopefully the new generation can avoid (Muller, 2009). The Prxs have the advantage of being endogenous and not requiring the introduction of probes that may alter cell physiology. Furthermore, by monitoring the redox state of the individual Prxs simultaneously, one can detect compartmentalized redox stress e.g., Prx 3 in mitochondria.

The most important step in the nonreducing immunoblot method for detecting reversible oxidation is the alkylation of Prx active site cysteines before they are exposed to adventitious H_2O_2. The most effective approach is to introduce high levels of NEM before cell lysis. This becomes more challenging with tissues, but as illustrated with reperfused hearts, immersion of the heart in buffer containing NEM before extraction was considerably more effective than just being present in the extract buffer (Kumar et al.,

2009). This approach has the potential to cause cell changes before blocking is complete, for example, differential alkylation through blocking of Trx before Prx might be expected to promote Prx oxidation within the cell, but in all of the systems we have investigated the majority of cellular Prxs are captured in the reduced state under normal conditions.

Quantification of immunoblots is reliant on the Prx antibodies detecting monomer and dimer with equal sensitivity. In some situations we have obtained gels where the dimer appears to label more intensely than the monomer. This could be caused by a variety of factors, including an inherent difference in recognition by the primary antibody (potentially amplified when antibody levels are too low), interference from colocalized proteins or alterations in transfer efficiency. We have undertaken experiments with purified Prxs, comparing protein staining with immunoblots, and found the AbFrontier antibodies detect monomer and dimer with similar efficiency. Where accurate quantification is vital, standards of prepared mixtures of fully reduced (and alkylated) and oxidized Prxs could be prepared to generate standard curves.

The sensitivity of the Prx hyperoxidation method would be limited by the formation of a Prx dimer where one of the pair had a hyperoxidized peroxidatic cysteine, because the complex would still run as a dimer under nonreducing conditions. Although such complexes have been observed with purified protein, the reducing environment of the cell appears to prevent these species from accumulating (Cox *et al.*, 2009a). Another possible limitation of the hyperoxidation method is that it does not differentiate between the sulfinic and sulfonic species. In some cases this may be important, as the sulfonic species is not reduced by sulfiredoxin and exhibits distinct properties (Lim *et al.*, 2008; Seo *et al.*, 2009).

There are several alternative methods that have been used to examine the oxidation of Prxs. We have already discussed one immunological approach, which detects hyperoxidized Prxs using an antibody raised against the hyperoxidized cysteine-containing sequence (Woo *et al.*, 2003b). There is also a hyperoxidized antibody that has been raised specifically against the sulfonic acid form of Prx in yeast (Lim *et al.*, 2008). Other investigators have used an immunochemical method that involves labeling the Prx sulfenic acid intermediate with dimedone and detecting the unique adduct with a specific antibody (Seo and Carroll, 2009). All of the other methods that have been used to detect Prx oxidation utilize aspects of 2D SDS–PAGE and mass spectrometry. Up until recently, the most commonly used method to detect Prx hyperoxidation was with 2D SDS–PAGE and immunoblotting, taking advantage of the fact that Prxs present in the sulfinic or sulfonic acid forms, migrate with a more acidic pI (Mitsumoto *et al.*, 2001; Rabilloud *et al.*, 2002; Woo *et al.*, 2003a). Although the acidic shift assays have provided a wealth of information regarding the hyperoxidation of Prxs, they become labor-intensive when comparing multiple samples and are prone to interference from other posttranslational modifications such as phosphorylation. Other proteomic approaches used to detect Prx oxidation include sequential nonreducing/reducing SDS–PAGE (diagonal gel electrophoresis) (Brennan et al., 2004) and the fluorescent labeling of reversibly oxidized protein thiols with 5-iodoacetamide (Baty et al., 2005). While these methods are not suitable for routine assessment of the redox state of Prxs, they place the Prx changes within the context of global thiol changes within cells. The nonreducing immunoblot methods described in this chapter also have an important place in validating the results obtained from the proteomic methods.

ACKNOWLEDGMENTS

The authors acknowledge the efforts and insight provided by past and present colleagues of the Free Radical Research Group, including Felicia Low, Alexander Peskin, James Baty, Andree Pearson, Kristin Brown, Sarah Cuddihy, Melissa Stacey, and Vikas Kumar. This project was supported by the Royal Society Marsden Fund, the Health Research Council of New Zealand, and the National Research Centre for Growth and Development. A. G. C. is a recipient of a Top Achiever Doctoral Scholarship from the Tertiary Education Commission.

REFERENCES

Baty, J. W., Hampton, M. B., and Winterbourn, C. C. (2005). Proteomic detection of hydrogen peroxide-sensitive thiol proteins in Jurkat cells. *Biochem. J.* **389,** 785–795.

Biteau, B., Labarre, J., and Toledano, M. B. (2003). ATP-dependent reduction of cysteine-sulphinic acid by *S. cerevisiae* sulphiredoxin. *Nature* **425**(6961), 980–984.

Bonini, M. G., Rota, C., Tomasi, A., and Mason, R. P. (2006). The oxidation of 2′,7′-dichlorofluorescin to reactive oxygen species: A self-fulfilling prophesy? *Free Radic. Biol. Med.* **40**(6), 968–975.

Brennan, J. P., Wait, R., Begum, S., Dunn, M. J., and Eaton, P. (2004). Detection and mapping of widespread intermolecular protein disulfide formation during cardiac oxidative stress using proteomics with diagonal electrophoresis. *J. Biol. Chem.* **279**(40), 41352–41360.

Brown, K. K., Eriksson, S. E., Arner, E. S., and Hampton, M. B. (2008). Mitochondrial peroxiredoxin 3 is rapidly oxidized in cells treated with isothiocyanates. *Free Radic. Biol. Med.* **45**(4), 494–502.

Burkitt, M. J., and Wardman, P. (2001). Cytochrome C is a potent catalyst of dichlorofluorescin oxidation: Implications for the role of reactive oxygen species in apoptosis. *Biochem. Biophys. Res. Commun.* **282**(1), 329–333.

Cox, A. G., and Hampton, M. B. (2007). Bcl-2 over-expression promotes genomic instability by inhibiting apoptosis of cells exposed to hydrogen peroxide. *Carcinogenesis* **28**(10), 2166–2171.

Cox, A. G., Brown, K. K., Arner, E. S., and Hampton, M. B. (2008a). The thioredoxin reductase inhibitor auranofin triggers apoptosis through a Bax/Bak-dependent process that involves peroxiredoxin 3 oxidation. *Biochem. Pharmacol.* **76,** 1097–1109.

Cox, A. G., Pullar, J. M., Hughes, G., Ledgerwood, E. C., and Hampton, M. B. (2008b). Oxidation of mitochondrial peroxiredoxin 3 during the initiation of receptor-mediated apoptosis. *Free Radic. Biol. Med.* **44**(6), 1001–1009.

Cox, A. G., Pearson, A. G., Pullar, J. M., Jonsson, T. J., Lowther, W. T., Winterbourn, C. C., and Hampton, M. B. (2009a). Mitochondrial peroxiredoxin 3 is more resilient to hyperoxidation than cytoplasmic peroxiredoxins. *Biochem. J.* **421**(1), 51–58.

Cox, A. G., Peskin, A. V., Patton, L. N., Winterbourn, C. C., and Hampton, M. B. (2009b). Redox potential and peroxide reactivity of peroxiredoxin 3. *Biochemistry* **48**(27), 6495–6501.

Hall, A., Karplus, P. A., and Poole, L. B. (2009). Typical 2-Cys peroxiredoxins—structures, mechanisms and functions. *FEBS J.* **276**(9), 2469–2477.

Jang, H. H., Lee, K. O., Chi, Y. H., Jung, B. G., Park, S. K., Park, J. H., Lee, J. R., Lee, S. S., Moon, J. C., Yun, J. W., Choi, Y. O., Kim, W. Y., et al. (2004). Two enzymes in one; Two yeast peroxiredoxins display oxidative stress-dependent switching from a peroxidase to a molecular chaperone function. *Cell* **117**(5), 625–635.

Jonsson, T. J., and Lowther, W. T. (2007). The peroxiredoxin repair proteins. *Subcell. Biochem.* **44**, 115–141.

Kumar, V., Kitaeff, N., Hampton, M. B., Cannell, M. B., and Winterbourn, C. C. (2009). Reversible oxidation of mitochondrial peroxiredoxin 3 in mouse heart subjected to ischemia and reperfusion. *FEBS Lett.* **583**(6), 997–1000.

Lim, J. C., Choi, H. I., Park, Y. S., Nam, H. W., Woo, H. A., Kwon, K. S., Kim, Y. S., Rhee, S. G., Kim, K., and Chae, H. Z. (2008). Irreversible oxidation of the active-site cysteine of peroxiredoxin to cysteine sulfonic acid for enhanced molecular chaperone activity. *J. Biol. Chem.* **283**(43), 28873–28880.

Low, F. M., Hampton, M. B., Peskin, A. V., and Winterbourn, C. C. (2007). Peroxiredoxin 2 functions as a noncatalytic scavenger of low-level hydrogen peroxide in the erythrocyte. *Blood* **109**(6), 2611–2617.

Manevich, Y., Feinstein, S. I., and Fisher, A. B. (2004). Activation of the antioxidant enzyme 1-CYS peroxiredoxin requires glutathionylation mediated by heterodimerization with pi GST. *Proc. Natl. Acad. Sci. USA* **101**(11), 3780–3785.

Mitsumoto, A., Takanezawa, Y., Okawa, K., Iwamatsu, A., and Nakagawa, Y. (2001). Variants of peroxiredoxins expression in response to hydroperoxide stress. *Free Radic. Biol. Med.* **30**(6), 625–635.

Muller, F. L. (2009). A critical evaluation of cpYFP as a probe for superoxide. *Free Radic. Biol. Med.* **47**(12), 1779–1780.

Myers, C. R., and Myers, J. M. (2009). The effects of hexavalent chromium on thioredoxin reductase and peroxiredoxins in human bronchial epithelial cells. *Free Radic. Biol. Med.* **47**(10), 1477–1485.

Ogusucu, R., Rettori, D., Munhoz, D. C., Netto, L. E., and Augusto, O. (2007). Reactions of yeast thioredoxin peroxidases I and II with hydrogen peroxide and peroxynitrite: Rate constants by competitive kinetics. *Free Radic. Biol. Med.* **42**(3), 326–334.

Parsonage, D., Karplus, P. A., and Poole, L. B. (2008). Substrate specificity and redox potential of AhpC, a bacterial peroxiredoxin. *Proc. Natl. Acad. Sci. USA* **105**(24), 8209–8214.

Peskin, A. V., Low, F. M., Paton, L. N., Maghzal, G. J., Hampton, M. B., and Winterbourn, C. C. (2007). The high reactivity peroxiredoxin 2 with H_2O_2 is not reflected in its reaction with other oxidants and thiol reagents. *J. Biol. Chem.* **282**(16), 11885–11892.

Rabilloud, T., Heller, M., Gasnier, F., Luche, S., Rey, C., Aebersold, R., Benahmed, M., Louisot, P., and Lunardi, J. (2002). Proteomics analysis of cellular response to oxidative stress—Evidence for in vivo overoxidation of peroxiredoxins at their active site. *J. Biol. Chem.* **277**(22), 19396–19401.

Requejo, R., Chouchani, E. T., Hurd, T. R., Menger, K. E., Hampton, M. B., and Murphy, M. P. (2010). Measuring mitochondrial protein thiol redox state. *Meth. Enzymol.* **474**, 123–147.

Rhee, S. G., Chae, H. Z., and Kim, K. (2005). Peroxiredoxins: A historical overview and speculative preview of novel mechanisms and emerging concepts in cell signaling. *Free Radic. Biol. Med.* **38**(12), 1543–1552.

Schroder, E., Brennan, J. P., and Eaton, P. (2008). Cardiac peroxiredoxins undergo complex modifications during cardiac oxidant stress. *Am. J. Physiol. Heart Circ. Physiol.* **295**(1), H425–H433.

Seo, Y. H., and Carroll, K. S. (2009). Profiling protein thiol oxidation in tumor cells using sulfenic acid-specific antibodies. *Proc. Natl. Acad. Sci. USA* **106**(38), 16163–16168.

Seo, M. S., Kang, S. W., Kim, K., Baines, I. C., Lee, T. H., and Rhee, S. G. (2000). Identification of a new type of mammalian peroxiredoxin that forms an intramolecular disulfide as a reaction intermediate. *J. Biol. Chem.* **275**(27), 20346–20354.

Seo, J. H., Lim, J. C., Lee, D. Y., Kim, K. S., Piszczek, G., Nam, H. W., Kim, Y. S., Ahn, T., Yun, C. H., Kim, K., Chock, P. B., and Chae, H. Z. (2009). Novel protective mechanism against irreversible hyperoxidation of peroxiredoxin: Nalpha-terminal acetylation of human peroxiredoxin II. *J. Biol. Chem.* **284**(20), 13455–13465.

Stacey, M. M., Peskin, A. V., Vissers, M. C., and Winterbourn, C. C. (2009). Chloramines and hypochlorous acid oxidize erythrocyte peroxiredoxin 2. *Free Radic. Biol. Med.* **47**(10), 1468–1476.

Winterbourn, C. C., and Hampton, M. B. (2008). Thiol chemistry and specificity in redox signaling. *Free Radic. Biol. Med.* **45**(5), 549–561.

Woo, H. A., Chae, H. Z., Hwang, S. C., Yang, K. S., Kang, S. W., Kim, K., and Rhee, S. G. (2003a). Reversing the inactivation of peroxiredoxins caused by cysteine sulfinic acid formation. *Science* **300**(5619), 653–656.

Woo, H. A., Kang, S. W., Kim, H. K., Yang, K. S., Chae, H. Z., and Rhee, S. G. (2003b). Reversible oxidation of the active site cysteine of peroxiredoxins to cysteine sulfinic acid—Immunoblot detection with antibodies specific for the hyperoxidized cysteine-containing sequence. *J. Biol. Chem.* **278**(48), 47361–47364.

Wood, Z. A., Poole, L. B., and Karplus, P. A. (2003). Peroxiredoxin evolution and the regulation of hydrogen peroxide signaling. *Science* **300**(5619), 650–653.

Zielonka, J., Vasquez-Vivar, J., and Kalyanaraman, B. (2008). Detection of 2-hydroxyethidium in cellular systems: A unique marker product of superoxide and hydroethidine. *Nat. Protoc.* **3**(1), 8–21.

CHAPTER FIVE

Thiol Redox Transitions by Thioredoxin and Thioredoxin-Binding Protein-2 in Cell Signaling

Eiji Yoshihara,*,† Zhe Chen,* Yoshiyuki Matsuo,* Hiroshi Masutani,* and Junji Yodoi*

Contents

1. Functional Regulation of Redox-Sensitive Proteins by Thiol Modification — 68
   1.1. Thioredoxin — 69
2. Thiol Reduction by the Thioredoxin Redox System — 70
   2.1. Thiol-redox regulation by thioredoxin in kinase-mediated cellular signal transduction — 71
   2.2. Thiol-redox regulation by thioredoxin in nuclear receptors and transcription factors-mediated cellular signal transduction — 72
3. Thioredoxin Superfamily — 73
4. Reversible Redox and Signal Regulation by Thioredoxin and Thioredoxin-binding Protein-2 (TBP-2) — 74
5. Conclusion — 76
References — 76

Abstract

The cellular thiol redox state is a crucial mediator of metabolic, signaling and transcriptional processes in cells, and an exquisite balance between the oxidizing and reducing states is essential for the normal function and survival of cells. Reactive oxygen species (ROS) are widely known to function as a kind of second messenger for intracellular signaling and to modulate the thiol redox state. Thiol reduction is mainly controlled by the thioredoxin (TRX) system and glutathione (GSH) systems as scavengers of ROS and regulators of the protein redox states. The thioredoxin system is composed of several related molecules interacting through the cysteine residues at the active site, including thioredoxin, thioredoxin-2, a mitochondrial thioredoxin family, and transmembrane

* Department of Biological Responses, Institute for Virus Research, Kyoto University, Kyoto, Japan
† Division of Systemic Life Science, Graduate School of Biostudies, Kyoto University, Kyoto, Japan

thioredoxin-related protein (TMX), an endoplasmic reticulum (ER)-specific thioredoxin family. Thioredoxin couples with thioredoxin-dependent peroxidases (peroxiredoxin) to scavenge hydrogen peroxide. In addition, thioredoxin does not simply act only as a scavenger of ROS but also as an important regulator of oxidative stress response through protein–protein interaction. The interaction of thioredoxin and thioredoxin-binding proteins such as thioredoxin-binding protein-2 (TBP-2, also called as Txnip or VDUP1), apoptosis signal kinase (ASK-1), redox factor 1 (Ref-1), Forkhead box class O 4 (FoxO4), and nod-like receptor proteins (NLRPs) suggested unconventional functions of thioredoxin and a novel mechanism of redox regulation. Here, we introduce the central mechanism of thiol redox transition in cell signaling regulated by thioredoxin and related molecules.

1. Functional Regulation of Redox-Sensitive Proteins by Thiol Modification

Phosphorylation and dephosphorylation of protein through the protein kinase and protein phosphatase kinase form the core of the signal transmission for the various physiological actions in the cell. Proteins receive oxidation on cysteine residues by reactive oxygen species (ROS) such as the H_2O_2, and the modulation causes protein functional change and physiological changes in cellular signaling. The protein function changes by thiol modification are a change in enzymatic activity, the induction of an allosteric effect and a modulation of protein–protein interaction. When cells are stimulated with ROS, many transcription factors (TFs) are also activated. However, higher doses of ROS may not activate these transcriptional factors, because they are highly susceptible to oxidation. Thus, the dosage of the redox signal is an important factor in the modulation of cell function. With respect to proteins, the thiol group of cysteinyl side chains is susceptible to a number of oxidative modifications, for instance, the formation of inter- or intramolecular disulfides between thiols ((P)–S–S–P) or between thiols and low-molecular-weight thiols such as glutathione (P–S–SG), the oxidation to sulfenic (P–SOH), sulfinic (P–SO_2H), and sulfonic (P–SO_3H) acid, and S-nitrosylation (Berndt et al., 2007). These modifications can alter the functions of numerous proteins that contain cysteines of structural importance, within their catalytic centers or at the interface of protein–protein interactions. These cysteinyl regulations affect cell signaling such as MAPKinase, Ca^{2+} signaling through PKCs and transcriptional factors (Table 5.1).

To a great extent, the redox state of these cysteinyl residues is controlled by the thioredoxin (TRX) and the glutathione (GRX) systems.

Table 5.1 Signaling molecules regulated by redox-thiol modification in the cells

Classification	Protein	Function
Phosphorylation enzyme	MEKK1, JNK	Inactivation
	Src, PKCs, PKA	Activation
Dephosphorylation enzyme	SHP-2, PTP1B	Inactivation
Acetylatic enzyme	FOXO4	Activation
Redox enzyme	Thioredoxin, glutathione	ASK1 regulation, signal complex (thioredoxin)
Small G protein	P21ras	Activation
Ca^{2+} signaling	Ins (1,4,5) P_3 receptor	Activation
Glycolytic enzyme	GAPDH	Bind with Siah1 and shuttle to nucleus
Protease	Caspase-3	Activation
	MMP-1, 8, 9	Inactivation
Metabolic enzyme	SUMO E1, E2	Inactivation
Transcriptional factors	NF-κB, AP-1 (c-Jun, c-Fos), p53, ATF, USF, Oct-2	Inactivation
	Keap1, Creb	Activation

Citation and modification from Berndt *et al.* (2007), Kamata and Hirata (1999), and Dansen *et al.* (2009).

1.1. Thioredoxin

Thioredoxin is a 12-kDa ubiquitous protein that has disulfide-reducing activity. Two cysteine residues (Cys32 and Cys35) of the active site -Cys-Gly-Pro-Cys- are responsible for this reducing activity. Thioredoxin was originally identified as a hydrogen donor for ribonucleotide reductase in *Escherichia Coli* (Holmgren, 1985). Decades of studies have indicated that thioredoxin is highly conserved in many organisms, including from bacterial organisms, plants, and mammals. We have identified human thioredoxin as an adult T-cell leukemia-derived factor (ADF) from the supernatant of HTLV-I infected T-cell line and named it the ADF, which was initially defined as an autocrine lymphokine and IL-2 receptor-inducing factor (Tagaya *et al.*, 1989). Given its nature to respond to oxidative stresses, thioredoxin expression is induced by a variety of physiochemical stimuli, including virus infection, mitogen, UV-irradiation, hydrogen peroxide, ischemia reperfusion, which we have broadly reviewed (Masutani *et al.*, 2005; Nakamura *et al.*, 1997, 2006). Natural metabolic or endocrine substances including hemin, estrogen, prostaglandins, sulforaphane, and cAMP can also induce the expression and secretion of TRX. Geranylgeranylacetone (GGA), an acyclic polyisoprenoid used as an antiulcer drug, or *tert*-butylhydroquinone (*t*BHQ), an electrophile stressor, can also induce TRX expression. A series of stress-responsive elements in the promoter region have been identified, including the oxidative stress response element (ORE), antioxidant responsive element (ARE), cAMP responsive element (CRE), xenobiotics responsive element (XRE), and Sp-1 (Bai *et al.*, 2003; Han *et al.*, 2003; Kim *et al.*, 2001; Tonissen and Wells, 1991; Yodoi *et al.*, 2002). Recently, we showed that fragrant unsaturated aldehydes from edible plants are novel thioredoxin inducers, through activation of the ARE in the promoter region and that they may be beneficial for protection against oxidative stress-induced cellular damage (Masutani *et al.*, 2009). In addition to ATL, an elevation of serum thioredoxin was detected in patients with HIV, HCV, HPV, or a variety of cancers (Fujii *et al.*, 1991; Kinnula *et al.*, 2004; Miyazaki *et al.*, 1998; Nakamura *et al.*, 2000, 2001a; Sumida *et al.*, 2000), all of which are involved with increased oxidative stresses. In addition to the mitogenic activity on lymphocytes, we broadly showed that extracellular thioredoxin can harbor cytoprotective effects, cytokine-stimulating effects, and antichemotactic effects in higher concentration ($>1\ \mu g$/ml in circulation) (Nakamura *et al.*, 2006).

Thioredoxin-knockout in mice is lethal for the early development and morphogenesis of the mouse embryo (Matsui *et al.*, 1996), whereas thioredoxin-transgenic mice are more resistant to oxidative stress with longer life spans, compared with wild-type mice (Mitsui *et al.*, 2002). These mice experiments suggested that thioredoxin is an essential gene for survival. Additional information is available in the other volume of this issue (Masutani and Yodoi, 2002; Nakamura *et al.*, 2002; Nishinaka *et al.*, 2002; Sato *et al.*, 1995).

2. Thiol Reduction by the Thioredoxin Redox System

Redox-sensitive cysteines have been shown to regulate several cellular signaling processes when cells are being exposed to oxidative stress. In the majority of cases, cysteine-dependent regulation of signaling by oxidative stress is due to the fact that many enzymes, such as lipids and protein phosphatases, deubiquitinating enzymes, have a critical cysteine residue within their catalytic center, oxidation of this cysteine residue results in impaired enzymatic function.

The thioredoxin system and the glutathione system constitute the major thiol-reducing systems (Holmgren, 1989). TRXs and GRXs comprise the TRX family of dithiol-disulfide oxidoreductases that are characterized by the TRX-like fold and a common dithiol/disulfide active site motif,

-Cys-X-X-Cys-, located at the end of β-strand and at the beginning of an α-helix (Martin, 1995).

2.1. Thiol-redox regulation by thioredoxin in kinase-mediated cellular signal transduction

Recent studies have revealed that cellular signaling pathways are regulated by the intracellular redox states. Reactive oxygen species (ROS) are widely known as regulators of cell signaling through regulation of the protein-thiol redox state. However, a high concentration of ROS induces cell dysfunction or cell death, a low physiological concentration of ROS activates cell proliferation, mobility, and gene expression. ROS acts as a second messenger in cells. Key molecules in the defense against oxidative stress (including ROS) are the members of the thioredoxin-fold family of proteins. Thioredoxins belong to corresponding systems containing NADP(H) and thioredoxin reductase, and they maintain a reduced intracellular redox state in mammalian cells by the reduction of protein thiols. It has been reported that thioredoxin negatively regulates cellular signal transduction of apoptosis signal-regulating kinase 1 (ASK1) through direct binding and inhibits the conformation change ASK1 for active formation (Saitoh et al., 1998). ASK1, also known as mitogen-activated protein kinase kinase kinase 5 (MAP3K5), is a part of mitogen-activated protein kinase (MAPK) cascade, which triggers the activation of c-Jun N-terminal kinase (JNK). ASK1 can bind to thioredoxin only in the reduced, but not oxidized form, since the dithiol-disulfide active site of thioredoxin is involved in the association. In the reducing environment, thioredoxin is associated with ASK1, giving rise to the inhibition of ASK1-mediated apoptosis. However, in the oxidizing environment, ASK1 is released from the ASK1–thioredoxin complex, giving rise to cell apoptosis. This association/dissociation process is a typical type of redox regulation conducted by thioredoxin. Thioredoxin NADP(H)-dependent reduction is required to maintain the activity of phosphatase and tensin homologue deleted from chromosome 10 (PTEN). The thioredoxin, NADP(H), thioredoxin reductase system reduced oxidized-form protein tyrosine phosphatase 1B (PTP1B), which is a member of the sensor protein family for ROS and suppressed H_2O_2 induced PTP1B inactivation (Lee et al., 1998). We also reported that thioredoxin is critical for inhibiting the p38 MAPK pathway and cytokine production (Hashimoto et al., 1999; Nakamura et al., 2001b). Recently, we have shown that thioredoxin overexpression showed elevation on the ERK1/2 phosphorylation levels over the entire range of S-nitrosoglutathione (GSNO) concentrations (Arai et al., 2008), and thioredoxin knockdown inactivated EGF-induced ERK phosphorylation in MCF-7 breast cancer and A549 lung cancer cells (Mochizuki et al., 2009). Further investigation is required to clarify the issue of thioredoxin-dependent redox regulation in ERK signaling.

2.2. Thiol-redox regulation by thioredoxin in nuclear receptors and transcription factors-mediated cellular signal transduction

Reducing condition is also very important for the regulation of DNA-binding activities of TFs. Redox-regulated TF or nuclear receptor (NR) with increased DNA-binding abilities, translocate from the cytoplasm to the nucleus in response to stress or ligand-bindings, for the transcriptional regulation. It was thought initially and demonstrated that thioredoxin shuttles between these two compartments, presumably providing a reducing environment for these targets when necessary.

Previously, we showed that thioredoxin enhances the DNA-binding activity of TFs such as NF-κB (Hirota et al., 1999a), AP-1 (Hirota et al., 1997), p53 (Ueno et al., 1999), and polyoma virus enhancer-binding protein-2 (PEBP-2) (Akamatsu et al., 1997). The NF-κB p50 and p65/RelA subunits have a well-conserved cysteine on its DNA-binding domain. A direct interaction between thioredoxin and NF-κB has been shown in *in vitro* cross-linking assay and nuclear magnetic resonance (NMR) studies (Hirota et al., 1999b; Qin et al., 1995). In the cytoplasm, TRX blocks the degradation of I-κB and suppresses NF-κB signaling, whereas in the nucleus, thioredoxin enhances NF-κB translocational activities by enhancing its DNA-binding capability (Hirota et al., 1999a). Thioredoxin enhances the DNA-binding activity and transactivation of activator protein-1 (AP-1) through Redox factor-1 (Ref-1), which was identified as a factor of restoring AP-1 DNA-binding activity and is identical to an AP-endonuclease (Xanthoudakis et al., 1992). The direct interaction between thioredoxin and Ref-1 has been shown in *in vitro* cross-linking assay and mammalian two-hybrid assay (Hirota et al., 1997). Thioredoxin is associated with Ref-1 through the redox active cysteines and coexpression of thioredoxin and Ref-1 enhances the transactivation of AP-1. Thioredoxin regulated Ref-1 activation also promotes the DNA binding of PEBP-2 and p53. We previously showed that thioredoxin and Ref-1 enhances p53-dependent p21 activation for DNA repair (Ueno et al., 1999). The thioredoxin-dependent redox regulation of p53 activity is coupled with the oxidative stress response and p53-dependent DNA repair mechanism. The interaction of hypoxia-inducible factor (HIF) and coactivators may also be a target of redox regulation. The C-terminal activation domain of HIF-1α and its related factors have a specific cysteine. The expression of thioredoxin and Ref-1 enhanced the interaction of these factors with a coactivator, CREB-binding protein (CBP)/p300 (Ema et al., 1999). Recently, thioredoxin was also reported to aid ERα in regulating the expression of estrogen responsive genes by direct protein–protein interaction (Rao et al., 2009). Moreover, reduced thioredoxin was reported to increase the XRE binding of Ah receptor heterodimers (Ireland et al., 1995). In fact, the thiol group associated with metals, including the highly conserved zinc-finger structure in the DBD domain of NRs, has been postulated to be one of strongest nucleophiles in cells (Beato *et al.*, 1995; Makino *et al.*, 1996; Mangelsdorf *et al.*, 1995), suggesting a broader significance of thioredoxin in the regulation of NR-mediated signal transduction. Of particular interest, as also described in the introduction, thioredoxin can also be upregulated upon high estrogen exposure (Maruyama *et al.*, 1999; Sahlin *et al.*, 1997), suggesting a putative positive feedback in the TRX-mediated ER signaling.

Redox regulation appears to be involved in various steps in the activation of TFs. In yeast, Yap1 and Skn7 cooperate on the yeast thioredoxin promoter to induce transcription in response to oxidative stress. Cysteine residues in the nuclear export signal (NES)-like sequence of the Yap1p regulatory domain serve as a sensor for the oxidative stress (Kuge *et al.*, 1998). In the thioredoxin-deficient mutant, Yap1p was constitutively concentrated in the nucleus and the level of expression of the Yap1 target genes was high (Izawa *et al.*, 1999). Taken together, redox regulation through thioredoxin is critical for the control of cell signaling and gene transcription.

3. Thioredoxin Superfamily

The thioredoxin-like active site motif (-Cys-X-X-Cys-) is found in several proteins classified in the thioredoxin superfamily (Matsuo *et al.*, 2002) and located in various cellular components, such as the mitochondria (Spyrou *et al.*, 1997; Tanaka *et al.*, 2002), the nucleus (Kurooka *et al.*, 1997), and the endoplasmic reticulum (ER) (Ferrari and Soling, 1999; Matsuo *et al.*, 2001). The mitochondrial thioredoxin family thioredoxin-2 (TRX-2) prevents apoptosis through the regulation of Bcl-xL protein and other mitochondrial proteins (Holmgren and Reichard, 1967; Ritz *et al.*, 2000; Tanaka *et al.*, 2002; Wang *et al.*, 2006; Zhu *et al.*, 2008) and nucleus thioredoxin family, nucleoredoxin controls cell polarity and development through Wnt-β-catenin signaling (Funato *et al.*, 2006; Hirota *et al.*, 2000). We previously identified the ER-localized thioredoxin family protein, transmembrane thioredoxin-related protein (TMX). TMX contains one catalytic thioredoxin-like domain with a unique active site motif, -Cys-Pro-Ala-Cys-, and a single transmembrane region. TMX orthologs have been found in other animal species, including mammals, *Drosophila melanogaster*, and Caenorhabditis (Ko and Chow, 2002) but not in plants, fungi, or prokaryotes. The thioredoxin-like domain of TMX is present in the ER lumen, and shows reductase and isomerase activity *in vitro* (Matsuo *et al.*, 2004). Recently, we reported TMX interacts with the MHC class I heavy chain and prevents the ER to cytosol retrotranslocation of misfolded class I HC (Heavy Chain) targeted for proteasome degradation (Matsuo *et al.*, 2009). These findings suggest a specific role of TMX in ER stress and its mechanism of action in redox-based ER quality control. Localization-specific thioredoxin family proteins may contribute to redox and the related signal homeostasis. Further investigation is required to clarify their physiological roles in the organelle-specific redox regulation.

4. Reversible Redox and Signal Regulation by Thioredoxin and Thioredoxin-binding Protein-2 (TBP-2)

As an endogenous negative regulator of thioredoxin, TBP-2 was identified in the yeast two-hybrid assay (Holmgren *et al.*, 2005; Masutani and Yodoi, 2002; Nishiyama *et al.*, 1999). TBP-2 was also given the name thioredoxin interacting protein (Txnip) (Bodnar *et al.*, 2002) or vitamin-D3 upregulating protein (VDUP1) (Chen and DeLuca, 1994). TBP-2 directly bind thioredoxin to inhibit thioredoxin-reducing activity (Nishiyama *et al.*, 1999). Two TBP-2 cysteines are important for thioredoxin binding through a disulfide exchange reaction between oxidized TBP-2 and reduced thioredoxin (Patwari *et al.*, 2006), suggesting that the thioreodxin–TBP-2 complex is important for redox dependant cell function.

Interestingly, TBP-2 is a member of the α-arrestin protein family, containing two characteristic arrestin-like domains with PxxP sequence, which is a known binding motif for SH3-domains containing proteins, and the PPXY sequence is a known binding motif for the WW domain (Alvarez, 2008; Oka *et al.*, 2006b; Patwari *et al.*, 2006, 2009). We also showed that TBP-2 interacts with importin-α (Nishinaka *et al.*, 2004a), while another group suggested its interaction with the SMRT-mSin3-HDAC corepressor complex or JAB1 (Han *et al.*, 2003; Jeon *et al.*, 2005). These results suggest that TBP-2 is an important molecule as a scaffold protein. Evidence suggesting that TBP-2 plays an important role in a wide variety of biological functions are growing, such as the regulation of cell death, growth, differentiation, and energy metabolism (Ahsan *et al.*, 2006; Aitken *et al.*, 2004; Chen *et al.*, 2008b; Corbett, 2008; Lee *et al.*, 2005; Nishinaka *et al.*, 2004b; Oka *et al.*, 2009). We reported that gene disruption of TBP-2 in mice results in a predisposition to death with severe bleeding, hypoglycemia, hyperinsulinemia, and liver steatosis during fasting (Oka *et al.*, 2006a). Other groups also showed the similar phenotype in Hcb-19 mice, which has a nonsense mutation of the TBP-2 gene (Hui *et al.*, 2004). These reports clearly suggest that TBP-2 is an important molecule in glucose and lipid metabolism *in vivo*. Furthermore, a recent reports have shown that TBP-2 regulates glucose uptake in human skeletal muscle (Parikh *et al.*, 2007), hepatic glucose production (Chutkow *et al.*, 2008),

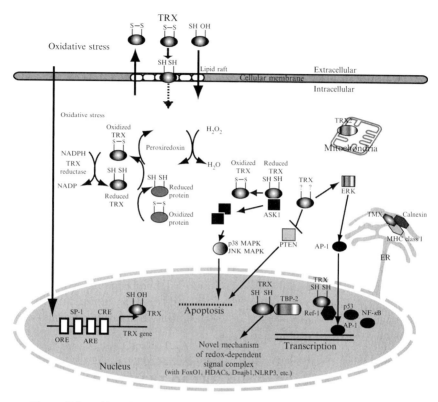

Figure 5.1 Thioredoxin and thioredoxin related molecules in cell signaling.

and pancreatic β-cell apoptosis (Chen et al., 2008a,b). We and other groups demonstrated that TBP-2-deficient mice showed increased insulin sensitivity (Chutkow et al., 2008; Hui et al., 2008; Oka et al., 2009) and insulin secretion (Hui et al., 2004; Oka et al., 2009). These reports raise the possibility that there is a fundamental novel redox regulatory mechanism occurring through thioredoxin and TBP-2 complex in cell and whole body metabolism. Very recently, with our collaborators, we elucidated the ROS induction of the formation of cysteine-thiol disulfide-dependent complexes of Forkhead box class O (FoxO) and the p300/CBP acetyltransferase, and showed that the modulation of FoxO biological activity by p300/CBP-mediated acetylation is fully dependent on the formation of this redox-dependent complex (Dansen et al., 2009). This complex contains thioredoxin and we showed coimmunoprecipitated complexes of FoxO4 and p300 with a buffer supplemented with active recombinant human thioredoxin almost completely abolished the interaction and that, in the presence of endogenous thioredoxin inhibitor, TBP-2, the interaction of

Foxo4Cys477 and p300 was stronger (Dansen et al., 2009). These results suggest that the redox-sensitive cysteines bridge p300/CBP-mediated acetylation regulated by thioredoxin and the TBP-2 complex link cellular redox status to the activity of the longevity protein FoxO.

It has been reported that the thioredoxin, TBP-2, and Dnajb5 complex modulates class II HDACs in response to oxidative stresses, regulating cardiac hypertrophy (Ago et al., 2008) in the heart. TBP-2 has been reported to dissociate the thioredoxin and nod-like receptor proteins (NLRP3) inflammasome complex in macrophage under oxidative stress when responding to ROS. Collectively, the complex formation of thioredoxin and TBP-2 axis may be a novel mechanism for biological function in the redox and cellular signal transition, related to responses to various factors, tissues structure, and type or timing.

5. Conclusion

Thioredoxin and the related molecules are important for the regulation of thiol-redox transition and related cellular signaling. The various functions of thioredoxin and related molecules for the redox transition and cellular signaling are summarized in Fig. 5.1. Thioredoxin regulates redox-dependent cell function through direct interaction with other proteins, including TBP-2. Further elucidation of the mechanism of the thioredoxin central protein complex under the various environments is required to understand how cells and the whole body integrate the various redox signals.

REFERENCES

Ago, T., Liu, T., Zhai, P., Chen, W., Li, H., Molkentin, J. D., Vatner, S. F., and Sadoshima, J. (2008). A redox-dependent pathway for regulating class II HDACs and cardiac hypertrophy. Cell **133**, 978–993.

Ahsan, M. K., Masutani, H., Yamaguchi, Y., Kim, Y. C., Nosaka, K., Matsuoka, M., Nishinaka, Y., Maeda, M., and Yodoi, J. (2006). Loss of interleukin-2-dependency in HTLV-I-infected T cells on gene silencing of thioredoxin-binding protein-2. Oncogene **25**, 2181–2191.

Aitken, C. J., Hodge, J. M., Nishinaka, Y., Vaughan, T., Yodoi, J., Day, C. J., Morrison, N. A., and Nicholson, G. C. (2004). Regulation of human osteoclast differentiation by thioredoxin binding protein-2 and redox-sensitive signaling. J. Bone Miner. Res. **19**, 2057–2064.

Akamatsu, Y., Ohno, T., Hirota, K., Kagoshima, H., Yodoi, J., and Shigesada, K. (1997). Redox regulation of the DNA binding activity in transcription factor PEBP2. The roles of two conserved cysteine residues. J. Biol. Chem. **272**, 14497–14500.

Alvarez, C. E. (2008). On the origins of arrestin and rhodopsin. BMC Evol. Biol. **8**, 222.

Arai, R. J., Ogata, F. T., Batista, W. L., Masutani, H., Yodoi, J., Debbas, V., Augusto, O., Stern, A., and Monteiro, H. P. (2008). Thioredoxin-1 promotes survival in cells exposed to S-nitrosoglutathione: Correlation with reduction of intracellular levels of nitrosothiols and up-regulation of the ERK1/2 MAP kinases. *Toxicol. Appl. Pharmacol.* **233**, 227–237.

Bai, J., Nakamura, H., Kwon, Y. W., Hattori, I., Yamaguchi, Y., Kim, Y. C., Kondo, N., Oka, S., Ueda, S., Masutani, H., and Yodoi, J. (2003). Critical roles of thioredoxin in nerve growth factor-mediated signal transduction and neurite outgrowth in PC12 cells. *J. Neurosci.* **23**, 503–509.

Beato, M., Herrlich, P., and Schutz, G. (1995). Steroid hormone receptors: Many actors in search of a plot. *Cell* **83**, 851–857.

Berndt, C., Lillig, C. H., and Holmgren, A. (2007). Thiol-based mechanisms of the thioredoxin and glutaredoxin systems: implications for diseases in the cardiovascular system. *Am. J. Physiol. Heart Circ. Physiol.* **292**, H1227–H1236.

Bodnar, J. S., Chatterjee, A., Castellani, L. W., Ross, D. A., Ohmen, J., Cavalcoli, J., Wu, C., Dains, K. M., Catanese, J., Chu, M., Sheth, S. S., Charugundla, K., et al. (2002). Positional cloning of the combined hyperlipidemia gene Hyplip1. *Nat. Genet.* **30**, 110–116.

Chen, K. S., and DeLuca, H. F. (1994). Isolation and characterization of a novel cDNA from HL-60 cells treated with 1, 25-dihydroxyvitamin D-3. *Biochim. Biophys. Acta* **1219**, 26–32.

Chen, J., Hui, S. T., Couto, F. M., Mungrue, I. N., Davis, D. B., Attie, A. D., Lusis, A. J., Davis, R. A., and Shalev, A. (2008a). Thioredoxin-interacting protein deficiency induces Akt/Bcl-xL signaling and pancreatic beta-cell mass and protects against diabetes. *FASEB J.* **22**, 3581–3594.

Chen, J., Saxena, G., Mungrue, I. N., Lusis, A. J., and Shalev, A. (2008b). Thioredoxin-interacting protein: a critical link between glucose toxicity and beta-cell apoptosis. *Diabetes* **57**, 938–944.

Chutkow, W. A., Patwari, P., Yoshioka, J., and Lee, R. T. (2008). Thioredoxin-interacting protein (Txnip) is a critical regulator of hepatic glucose production. *J. Biol. Chem.* **283**, 2397–2406.

Corbett, J. A. (2008). Thioredoxin-interacting protein is killing my beta-cells!. *Diabetes* **57**, 797–798.

Dansen, T. B., Smits, L. M., van Triest, M. H., de Keizer, P. L., van Leenen, D., Koerkamp, M. G., Szypowska, A., Meppelink, A., Brenkman, A. B., Yodoi, J., Holstege, F. C., and Burgering, B. M. (2009). Redox-sensitive cysteines bridge p300/CBP-mediated acetylation and FoxO4 activity. *Nat. Chem. Biol.* **5**, 664–672.

Ema, M., Hirota, K., Mimura, J., Abe, H., Yodoi, J., Sogawa, K., Poellinger, L., and Fujii-Kuriyama, Y. (1999). Molecular mechanisms of transcription activation by HLF and HIF1alpha in response to hypoxia: Their stabilization and redox signal-induced interaction with CBP/p300. *EMBO J.* **18**, 1905–1914.

Ferrari, D. M., and Soling, H. D. (1999). The protein disulphide-isomerase family: Unravelling a string of folds. *Biochem. J.* **339**(Pt 1), 1–10.

Fujii, S., Nanbu, Y., Nonogaki, H., Konishi, I., Mori, T., Masutani, H., and Yodoi, J. (1991). Coexpression of adult T-cell leukemia-derived factor, a human thioredoxin homologue, and human papillomavirus DNA in neoplastic cervical squamous epithelium. *Cancer* **68**, 1583–1591.

Funato, Y., Michiue, T., Asashima, M., and Miki, H. (2006). The thioredoxin-related redox-regulating protein nucleoredoxin inhibits Wnt-beta-catenin signalling through dishevelled. *Nat. Cell Biol.* **8**, 501–508.

Han, S. H., Jeon, J. H., Ju, H. R., Jung, U., Kim, K. Y., Yoo, H. S., Lee, Y. H., Song, K. S., Hwang, H. M., Na, Y. S., Yang, Y., Lee, K. N., and Choi, I. (2003). VDUP1 upregulated by TGF-beta1 and 1, 25-dihydorxyvitamin D3 inhibits tumor cell growth by blocking cell-cycle progression. *Oncogene* **22,** 4035–4046.

Hashimoto, S., Matsumoto, K., Gon, Y., Furuichi, S., Maruoka, S., Takeshita, I., Hirota, K., Yodoi, J., and Horie, T. (1999). Thioredoxin negatively regulates p38 MAP kinase activation and IL-6 production by tumor necrosis factor-alpha. *Biochem. Biophys. Res. Commun.* **258,** 443–447.

Hirota, K., Matsui, M., Iwata, S., Nishiyama, A., Mori, K., and Yodoi, J. (1997). AP-1 transcriptional activity is regulated by a direct association between thioredoxin and Ref-1. *Proc. Natl. Acad. Sci. USA* **94,** 3633–3638.

Hirota, K., Murata, M., Sachi, Y., Nakamura, H., Takeuchi, J., Mori, K., and Yodoi, J. (1999a). Distinct roles of thioredoxin in the cytoplasm and in the nucleus. A two-step mechanism of redox regulation of transcription factor NF-kappaB. *J. Biol. Chem.* **274,** 27891–27897.

Hirota, K., Nishiyama, A., and Yodoi, J. (1999b). Reactive oxygen intermediates, thioredoxin, and Ref-1 as effector molecules in cellular signal transduction. *Tanpakushitsu Kakusan Koso* **44,** 2414–2419.

Hirota, K., Matsui, M., Murata, M., Takashima, Y., Cheng, F. S., Itoh, T., Fukuda, K., and Yodoi, J. (2000). Nucleoredoxin, glutaredoxin, and thioredoxin differentially regulate NF-kappaB, AP-1, and CREB activation in HEK293 cells. *Biochem. Biophys. Res. Commun.* **274,** 177–182.

Holmgren, A. (1985). Thioredoxin. *Annu. Rev. Biochem.* **54,** 237–271.

Holmgren, A. (1989). Thioredoxin and glutaredoxin systems. *J. Biol. Chem.* **264,** 13963–13966.

Holmgren, A., and Reichard, P. (1967). Thioredoxin 2: Cleavage with cyanogen bromide. *Eur. J. Biochem.* **2,** 187–196.

Holmgren, A., Johansson, C., Berndt, C., Lonn, M. E., Hudemann, C., and Lillig, C. H. (2005). Thiol redox control via thioredoxin and glutaredoxin systems. *Biochem. Soc. Trans.* **33,** 1375–1377.

Hui, T. Y., Sheth, S. S., Diffley, J. M., Potter, D. W., Lusis, A. J., Attie, A. D., and Davis, R. A. (2004). Mice lacking thioredoxin-interacting protein provide evidence linking cellular redox state to appropriate response to nutritional signals. *J. Biol. Chem.* **279,** 24387–24393.

Hui, S. T., Andres, A. M., Miller, A. K., Spann, N. J., Potter, D. W., Post, N. M., Chen, A. Z., Sachithanantham, S., Jung, D. Y., Kim, J. K., and Davis, R. A. (2008). Txnip balances metabolic and growth signaling via PTEN disulfide reduction. *Proc. Natl. Acad. Sci. USA* **105,** 3921–3926.

Ireland, R. C., Li, S. Y., and Dougherty, J. J. (1995). The DNA binding of purified Ah receptor heterodimer is regulated by redox conditions. *Arch. Biochem. Biophys.* **319,** 470–480.

Izawa, S., Maeda, K., Sugiyama, K., Mano, J., Inoue, Y., and Kimura, A. (1999). Thioredoxin deficiency causes the constitutive activation of Yap1, an AP-1-like transcription factor in *Saccharomyces cerevisiae*. *J. Biol. Chem.* **274,** 28459–28465.

Jeon, J. H., Lee, K. N., Hwang, C. Y., Kwon, K. S., You, K. H., and Choi, I. (2005). Tumor suppressor VDUP1 increases p27(kip1) stability by inhibiting JAB1. *Cancer Res.* **65,** 4485–4489.

Kamata, H., and Hirata, H. (1999). *Cell Signal.* **11,** 1–14.

Kim, Y. C., Masutani, H., Yamaguchi, Y., Itoh, K., Yamamoto, M., and Yodoi, J. (2001). Hemin-induced activation of the thioredoxin gene by Nrf2. A differential regulation of the antioxidant responsive element by a switch of its binding factors. *J. Biol. Chem.* **276,** 18399–18406.

Kinnula, V. L., Paakko, P., and Soini, Y. (2004). Antioxidant enzymes and redox regulating thiol proteins in malignancies of human lung. *FEBS Lett.* **569,** 1–6.

Ko, F. C., and Chow, K. L. (2002). A novel thioredoxin-like protein encoded by the *C. elegans* dpy-11 gene is required for body and sensory organ morphogenesis. *Development* **129**, 1185–1194.

Kuge, S., Toda, T., Iizuka, N., and Nomoto, A. (1998). Crm1 (XpoI) dependent nuclear export of the budding yeast transcription factor yAP-1 is sensitive to oxidative stress. *Genes Cells* **3**, 521–532.

Kurooka, H., Kato, K., Minoguchi, S., Takahashi, Y., Ikeda, J., Habu, S., Osawa, N., Buchberg, A. M., Moriwaki, K., Shisa, H., and Honjo, T. (1997). Cloning and characterization of the nucleoredoxin gene that encodes a novel nuclear protein related to thioredoxin. *Genomics* **39**, 331–339.

Lee, S. R., Kwon, K. S., Kim, S. R., and Rhee, S. G. (1998). Reversible inactivation of protein-tyrosine phosphatase 1B in A431 cells stimulated with epidermal growth factor. *J. Biol. Chem.* **273**, 15366–15372.

Lee, K. N., Kang, H. S., Jeon, J. H., Kim, E. M., Yoon, S. R., Song, H., Lyu, C. Y., Piao, Z. H., Kim, S. U., Han, Y. H., Song, S. S., Lee, Y. H., Song, K. S., Kim, Y. M., Yu, D. Y., and Choi, I. (2005). VDUP1 is required for the development of natural killer cells. *Immunity* **22**, 195–208.

Makino, Y., Okamoto, K., Yoshikawa, N., Aoshima, M., Hirota, K., Yodoi, J., Umesono, K., Makino, I., and Tanaka, H. (1996). Thioredoxin: A redox-regulating cellular cofactor for glucocorticoid hormone action. Cross talk between endocrine control of stress response and cellular antioxidant defense system. *J. Clin. Invest.* **98**, 2469–2477.

Mangelsdorf, D. J., Thummel, C., Beato, M., Herrlich, P., Schutz, G., Umesono, K., Blumberg, B., Kastner, P., Mark, M., Chambon, P., and Evans, R. M. (1995). The nuclear receptor superfamily: The second decade. *Cell* **83**, 835–839.

Martin, J. L. (1995). Thioredoxin—A fold for all reasons. *Structure* **3**, 245–250.

Maruyama, T., Sachi, Y., Furuke, K., Kitaoka, Y., Kanzaki, H., Yoshimura, Y., and Yodoi, J. (1999). Induction of thioredoxin, a redox-active protein, by ovarian steroid hormones during growth and differentiation of endometrial stromal cells in vitro. *Endocrinology* **140**, 365–372.

Masutani, H., and Yodoi, J. (2002). Thioredoxin. Overview. *Methods Enzymol.* **347**, 279–286.

Masutani, H., Ueda, S., and Yodoi, J. (2005). The thioredoxin system in retroviral infection and apoptosis. *Cell Death Differ.* **12**(Suppl. 1), 991–998.

Masutani, H., Otsuki, R., Yamaguchi, Y., Takenaka, M., Kanoh, N., Takatera, K., Kunimoto, Y., and Yodoi, J. (2009). Fragrant unsaturated aldehydes elicit activation of the Keap1/Nrf2 system leading to the upregulation of thioredoxin expression and protection against oxidative stress. *Antioxid. Redox Signal.* **11**, 949–962.

Matsui, M., Oshima, M., Oshima, H., Takaku, K., Maruyama, T., Yodoi, J., and Taketo, M. M. (1996). Early embryonic lethality caused by targeted disruption of the mouse thioredoxin gene. *Dev. Biol.* **178**, 179–185.

Matsuo, Y., Akiyama, N., Nakamura, H., Yodoi, J., Noda, M., and Kizaka-Kondoh, S. (2001). Identification of a novel thioredoxin-related transmembrane protein. *J. Biol. Chem.* **276**, 10032–10038.

Matsuo, Y., Hirota, K., Nakamura, H., and Yodoi, J. (2002). Redox regulation by thioredoxin and its related molecules. *Drug News Perspect.* **15**, 575–580.

Matsuo, Y., Nishinaka, Y., Suzuki, S., Kojima, M., Kizaka-Kondoh, S., Kondo, N., Son, A., Sakakura-Nishiyama, J., Yamaguchi, Y., Masutani, H., Ishii, Y., and Yodoi, J. (2004). TMX, a human transmembrane oxidoreductase of the thioredoxin family: The possible role in disulfide-linked protein folding in the endoplasmic reticulum. *Arch. Biochem. Biophys.* **423**, 81–87.

Matsuo, Y., Masutani, H., Son, A., Kizaka-Kondoh, S., and Yodoi, J. (2009). Physical and functional interaction of transmembrane thioredoxin-related protein with major histocompatibility complex class I heavy chain: Redox-based protein quality control and its potential relevance to immune responses. *Mol. Biol. Cell* **20**, 4552–4562.

Mitsui, A., Hamuro, J., Nakamura, H., Kondo, N., Hirabayashi, Y., Ishizaki-Koizumi, S., Hirakawa, T., Inoue, T., and Yodoi, J. (2002). Overexpression of human thioredoxin in transgenic mice controls oxidative stress and life span. *Antioxid. Redox Signal.* **4**, 693–696.

Miyazaki, K., Noda, N., Okada, S., Hagiwara, Y., Miyata, M., Sakurabayashi, I., Yamaguchi, N., Sugimura, T., Terada, M., and Wakasugi, H. (1998). Elevated serum level of thioredoxin in patients with hepatocellular carcinoma. *Biotherapy* **11**, 277–288.

Mochizuki, M., Kwon, Y. W., Yodoi, J., and Masutani, H. (2009). Thioredoxin regulates cell cycle via the ERK1/2-cyclin D1 pathway. *Antioxid. Redox Signal.* **11**, 2957–2971.

Nakamura, H., Nakamura, K., and Yodoi, J. (1997). Redox regulation of cellular activation. *Annu. Rev. Immunol.* **15**, 351–369.

Nakamura, H., Bai, J., Nishinaka, Y., Ueda, S., Sasada, T., Ohshio, G., Imamura, M., Takabayashi, A., Yamaoka, Y., and Yodoi, J. (2000). Expression of thioredoxin and glutaredoxin, redox-regulating proteins, in pancreatic cancer. *Cancer Detect. Prev.* **24**, 53–60.

Nakamura, H., De Rosa, S. C., Yodoi, J., Holmgren, A., Ghezzi, P., and Herzenberg, L. A. (2001a). Chronic elevation of plasma thioredoxin: inhibition of chemotaxis and curtailment of life expectancy in AIDS. *Proc. Natl. Acad. Sci. USA* **98**, 2688–2693.

Nakamura, H., Herzenberg, L. A., Bai, J., Araya, S., Kondo, N., Nishinaka, Y., and Yodoi, J. (2001b). Circulating thioredoxin suppresses lipopolysaccharide-induced neutrophil chemotaxis. *Proc. Natl. Acad. Sci. USA* **98**, 15143–15148.

Nakamura, H., Mitsui, A., and Yodoi, J. (2002). Thioredoxin overexpression in transgenic mice. *Methods Enzymol.* **347**, 436–440.

Nakamura, H., Masutani, H., and Yodoi, J. (2006). Extracellular thioredoxin and thioredoxin-binding protein 2 in control of cancer. *Semin. Cancer Biol.* **16**, 444–451.

Nishinaka, Y., Nakamura, H., and Yodoi, J. (2002). Thioredoxin cytokine action. *Methods Enzymol.* **347**, 332–338.

Nishinaka, Y., Masutani, H., Oka, S., Matsuo, Y., Yamaguchi, Y., Nishio, K., Ishii, Y., and Yodoi, J. (2004a). Importin alpha1 (Rch1) mediates nuclear translocation of thioredoxin-binding protein-2/vitamin D(3)-up-regulated protein 1. *J. Biol. Chem.* **279**, 37559–37565.

Nishinaka, Y., Nishiyama, A., Masutani, H., Oka, S., Ahsan, K. M., Nakayama, Y., Ishii, Y., Nakamura, H., Maeda, M., and Yodoi, J. (2004b). Loss of thioredoxin-binding protein-2/vitamin D3 up-regulated protein 1 in human T-cell leukemia virus type I-dependent T-cell transformation: implications for adult T-cell leukemia leukemogenesis. *Cancer Res.* **64**, 1287–1292.

Nishiyama, A., Matsui, M., Iwata, S., Hirota, K., Masutani, H., Nakamura, H., Takagi, Y., Sono, H., Gon, Y., and Yodoi, J. (1999). Identification of thioredoxin-binding protein-2/vitamin D(3) up-regulated protein 1 as a negative regulator of thioredoxin function and expression. *J. Biol. Chem.* **274**, 21645–21650.

Oka, S., Liu, W., Masutani, H., Hirata, H., Shinkai, Y., Yamada, S., Yoshida, T., Nakamura, H., and Yodoi, J. (2006a). Impaired fatty acid utilization in thioredoxin binding protein-2 (TBP-2)-deficient mice: A unique animal model of Reye syndrome. *FASEB J.* **20**, 121–123.

Oka, S., Masutani, H., Liu, W., Horita, H., Wang, D., Kizaka-Kondoh, S., and Yodoi, J. (2006b). Thioredoxin-binding protein-2-like inducible membrane protein is a novel vitamin D3 and peroxisome proliferator-activated receptor (PPAR)gamma ligand target protein that regulates PPARgamma signaling. *Endocrinology* **147**, 733–743.

Oka, S., Yoshihara, E., Bizen-Abe, A., Liu, W., Watanabe, M., Yodoi, J., and Masutani, H. (2009). Thioredoxin binding protein-2/thioredoxin-interacting protein is a critical regulator of insulin secretion and peroxisome proliferator-activated receptor function. *Endocrinology* **150,** 1225–1234.

Parikh, H., Carlsson, E., Chutkow, W. A., Johansson, L. E., Storgaard, H., Poulsen, P., Saxena, R., Ladd, C., Schulze, P. C., Mazzini, M. J., Jensen, C. B., Krook, A., *et al.* (2007). TXNIP regulates peripheral glucose metabolism in humans. *PLoS Med.* **4,** e158.

Patwari, P., Higgins, L. J., Chutkow, W. A., Yoshioka, J., and Lee, R. T. (2006). The interaction of thioredoxin with Txnip. Evidence for formation of a mixed disulfide by disulfide exchange. *J. Biol. Chem.* **281,** 21884–21891.

Patwari, P., Chutkow, W. A., Cummings, K., Verstraeten, V. L., Lammerding, J., Schreiter, E. R., and Lee, R. T. (2009). Thioredoxin-independent regulation of metabolism by the alpha-arrestin proteins. *J. Biol. Chem.* **284,** 24996–25003.

Qin, J., Clore, G. M., Kennedy, W. M., Huth, J. R., and Gronenborn, A. M. (1995). Solution structure of human thioredoxin in a mixed disulfide intermediate complex with its target peptide from the transcription factor NF kappa B. *Structure* **3,** 289–297.

Rao, A. K., Ziegler, Y. S., McLeod, I. X., Yates, J. R., and Nardulli, A. M. (2009). Thioredoxin and thioredoxin reductase influence estrogen receptor alpha-mediated gene expression in human breast cancer cells. *J. Mol. Endocrinol.* **43,** 251–261.

Ritz, D., Patel, H., Doan, B., Zheng, M., Aslund, F., Storz, G., and Beckwith, J. (2000). Thioredoxin 2 is involved in the oxidative stress response in *Escherichia coli*. *J. Biol. Chem.* **275,** 2505–2512.

Sahlin, L., Stjernholm, Y., Holmgren, A., Ekman, G., and Eriksson, H. (1997). The expression of thioredoxin mRNA is increased in the human cervix during pregnancy. *Mol. Hum. Reprod.* **3,** 1113–1117.

Saitoh, M., Nishitoh, H., Fujii, M., Takeda, K., Tobiume, K., Sawada, Y., Kawabata, M., Miyazono, K., and Ichijo, H. (1998). Mammalian thioredoxin is a direct inhibitor of apoptosis signal-regulating kinase (ASK) 1. *EMBO J.* **17,** 2596–2606.

Sato, N., Iwata, S., Yamauchi, A., Hori, T., and Yodoi, J. (1995). Thiol compounds and adult T-cell leukemia virus infection: A potential therapeutic approach. *Methods Enzymol.* **252,** 343–348.

Spyrou, G., Enmark, E., Miranda-Vizuete, A., and Gustafsson, J. (1997). Cloning and expression of a novel mammalian thioredoxin. *J. Biol. Chem.* **272,** 2936–2941.

Sumida, Y., Nakashima, T., Yoh, T., Nakajima, Y., Ishikawa, H., Mitsuyoshi, H., Sakamoto, Y., Okanoue, T., Kashima, K., Nakamura, H., and Yodoi, J. (2000). Serum thioredoxin levels as an indicator of oxidative stress in patients with hepatitis C virus infection. *J. Hepatol.* **33,** 616–622.

Tagaya, Y., Maeda, Y., Mitsui, A., Kondo, N., Matsui, H., Hamuro, J., Brown, N., Arai, K., Yokota, T., Wakasugi, H., *et al.* (1989). ATL-derived factor (ADF), an IL-2 receptor/Tac inducer homologous to thioredoxin; Possible involvement of dithiol-reduction in the IL-2 receptor induction. *EMBO J.* **8,** 757–764.

Tanaka, T., Hosoi, F., Yamaguchi-Iwai, Y., Nakamura, H., Masutani, H., Ueda, S., Nishiyama, A., Takeda, S., Wada, H., Spyrou, G., and Yodoi, J. (2002). Thioredoxin-2 (TRX-2) is an essential gene regulating mitochondria-dependent apoptosis. *EMBO J.* **21,** 1695–1703.

Tonissen, K. F., and Wells, J. R. (1991). Isolation and characterization of human thioredoxin-encoding genes. *Gene* **102,** 221–228.

Ueno, M., Masutani, H., Arai, R. J., Yamauchi, A., Hirota, K., Sakai, T., Inamoto, T., Yamaoka, Y., Yodoi, J., and Nikaido, T. (1999). Thioredoxin-dependent redox regulation of p53-mediated p21 activation. *J. Biol. Chem.* **274,** 35809–35815.

Wang, D., Masutani, H., Oka, S., Tanaka, T., Yamaguchi-Iwai, Y., Nakamura, H., and Yodoi, J. (2006). Control of mitochondrial outer membrane permeabilization and Bcl-xL levels by thioredoxin 2 in DT40 cells. *J. Biol. Chem.* **281,** 7384–7391.

Xanthoudakis, S., Miao, G., Wang, F., Pan, Y. C., and Curran, T. (1992). Redox activation of Fos-Jun DNA binding activity is mediated by a DNA repair enzyme. *EMBO J.* **11,** 3323–3335.

Yodoi, J., Nakamura, H., and Masutani, H. (2002). Redox regulation of stress signals: possible roles of dendritic stellate TRX producer cells (DST cell types). *Biol. Chem.* **383,** 585–590.

Zhu, X., Sato, E. F., Wang, Y., Nakamura, H., Yodoi, J., and Inoue, M. (2008). Acetyl-L-carnitine suppresses apoptosis of thioredoxin 2-deficient DT40 cells. *Arch. Biochem. Biophys.* **478,** 154–160.

CHAPTER SIX

DETECTION OF PROTEIN THIOLS IN MITOCHONDRIAL OXIDATIVE PHOSPHORYLATION COMPLEXES AND ASSOCIATED PROTEINS

Kelly K. Andringa *and* Shannon M. Bailey

Contents

1. Introduction	84
2. Mitochondria Isolation and Protein Thiol Labeling	86
2.1. Preparation of mitochondria from liver	86
2.2. Labeling of reduced (i.e., unmodified) mitochondrial protein thiols using IBTP	87
3. Application of Blue Native-PAGE for the Isolation of Oxidative Phosphorylation Protein Subunits and Other Proteins Associated with the Complexes	89
3.1. Preparation and running of 1D BN-PAGE gels	89
3.2. Separation of the protein complexes into their individual protein subunits using 2D BN-PAGE	92
4. Detection of IBTP-Labeled Protein Thiols in Protein Complexes	94
4.1. Immunoblotting protocol for gels	94
5. Analysis and Mass Spectrometry Identification of Protein	95
5.1. Imaging and analysis of gels and blots	95
5.2. Mass spectrometry identification of proteins	97
6. Other Considerations	104
7. Conclusion	105
Acknowledgments	106
References	106

Abstract

The ability to detect and identify mitochondrial proteins that are sensitive to oxidative modification and inactivation by reactive species is important in understanding the molecular mechanisms responsible for mitochondrial

Department of Environmental Health Sciences, Center for Free Radical Biology, University of Alabama at Birmingham, Birmingham, Alabama, USA

Methods in Enzymology, Volume 474 © 2010 Elsevier Inc.
ISSN 0076-6879, DOI: 10.1016/S0076-6879(10)74006-4 All rights reserved.

dysfunction and tissue injury. In particular, cysteine residues play critical roles in maintaining the functional and structural integrity of numerous proteins in the mitochondrion and throughout the cell. To define changes in mitochondrial protein thiol status, proteomic approaches have been developed in which unmodified, reduced thiols (i.e., R–SH or thiolate species R–S$^-$) are tagged with thiol-labeling reagents that can be visualized following gel electrophoresis and immunoblotting techniques. Herein, we describe the use of one thiol-labeling approach in combination with blue native gel electrophoresis (BN-PAGE) to detect reactive thiol groups within mitochondrial proteins including those of the oxidative phosphorylation (OxPhos) system. Labeling or "tagging" of protein thiol groups in combination with various gel electrophoresis and proteomics techniques is a valuable way to measure alterations in cellular or organelle thiol proteomes in response to drug treatment, disease state, or metabolic/oxidative stress.

1. INTRODUCTION

Mitochondrial dysfunction, manifested by the inability to maintain cellular ATP levels, has been linked to a number of pathologies and diseases. As both a source for the formation and target of modifications mediated by reactive oxygen, nitrogen, and lipid species (ROS, RNS, and RLS), the mitochondrion is also recognized as a critical site in cellular stress responses. An emerging and novel aspect of mitochondrial function is the role this organelle plays in regulating key signaling pathways, which impact cellular responses to metabolic, hypoxic, and oxidative stress. While the sequence of events leading to toxicant and disease induced mitochondrial dysfunction are not fully known, accumulating evidence suggests that posttranslational modification of critical functional groups within mitochondrial proteins by reactive species may alter mitochondrial and cellular systems including energy metabolism and redox signaling. Specifically, mitochondrial protein thiols have received a significant amount of attention in the posttranslational modification arena as these alterations can regulate several mitochondrial functions (Costa *et al.*, 2003; Dahm *et al.*, 2006). Therefore, the ability to detect, identify, and structurally characterize mitochondrial protein thiols that might be susceptible to oxidative modifications is critical to understanding the molecular mechanisms contributing to oxidative damage and mitochondrial dysfunction in pathobiology.

Studies show that reactive species (i.e., ROS, RNS, and RLS) modify critical amino acid residues thereby disrupting the structure and catalytic function of proteins. These toxic effects in mitochondria largely stem from the diffusion of nitric oxide (NO) into mitochondria and its reaction with superoxide ($O_2^{\bullet-}$) to generate more reactive species such as peroxynitrite (ONOO$^-$), other secondary RNS, and RLS. A number of reversible and irreversible modifications to cysteine residues are known to occur upon the interaction of the free sulfhydryl groups with reactive species (Ying et al., 2007). Reversible modifications to thiols include the formation of nitrosothiols, sulfenic acids, and protein mixed disulfides. Similarly, cysteines can be irreversibly oxidized to higher oxidation states such as sulfinic and sulfonic acids, as well as modified via the adduction of electrophilic lipids like 4-hydroxynonenal. Each of these modifications has the potential to elicit a unique biologic response that may disrupt mitochondrial function. Reversible modifications may also be important as they may "protect" vulnerable thiol groups from permanent modification. Emerging evidence has linked the oxidation and modification of mitochondrial protein thiols with alterations in respiration, energy metabolism, oxidant production, and permeability transition (Costa et al., 2003; Dahm et al., 2006; Lemasters et al., 2002). Previously, we reported a chronic alcohol-dependent loss of function of the mitochondrial low K_m aldehyde dehydrogenase from cysteine modifications (Venkatraman et al., 2004b). This is noteworthy because inactivation of a key detoxification enzyme like aldehyde dehydrogenase would lead to the accumulation of acetaldehyde and other RLS, which themselves have been shown to inactivate proteins through cysteine modifications (Doorn et al., 2006).

To determine the role of posttranslational protein thiol modifications that control mitochondrial functions, methods are needed to detect and analyze changes in mitochondrial protein thiol status. With the upsurge in proteomics technology advancements, the ability to detect and identify specific mitochondrial protein thiols susceptible to posttranslational modifications has become more realistic (Torta et al., 2008; Witze et al., 2007). Typically, these approaches incorporate methodologies using thiol "labeling" or "tagging" reagents in combination with gel electrophoresis techniques (Hill et al., 2009; Kettenhofen et al., 2008); however, some recent studies are now forgoing the electrophoresis step and going straight to mass spectrometry identification (Sethuraman et al., 2004). One weakness of some of these approaches is that because many of the thiol-labeling reagents are membrane impermeable they cannot be used in fully functional mitochondria but only in mitochondrial extracts. Therefore, information about the redox status of protein thiols in the intact respiring mitochondrion cannot be experimentally gathered. To address this problem, Murphy and colleagues developed an approach using novel molecules to selectively label reduced (unmodified) protein thiol groups within mitochondria (Lin et al., 2002). Specifically, a lipophilic, cationic compound 4-iodobutyl triphenylphosphonium (IBTP) was synthesized and accumulates in mitochondria as a function of the large membrane potential across the mitochondrial inner membrane and the positive charge on the triphenylphosphonium (TPP) moiety. Thiolate groups ($R-S^-$) on mitochondrial proteins will displace the iodo group from the TPP moiety resulting in a TPP-tagged protein thiol

(Lin et al., 2002). These "tagged" proteins are then easily detected following gel electrophoresis and immunoblotting with an antibody developed against the TTP group. Therefore, protein thiol groups that have been oxidized or modified as a consequence of oxidative stress can be detected and identified using immunoblotting where decreased labeling with IBTP is shown by decreased anti-TPP immunoreactivity (i.e., signal) on the blot (Lin et al., 2002; Venkatraman et al., 2004b). Therefore, this approach has the distinct advantage of enabling detection of the thiol redox state of specific mitochondrial proteins in functional isolated mitochondria and intact cells.

To address the impacts of hepatotoxicants and disease on liver mitochondria protein thiols, we have used the IBTP thiol tagging strategy developed by Murphy and colleagues in combination with blue native gel electrophoresis (BN-PAGE). BN-PAGE was chosen for these studies in an attempt to focus analyses of thiols on those proteins that comprise the oxidative phosphorylation (OxPhos) system. While BN-PAGE offers advantages over other gel electrophoresis techniques, some limitations were identified and discussed. This chapter describes the approaches and techniques used to examine the mitochondrial protein thiol proteome in liver mitochondria with a goal to apply these methods to pathologies associated with mitochondrial dysfunction in liver.

2. Mitochondria Isolation and Protein Thiol Labeling

2.1. Preparation of mitochondria from liver

Mitochondria are prepared from fresh liver tissue using standard differential centrifugation techniques as described in Bailey et al. (2001, 2006). Briefly, fresh liver tissue is chopped into tiny pieces and then gently disrupted (i.e., homogenized) in 10 vol/g liver ice-cold isolation buffer (0.25 M sucrose, 5 mM Tris–HCl, 1 mM EDTA, pH 7.4) using a motor-driven teflon-coated pestle (serrated bottom) and glass homogenizer. A cocktail of protease inhibitors is also added to the isolation buffer to prevent protein degradation. The liver homogenate is centrifuged at $560 \times g$ (10 min, 4 °C) to remove the nucleus and other cellular debris. The postnuclear supernatant is collected and centrifuged at $8500 \times g$ (10 min, 4 °C) to isolate the mitochondrial fraction. The postmitochondrial supernatant can be discarded or saved to prepare cytosol and microsomal fractions. It is important at this stage that the mitochondrial fraction (i.e., pellet) is subjected to three wash steps (i.e., gentle resuspension/centrifugation at $8500 \times g$) to obtain a more pure and highly functional aliquot of mitochondria. After the last wash step the mitochondrial pellet is resuspended in a small volume of buffer to obtain a protein concentration of approximately 30–50 mg protein/ml.

Note that it is important that mitochondria are kept cold at all times during liver harvest, homogenization, centrifugation, and wash steps. This is easily facilitated by performing mitochondrial isolation in a cold room at 4 °C. Immediately after isolation, mitochondrial functionality (i.e., respiration) must be measured to demonstrate high-respiratory capacity and coupling. Liver mitochondria isolated from control animals (rats or mice) should have respiratory control ratios [state 3 (ADP-dependent)/state 4 (ADP-independent) respiration] in the range of 5–8 using glutamate/malate or succinate as oxidizable substrates. Using this approach, mitochondrial protein yield is typically 25–30 mg mitochondrial protein/g wet liver and cytochrome c oxidase and citrate synthase activities should be determined as additional markers of mitochondrial purity and yield (Bailey *et al.*, 2006; Venkatraman *et al.*, 2004a). It is vitally important that mitochondria used for thiol labeling and proteomics studies are functional (i.e., coupled) and pure (Zhang *et al.*, 2008). Mitochondria kept on ice remain functional and "viable" for 2–3 h allowing sufficient time to perform a number of functional assays including protein thiol-labeling studies. Readers should refer to other methodologies for preparation of mitochondria from other tissues.

2.2. Labeling of reduced (i.e., unmodified) mitochondrial protein thiols using IBTP

For IBTP labeling, 0.5–1.0 mg of mitochondrial protein is pipetted into a test tube or microcentrifuge tube containing 1 ml mitochondrial respiration buffer (130 mM KCl, 2 mM KH$_2$PO$_4$, 3 mM HEPES, 2 mM MgCl$_2$, and 1 mM EGTA, pH 7.2) and 10 µl 1 M succinate is added (Venkatraman *et al.*, 2004b). The mitochondrial suspension is allowed to incubate for 3 min at 37 °C to allow mitochondria to respire and establish a membrane potential. After this initial 3 min incubation, 5 µl of a 5 mM IBTP solution (prepared in ethanol) is added to the mitochondrial suspension, gently mixed by rotation of the tube, and allowed to incubate for an additional 10 min at 37 °C. The labeling reaction is stopped by the addition of 2 µl of 1 mM FCCP, a mitochondrial protonophore that leads to uncoupling and dissipates the mitochondrial membrane potential. As the uptake of IBTP into mitochondria is dependent on the membrane potential, the addition of FCCP prevents further accumulation of IBTP, which essentially stops the labeling reaction. Mitochondria are then centrifuged at 10,000×g for 10 min at 4 °C. The supernatant is removed and the mitochondrial pellet is stored at −80 °C before being used in gel electrophoresis and immunoblotting experiments.

A scheme illustrating the IBTP-labeling protocol is provided in Fig. 6.1A. Again, this methodology is essentially the same as that used in Lin *et al.* (2002) and Venkatraman *et al.* (2004b). Moreover, an example showing the extent of IBTP labeling is provided in Fig. 6.1B using a low-resolution, global

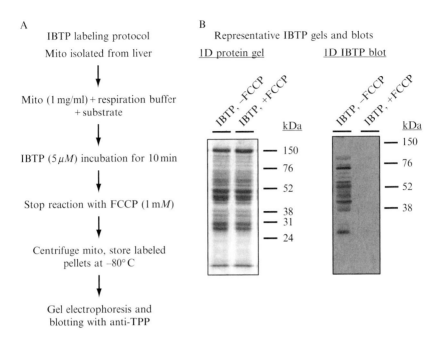

Figure 6.1 Scheme illustrating IBTP-labeling protocol of protein thiols with representative 1D SDS–PAGE gels and immunoblots of IBTP-labeled liver mitochondria. *Panel A*: a brief step-by-step description of the procedure used to label mitochondrial protein thiol groups with the reagent IBTP. *Panel B*: representative 1D SDS–PAGE gels and immunoblots of liver mitochondria subjected to IBTP-labeling procedure described in text. Mitochondrial protein (10 μg) was separated on 10% gel by SDS–PAGE followed by immunoblotting with anti-TPP antiserum. Note that IBTP labeling with and without pretreatment with FCCP had no effect on protein resolution (see gel image).

separation of mitochondrial proteins by SDS–PAGE and immunoblotting against IBTP-labeled proteins. The authors highly recommend that a low-resolution SDS–PAGE approach be used before attempting more high-resolution "2D" proteomics experiments to demonstrate first whether there are large differences in labeling among treatments and second whether there are labeling inconsistencies in replicates within treatments. This helps validate the reproducibility of the labeling reaction. As shown in the IBTP blot panel (Fig. 6.1B, right panel), there are several IBTP-reactive protein bands within the liver mitochondrial compartment as was shown previously (Venkatraman *et al.*, 2004b), and as expected, labeling with IBTP was prevented by pretreating mitochondria with FCCP; that is, no IBTP uptake demonstrated by an absence of labeling in uncoupled mitochondria. Previously, we have shown that IBTP labeling is also specific for protein thiol groups because pretreatment of mitochondria with the thiol-alkylating agent iodoacetamide blocks labeling (Venkatraman *et al.*, 2004b). Taken together, these results and others (Lin *et al.*, 2002; Sethuraman *et al.*, 2004; Venkatraman *et al.*, 2004b) show that the labeling of mitochondrial proteins by IBTP is membrane potential-dependent and specific for protein thiol groups.

3. Application of Blue Native-PAGE for the Isolation of Oxidative Phosphorylation Protein Subunits and Other Proteins Associated with the Complexes

In the low-resolution SDS–PAGE gel and blot shown in Fig. 6.1, the detection and identification of individual proteins labeled by IBTP is likely to be masked by comigration of proteins with identical molecular weights. In light of this, we have more routinely used the high-resolution 2D proteomic approaches (e.g., 2D isoelectric focusing (IEF)/SDS–PAGE) for identifying changes to the mitochondrial proteome with regards to alterations in protein abundance or posttranslational modifications (Venkatraman *et al.*, 2004a). While 2D IEF/SDS–PAGE proteomics is well-suited for the separation of the more hydrophilic, matrix proteins of the mitochondrion, analyses of the more basic and hydrophobic inner membrane proteins is hampered because these proteins precipitate at the basic end of the IEF gel and hence cannot be separated on gels (Bailey *et al.*, 2005). In attempts to circumvent this problem when studying the mitochondrial proteome, many laboratories have adapted the BN-PAGE approach to aid high-resolution analyses of the OxPhos complexes and other associated proteins. Additional information regarding history, development, and theory behind the development of BN-PAGE is presented in papers of Schägger and von Jagow (Schagger and von Jagow, 1991; Schagger *et al.*, 1994). In this approach, the five OxPhos complexes and other proteins associated with the inner membrane are maintained intact in their "native" and active state during the first dimension gel electrophoresis step (1D BN-PAGE). The protein complexes can then be cut from the 1D gel and transferred to a denaturing second dimension gel (2D BN-PAGE) to separate the OxPhos complexes into their individual protein subunits. A scheme illustrating this procedure is provided in Fig. 6.2.

3.1. Preparation and running of 1D BN-PAGE gels

For 1D BN-PAGE proteomics, remove the needed number of 1.0 mg mitochondrial pellets from the -80 °C freezer, place on ice, and prepare samples as described in Andringa *et al.* (2009) and Brookes *et al.* (2002). Briefly, each 1.0 mg mitochondria sample is resuspended in 100 μl

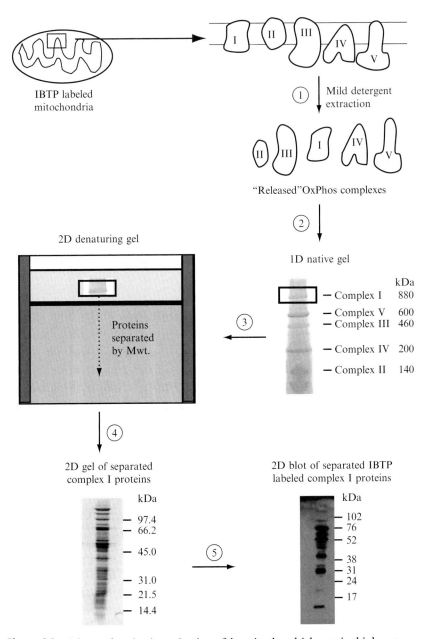

Figure 6.2 Scheme showing investigation of the mitochondrial protein thiol proteome by BN-PAGE proteomics. This diagram illustrates the five major steps used to detect thiols in proteins that makeup and are associated with the OxPhos system complexes. In this technique, mitochondria subjected to the IBTP-labeling protocol described in Fig. 6.1 are exposed to a mixture of mild detergents and buffers to extract the protein

BN-PAGE extraction buffer (0.75 M aminocaproic acid, 50 mM BisTris, pH 7.0) and 12.5 μl 10% (w/v) N-dodecyl-β-D-maltoside. Mitochondria are subjected to gentle up-and-down pipetting (on ice) to dissociate protein complexes from the inner membrane. Samples should also be gently vortexed every 5 min for 30–45 min in total to aid in the extraction procedure while keeping samples on ice between vortex steps. Note that incubation time in the extraction buffer should be optimized per sample type to ensure maximal release of intact OxPhos complexes. The mitochondrial extracts are centrifuged at 17,500×g for 5 min at 4 °C to pellet any nondissolved material/debris with the supernatant removed to a fresh tube. After the protein concentration of the mitochondrial extract has been determined, 6.3 μl of cold Coomassie Brilliant Blue buffer (0.5 M aminocaproic acid and 5% Coomassie Brilliant Blue G-250, pH 7.0) is added to each tube of ~100 μl of mitochondrial extract and gently mix. Samples (75–250 μg/sample well) should be loaded immediately onto the 1D BN-PAGE gel after the addition of Coomassie Blue G-250. The sample wells within the stacking gel should be filled with cold "Hi-Blue" cathode buffer (50 mM Tricine, 15 mM BisTris, and 0.02% Coomassie Brilliant Blue G-250, pH 7.0). The recipes needed for the preparation of 1D BN-PAGE gels are shown in Table 6.1.

For electrophoresis, most commercially available units can be adapted and used to perform BN-PAGE. Fill the inner (i.e., upper) buffer chamber with cold "Hi-Blue" cathode buffer and the outer chamber with cold anode buffer (50 mM BisTris, pH 7.0). Electrophoresis is performed in the cold (4 °C) at 40 V for 1 h or until samples have migrated from the stacking gel into the resolving gel. Once the sample (dye front) has migrated into the resolving gel, the cathode buffer is changed to cold "Low-Blue" cathode buffer (50 mM Tricine, 15 mM BisTris, and 0.002% Coomassie Brilliant Blue G-250, pH 7.0) and electrophoresis is continued for an additional 3–5 h at 80 V or until the dye front reaches the bottom of the gel. If duplicate gels are run (one for protein stain and one for immunoblotting)

---

complexes from the inner membrane (step 1). The extracted or "released" protein complexes are subjected to 1D native gel electrophoresis (step 2). During the 1D native electrophoresis step, the five OxPhos complexes remain intact and functional. After 1D BN-PAGE, each individual protein complex band is cut from the gel and placed on a 2D denaturing gel (step 3) where the individual polypeptides that comprise, and are associated with, each complex are separated vertically by molecular weight (step 4). At this stage in the experimental procedure, the protein gels can be stained for protein or be subjected to standard immunoblotting techniques so that IBTP-labeled protein can be detected (step 5). Both protein gels and blots can undergo image analysis to determine protein abundance and thiol-labeling intensity in mitochondrial proteins from tissue of interest. Remember that a decrease in IBTP-labeling intensity can be interpreted to mean that there has been an increase in oxidized/modified thiols within proteins as only the reduced thiol group binds the IBTP reagent.

Table 6.1 Recipe for 1D BN-PAGE gels

	Resolving gel solutions		Stacking gel solution
	Light—5%	Heavy—12%	4%
Protogel (ml)	0.66	1.60	0.60
Water (ml)	1.97	0.56	2.40
3× Gel Buffer[a] (ml)	1.34	1.34	1.50
Glycerol (ml)	–	0.47	–
10% AMPS (μl)	26.0	26.0	70.0
TEMED (μl)	4.0	4.0	9.0

[a] The 3× Gel Buffer contains 1.5 M aminocaproic acid, 150 mM BisTris. Adjust pH to 7.0 at 4 °C and store at room temperature. Protogel is a 37.5:1 acrylamide to bisacrylamide stabilized solution (30%) (National Diagnostics, Atlanta, GA) and AMPS is ammonium persulfate solution. Gradient gels can be prepared using a standard gradient-maker apparatus.

the gel being used for protein staining can be incubated with a Coomassie blue stain (0.3 g Coomassie Blue R-250, 100 ml glacial acetic acid, 250 ml isopropanol, and 650 ml ultrapure H_2O) overnight to visualize the intact complexes in the 1D BN-PAGE gel. Destaining is typically performed with 10% glacial acetic acid. Gels processed for 2D BN-PAGE and immunoblotting are subjected to the procedures described in the following sections. An example of a 1D BN-PAGE from liver mitochondria is shown in Fig. 6.2, step 2.

3.2. Separation of the protein complexes into their individual protein subunits using 2D BN-PAGE

The second dimension (2D) BN-PAGE gel step can be performed immediately after 1D gels or at a later time. After electrophoresis, remove the gel from the apparatus and place gel on a fresh sheet of laboratory Parafilm or clean glass. Separate each gel lane from the rest of the gel and cut the individual complexes (I, II, III, IV, and V) from each lane. Note that we typically cannot adequately visualize and retrieve Complex II from gels. Store each complex gel piece in a microcentrifuge tube at -80 °C until ready to perform the 2D gel step. Table 6.2 shows the recipe needed to prepare one 1.5-mm-thick 2D BN-PAGE gel. Make sure to leave a 1.0 cm gap at the top of the resolving gel for placement of gel pieces.

Once the 2D gel has polymerized, carefully unclamp the gel, raise the gel plates up in the clamps, refasten the clamps, and then place the top, that is, 1–2 cm gap of the gel plate sandwich onto an angled hot plate or heat block. Pour approximately 4–5 ml of freshly made warm agarose solution (100 mg low melting temperature agarose, 1 ml 10% SDS, 100 μl β-mercaptoethanol into 9.0 ml ultrapure H_2O) into the gap between the gel plates. Using the

Table 6.2 Recipe for 2D BN-PAGE gels

2D BN-PAGE Gel Buffer[a] (ml)	2.98
Protogel (ml)	2.98
H_2O (ml)	2.31
Glycerol (ml)	0.72
10% AMPS (μl)	60.0
TEMED (μl)	6.00

[a] The 2D BN-PAGE Gel Buffer contains 3 M Tris–HCl and 0.3% SDS, pH 8.45. Store buffer at room temperature.

back gel plate as a "staging area" and the agarose as a "lubricant," gently slide the gel slice down between the gel plates until it is in place on top of the SDS–PAGE resolving gel. Once the gel slice is in place, remove the gel assembly from the hot plate, insert a single "tooth" of a gel comb on the end to serve as a well for molecular weight markers and allow agarose to set (see panel 2D denaturing gel, Fig. 6.2, step 3). After the agarose solution has cooled and hardened, remove excess agarose from the top of the gel plate with a razor blade and then overlay the gel with a thin layer of freshly made SDS/β-mercaptoethanol solution (20 μl β-mercaptoethanol, 200 μl 10% SDS into 1780 μl ultrapure H_2O) every 5–10 min for 30 min to denature proteins present in gel slice. When comparing multiple treatment groups, several individual slices (i.e., individual complex pieces) can be placed side-by-side on the same gel to be run and analyzed together. This allows for minimization of gel-to-gel variability and also allows for direct comparison of effects of treatment on that complex. For electrophoresis, the gel electrophoresis apparatus is assembled and the inner (i.e., upper) buffer chamber is filled with 2D cathode buffer (100 mM Tris, 100 mM Tricine, and 0.1% SDS, pH 8.25) and the outer chamber is filled with 2D anode buffer (200 mM Tris, pH 8.9 with HCl). Molecular weight markers are placed in the "homemade" well and the gels are run at 30 V for 45 min followed by 110 V for 1.5–2 h, or until the blue dye front reaches the bottom of the gel. At this point gels can now be stained for protein content using any protein dye (Coomassie Blue or SYPRO® Ruby, Invitrogen, Carlsbad, CA) or processed for immunoblotting analysis. For our studies, we typically stain for total protein content using SYPRO® Ruby by first fixing gels (40% methanol, 10% glacial acetic acid, H_2O) for 1 h at room temperature. Gels are then stained overnight at room temperature with shaking in approximately 30 ml SYPRO® Ruby stain followed by destaining (10% methanol, 7% glacial acetic acid, H_2O) with multiple changes (approximately once per hour for up to 4 h) while shaking at room temperature. A representative 2D BN-PAGE gel is shown for Complex I in Fig. 6.2 along with an immunoblot of IBTP immunoreactive protein thiol groups (steps 4 and 5).

4. Detection of IBTP-Labeled Protein Thiols in Protein Complexes

4.1. Immunoblotting protocol for gels

After electrophoresis is completed, the gel is removed from between the plates, the agarose stacking gel is discarded, and the gel is soaked in transfer buffer (24 mM Tris base, 194 mM glycine, 20% methanol in H_2O, pH 8–8.5) for 15 min. Typically, we use nitrocellulose membrane for immunoblots; however, other membrane media can be used. Membranes are incubated in transfer buffer for 5 min before assembling the immunoblot apparatus. Again, most commercially available "wet/tank" immunoblotting units can be used at this step in the experimental protocol. Immunoblot transfers are set to run at 350 mA for 1 h. During the transfer time period a 5% (w/v) nonfat milk blocking buffer is prepared in 1× TBS-T (10× TBS-T is prepared: 0.2 M Tris pH 7.4, 9% (w/v) NaCl, H_2O, 0.5% Tween-20 and then diluted 1:10 with H_2O for use in washes). After transfer is complete, blots are removed and blocked overnight at 4 °C with rotation in the nonfat milk blocking buffer. On the next day blots are removed from blocking buffer and washed 2 × 5 min with 1× TBS-T to remove excess blocking buffer. The anti-TPP antibody (kindly provided by Dr. M. P. Murphy) is diluted in 1% (w/v) BSA in 1× TBS-T to 1:10,000 (e.g., 1 μl antibody into 10 ml buffer) and blots are incubated with shaking for 1 h at room temperature. After primary antibody incubation, blots are washed for 15 min in 1× TBS-T, followed by 4 × 5 min washes in 1× TBS-T. The secondary antibody (goat antirabbit) is prepared in 1% BSA blocking buffer at a dilution of 1:5000 (i.e., 1 μl antibody into 5 ml buffer) and blots are incubated with shaking for 1 h at room temperature. After incubation with the secondary antibody, blots are washed in 1× TBS-T for 15 min followed by 6 × 5 min washes. Washed blots are exposed to chemiluminescent substrate (Super-Signal West Pico Chemiluminescent substrate, Pierce Biotechnology) for 1–2 min. After the blots have been exposed to chemiluminescent substrates they can then be imaged using film with increasing exposure time points or using a chemiluminescent imager, for example, ChemiDoc XRS (Bio-Rad Laboratories, Hercules CA). The ChemiDoc XRS system and comparable systems allow immunoblots to be imaged at multiple time points like with film; however, images can be captured using an image acquisition setting that allows for "movies" of exposure to be created. These "movies" build the acquisition of the chemiluminescent image on top of one another to allow for accumulated images of each band to be measured. This may facilitate analysis and help to ensure that the image closest to saturation, but not oversaturation of the camera pixels, is acquired and used for image analysis of blots.

5. ANALYSIS AND MASS SPECTROMETRY IDENTIFICATION OF PROTEIN

5.1. Imaging and analysis of gels and blots

Analysis of 2D protein gels and IBTP blots is carried out using procedures described more fully in Bailey *et al.* (2006) and Venkatraman *et al.* (2004a,b). Briefly, images of 2D BN-PAGE gels can be exposed on an imager instrument, for example, Bio-Rad ChemiDoc XRS imager, using the appropriate filters for either Coomassie Blue or SYPRO® Ruby stain. The total protein gel images for each complex band (Fig. 6.3) can be analyzed using Quantity One software (Bio-Rad Laboratories, Hercules, CA) or any other comparable imaging software package. For imaging, the density of individual protein bands is determined and used to calculate the total protein density for each gel lane (data not shown). This type of analysis is needed when comparing control mitochondria to treatment mitochondria. Thus, any potential differences in total protein or individual bands will be known. Similarly, immunoblots are also imaged (Fig. 6.3) and analyzed using the same approach. Again, this type of analysis is needed when trying to determine differences in IBTP thiol-labeling intensity between treatment and control mitochondria. Using this approach, we detected several IBTP-reactive bands within each complex band (Fig. 6.3). For example, 12 IBTP-reactive bands were detected in the gel band for Complex I (panel A), 9 IBTP-reactive bands were detected in the gel band for Complex III (panel B), 11 IBTP-reactive bands were detected in the gel band for Complex IV (panel C), and 9 IBTP-reactive bands were detected in the gel band for Complex V (panel D). Currently, the identity of these IBTP-reactive protein(s) is unknown. Studies are underway to identify these proteins using the mass spectrometry methods described in the following section (Section 5.2). One nontrivial aspect of this experiment is the ability to match protein bands between gels and blots with 100% certainty. To help in this matching process, it might be important to consider a more high-resolution separation of the proteins—a "3D" approach—as described by Brookes and colleagues (Tompkins *et al.*, 2006), especially as multiple proteins of similar molecular weights are likely to be present in each individual protein band. In this "3D" approach, the protein band for each complex is cut from the 1D BN-PAGE gel, homogenized, and proteins are extracted from the gel band in a buffer containing urea, thiourea, detergents, and ampholytes used for IEF. The extracted proteins are then applied to 2D IEF gel strips (pH = 3–10); the strips are allowed to rehydrate overnight with IEF performed the following morning by standard protocols. After IEF, the gel strips are then applied to the top of an SDS–PAGE gel with proteins separated based on molecular weight. Thus, complex and associated complex proteins are separated in this technique in three dimensions: BN-PAGE → IEF → SDS–PAGE; resulting in a "3D"

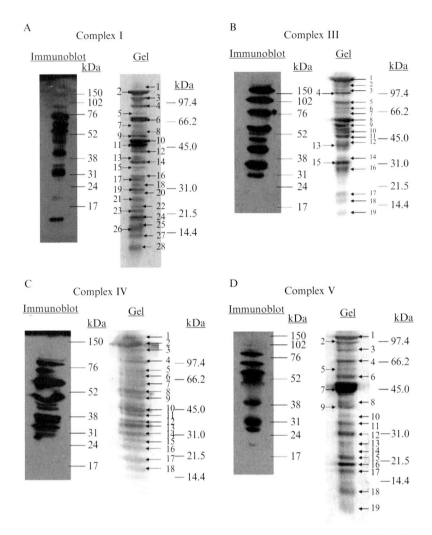

Figure 6.3 Detection of IBTP-labeled mitochondrial proteins using 2D BN-PAGE proteomics. Representative gels and IBTP immunoblots are shown for proteins within the major protein bands (Complexes I, III, IV, and V) resolved using 2D BN-PAGE as described in Fig. 6.2. Proteins identified within each complex band are labeled numerically and listed in their corresponding protein ID table (Tables 6.3–6.6). *Panel A*: proteins present within Complex I band; protein IDs included in Table 6.3. *Panel B*: proteins present within Complex III band; protein IDs included in Table 6.4. *Panel C*: proteins present within Complex IV band; protein IDs included in Table 6.5. *Panel D*: proteins present within Complex V band, proteins IDs included in Table 6.6. Note that two different types of molecular weight markers were included on gels and blots. Prestained markers were used for immunoblots (Full Range Molecular Weight, GE Healthcare and Life Sciences, Piscataway, NJ) whereas molecular weight markers compatible with protein staining techniques (SYPRO® Ruby) were used on protein gels (SDS–PAGE standards, Bio-Rad).

high-resolution protein map. And, like the experiments described in the current method, gels can be subjected to immunoblotting to probe for changes in thiol status using the same IBTP-labeled mitochondrial samples (Tompkins et al., 2006). However, as shown by Brookes and colleagues, this "3D" separation of Complex I resulted in only a few proteins being resolved on conventional 2D IEF/SDS–PAGE. Again, this is probably due to the incompatibility of these hydrophobic and basic proteins on IEF gels. Clearly, future studies need to be aimed at improving methodologies for proteomics analyses of OxPhos proteins with specific emphasis on posttranslational modifications identification.

5.2. Mass spectrometry identification of proteins

Proteins can be identified using standard mass spectrometry techniques available in most mass spectrometry core facilities (www.uab.edu/proteomics). Herein, proteins were identified by cutting bands from gels and performing matrix-assisted laser desorption ionization time-of-flight (MALDI-TOF) mass spectrometry by standard methods (Andringa et al., 2009; Brookes et al., 2002; Venkatraman et al., 2004a). Briefly, trypsin digestion of protein samples was performed at 37 °C with the resulting solution extracted in a 50/50 solution of 5% formic acid and acetonitrile. The sample supernatants were collected and SpeedVac dried, resuspended in 0.1% formic acid, desalted, and diluted 1:10 with saturated α-cyano-4-hydroxycinnamic acid matrix before being applied to the MALDI-TOF target plates. Peptides were analyzed with a Voyager De-Pro mass spectrometer in positive mode. Spectra were analyzed using Voyage Explorer software and the peptide masses of identified proteins were entered into MASCOT database (www.matrixscience.com) and the NCBI database was searched to match the peptide fragments with parent proteins. Proteins identified from these specific gels were classified using the Universal Protein Resource website (www.uniprot.org) maintained by UniProt Consortium. Proteins identified from gels shown in Fig. 6.3 are presented in Tables 6.3–6.6 for each of the complex bands.

It is important to identify the proteins that are present within the complex bands so that attempts can be made to match proteins in gels to IBTP-labeled proteins present in blots. Moreover, it should not be assumed that the proteins present within the complex band will be exclusively those known polypeptides that comprise the OxPhos complexes. Indeed, this is the case in this experimental method as many of the proteins identified within the individual complex bands are not complex proteins (Tables 6.3–6.6). For example, only 10 of 45+ Complex I proteins, 5 of 11 Complex III proteins, 2 of 13 Complex IV proteins, and 6 of 16+ Complex V proteins were identified using this BN-PAGE approach. This finding is not unlike what we and others have seen when using other modified BN-PAGE methods (Andringa et al., 2009;

Table 6.3 Proteins identified within the Complex I band

Band number	Protein designation	MOWSE	MW (kDa)	Accession number
1	Carbamoyl-phosphate synthetase 1	132	164.5	gi|8393186
2	Carbamoyl-phosphate synthetase 1	254	164.5	gi|8393186
3	Pyruvate carboxylase	160	129.7	gi|31543464
4	Aminoadipate-semialdehyde synthase precursor	128	103.4	gi|109473207
5	Aconitase 2	104	85.4	gi|40538860
6	NADH dehydrogenase (ubiquinone) Fe–S protein 1, 75 kDa	215	79.4	gi|53850628
	Hydroxyacyl-coenzyme A dehydrogenase/3-ketoacyl-coenzyme A thiolase/enoyl-coenzyme A hydratase	215	82.6	gi|60688124
7	Calcium-binding mitochondrial carrier protein Aralar2	123	74.4	gi|62646841
8	ND			
9	Propionyl coenzyme A carboxylase, beta polypeptide	167	58.6	gi|51260066
10	ATP synthase alpha subunit	126	58.8	gi|203055
11	ATP synthase beta subunit	138	51.2	gi|1374715
12	NADH dehydrogenase (ubiquinone) Fe–S protein 2	143	52.5	gi|58865384
13	Ubiquinol-cytochrome-*c* reductase complex core protein 2	89	48.4	gi|418146
14	Glutamate oxaloacetate transaminase 2	121	47.3	gi|6980972

Table 6.3 (continued)

Band number	Protein designation	MOWSE	MW (kDa)	Accession number
15	Glutamate oxaloacetate transaminase 2	121	47.3	gi|6980972
16	NADH dehydrogenase (ubiquinone) 1 alpha subcomplex, 9	95	41.8	gi|60688426
17	Malate dehydrogenase	81	35.6	gi|42476181
18	ND			
19	ATP synthase gamma chain	97	30.2	gi|728931
20	ND			
21	NADH dehydrogenase (ubiquinone) Fe–S protein 3	122	30.2	gi|27702072
22	24-kDa mitochondrial NADH dehydrogenase precursor	97	26.5	gi|205628
23	NADH-ubiquinone oxidoreductase PDSW subunit	89	21.0	gi|109487851
24	Mitochondrial ATP synthase, O subunit	102	23.4	gi|20302061
25	NADH dehydrogenase (ubiquinone) 1 beta subcomplex, 5	85	21.6	gi|34856800
26	NADH dehydrogenase (ubiquinone) 1 beta subcomplex, 4, 15 kDa	85	14.9	gi|62660147
27	NADH dehydrogenase (ubiquinone) 1, subunit C2	88	14.4	gi|57164133
28	NADH dehydrogenase (ubiquinone) 1 alpha subcomplex, 2	85	10.8	gi|27682913

Proteins listed within table were identified using MALDI-TOF as described in text. The band number is the same used to denote the protein bands shown in Fig. 6.3, panel A. The mass is measured in kDa and the MOWSE score is an algorithmic calculation used to assign a statistical weight to each peptide match; thus, a higher MOWSE score implies higher statistical certainty of the match being correct. Accession number is a unique NCBI label given to proteins.

Table 6.4 Proteins identified within the Complex III band

Band number	Protein	MOWSE	MW (kDa)	Accession number
1	Carbamoyl-phosphate synthetase 1	1054	164.5	gi|8393186
2	Carbamoyl-phosphate synthetase 1	1206	164.5	gi|8393186
3	Carbamoyl-phosphate synthetase 1	559	164.5	gi|8393186
4	Aldehyde dehydrogenase 1 family, member L2	1093	101.6	gi|109480409
5	Mitochondrial trifunctional protein, alpha subunit	843	82.6	gi|148747393
6	Stress-70 protein, mitochondrial precursor	607	73.7	gi|3122170
7	Acyl-coenzyme A dehydrogenase, very long chain	640	70.7	gi|6978435
8	Heat shock protein 1	783	60.9	gi|11560024
9	Glutamate dehydrogenase 1	867	61.3	gi|6980956
10	Aldehyde dehydrogenase 2	833	56.4	gi|14192933
11	Alanine-glyoxylate aminotransferase 2	529	57.2	gi|13929196
	3-Hydroxy-3-methylglutaryl-coenzyme A synthase 2	527	56.8	gi|27465521
12	Ubiquinol-cytochrome-c reductase core protein 1	476	52.8	gi|51948476
13	Ubiquinol-cytochrome-c reductase core protein 2	511	48.4	gi|55741544
14	Malate dehydrogenase	368	35.6	gi|42476181
15	Cytochrome c-1	166	35.4	gi|34866853

Table 6.4 (continued)

Band number	Protein	MOWSE	MW (kDa)	Accession number
16	Ubiquinol-cytochrome-c reductase, Rieske iron–sulfur polypeptide 1	167	29.4	gi|57114330
17	Microsomal glutathione S-transferase 1	58	17.4	gi|19705453
18	Ubiquinol-cytochrome-c reductase binding protein	74	13.5	gi|34866011
19	ND			

Proteins listed within table were identified using MALDI-TOF as described in text. The band number is the same used to denote the protein bands shown in Fig. 6.3, panel B. The mass is measured in kDa and the MOWSE score is an algorithmic calculation used to assign a statistical weight to each peptide match; thus, a higher MOWSE score implies higher statistical certainty of the match being correct. Accession number is a unique NCBI label given to proteins.

Table 6.5 Proteins identified within the Complex IV band

Band number	Protein	MOWSE	MW (kDa)	Accession number
1	Carbamoyl-phosphate synthetase 1	1400	164.5	gi|8393186
2	Carbamoyl-phosphate synthetase 1	2169	164.5	gi|8393186
3	Carbamoyl-phosphate synthetase 1	1382	164.5	gi|8393186
4	Dimethylglycine dehydrogenase	923	95.9	gi|55742723
5	Acyl-CoA synthetase long-chain family member 1	1027	78.2	gi|25742739
6	Succinate dehydrogenase complex, subunit A, flavoprotein	1054	71.6	gi|18426858
7	Heat shock protein 1	683	60.9	gi|11560024

(continued)

Table 6.5 (continued)

Band number	Protein	MOWSE	MW (kDa)	Accession number
8	Aldehyde dehydrogenase	803	48.2	gi|16073616
9	3-Hydroxy-3-methylglutaryl-coenzyme A synthase 2	840	56.9	gi|27465521
10	Acetyl-coenzyme A acyltransferase 2	858	41.8	gi|149027156
11	Acetyl-coenzyme A acyltransferase 2	698	41.8	gi|149027156
12	Ornithine transcarbamylase	461	39.9	gi|6981312
13	Electron transferring flavoprotein, alpha polypeptide	583	34.9	gi|57527204
	Malate dehydrogenase	577	35.6	gi|42476181
14	Electron-transfer-flavoprotein, beta polypeptide	550	27.7	gi|51948412
15	Ornithine transcarbamylase	304	39.9	gi|205892
		304	39.9	gi|6981312
16	Peptidylprolyl isomerase F	264	21.8	gi|26892289
17	Cytochrome-*c*-oxidase subunit IV isoform 1	244	19.5	gi|8393180
18	Cytochrome-*c*-oxidase subunit 5A	216	16.0	gi|117099

Proteins listed within table were identified using MALDI-TOF as described in text. The band number is the same used to denote the protein bands shown in Fig. 6.3, panel C. The mass is measured in kDa and the MOWSE score is an algorithmic calculation used to assign a statistical weight to each peptide match; thus, a higher MOWSE score implies higher statistical certainty of the match being correct. Accession number is a unique NCBI label given to proteins.

Brookes *et al.*, 2002) in which many non-OxPhos proteins are present in gels. To circumvent this problem, one might choose to use an alternative approach in which the OxPhos complexes are separated by liquid phase high-throughput chromatography. This approach has been used successfully to isolate Complex I on a Superose 6 column (Burwell *et al.*, 2006). Moreover, improved isolation of complex proteins might be achieved by using the recently developed "Immunocapture" methodology of Dr. Roderick Capaldi and colleagues to purify OxPhos complexes proteins from mitochondrial samples (Keeney *et al.*, 2006;

Table 6.6 Proteins identified within the Complex V band

Band number	Protein	MOWSE	MW (kDa)	Accession number
1	Carbamoyl-phosphate synthetase 1	2065	164.5	gi|8393186
2	Carbamoyl-phosphate synthetase 1	1235	164.5	gi|8393186
3	Aldehyde dehydrogenase 1 family, member L2	946	101.7	gi|109480409
4	Mitochondrial trifunctional protein, alpha subunit	798	82.6	gi|148747393
5	Calcium-binding mitochondrial carrier protein Aralar2	627	74.4	gi|62646841
6	Chaperonin 60	757	60.8	gi|1778213
7	Chain A, Rat Liver F1-ATPase	978	55.2	gi|93279422
8	Ubiquinol-cytochrome-*c* reductase core protein 2	685	48.4	gi|55741544
9	Glutamate oxaloacetate transaminase 2	622	47.3	gi|6980972
10	ND			
11	Malate dehydrogenase	778	35.7	gi|42476181
12	ATP synthase gamma chain	449	30.2	gi|728931
13	Electron-transfer-flavoprotein, beta polypeptide	222	27.7	gi|51948412
14	3-Hydroxyacyl-CoA dehydrogenase type-2	375	27.3	gi|7387724
15	ATP synthase, H+ transporting, mitochondrial F0 complex, subunit b, isoform 1	345	28.9	gi|19705465

(*continued*)

Table 6.6 (continued)

Band number	Protein	MOWSE	MW (kDa)	Accession number
16	ATP synthase, H+ transporting, mitochondrial F0 complex, subunit d	464	18.8	gi|9506411
17	ND			
18	ATP synthase, H+ transporting, mitochondrial F0 complex, subunit E	253	8.3	gi|17978459
19	ATP synthase, H+ transporting, subunit-epsilon	112	5.6	gi|258789

Proteins listed within table were identified using MALDI-TOF as described in text. The band number is the same used to denote the protein bands shown in Fig. 6.3, panel D. The mass is measured in kDa and the MOWSE score is an algorithmic calculation used to assign a statistical weight to each peptide match; thus, a higher MOWSE score implies higher statistical certainty of the match being correct. Accession number is a unique NCBI label given to proteins.

Schilling et al., 2005). This "pull-down" approach has been used to successfully resolve up to 42 proteins of the Complex I subproteome. With this said, however, the interesting feature of the BN-PAGE approach is that because the first dimension is done under native (nondenaturing) conditions using mild detergent extraction, this allows for protein:protein interactions to remain intact so that unique binding partners among the OxPhos complexes and other matrix proteins can be identified. Identifying these interactions will likely be important for increased understanding of alterations in mitochondria in diseases with different mitochondrial protein-interaction "fingerprints" being identified in diseased versus healthy tissue.

6. OTHER CONSIDERATIONS

In addition to some of the caveats presented in previous sections, it is important to briefly elaborate on the alternative mechanisms that might be responsible for decrease in the IBTP signal in immunoblots. A decrease in the IBTP signal could be due to: (1) a modification of a specific protein thiol by a redox-dependent mechanism. In this case, one would then need to identify the protein showing decreased thiol labeling by mass spectrometry. Ideally, the specific site and type of modification would be determined, but this is likely to be only possible for the more abundant mitochondrial proteins. Furthermore, it should be emphasized that this approach only allows one to determine that a protein thiol has been modified; it does not provide any information on the type of modification; (2) a decrease in the total amount of the specific thiol-labeled protein. This issue can be addressed by measuring the protein amount in the corresponding protein band or spot on the SYPRO® Ruby stained gel. For example, because we observed no alcohol-dependent change in total aldehyde dehydrogenase protein, the decrease in IBTP signal in the protein could then be attributed to oxidation/modification of protein cysteinyl groups (Venkatraman *et al.*, 2004b); and (3) a change in the mitochondrial membrane potential. Because IBTP accumulation into mitochondria is dependent on the mitochondria membrane potential, it could be argued that decreased labeling of proteins might be due to decreased uptake of IBTP. Therefore, it is important to measure whether differences in mitochondrial membrane potential exist between control and experimental groups and to test global thiol labeling between control and experimental groups by low-resolution gel electrophoresis (Fig. 6.1). In addition to these changes, it is also possible that one might see increased IBTP labeling in treatment groups due to increased reactivity of thiols after exposure to oxidants or toxicants (Andringa *et al.*, 2008). This might result as a consequence of a conformational change in the protein resulting in the exposure of a thiol group rendering it more reactive either by increasing its accessibility or decreasing the pK_a by moving the cysteine closer to charged amino acids.

As the OxPhos system is comprised of a discrete number of proteins (approximately 90), these types of analyses are focused on a limited set of known proteins. In light of this, and the fact that many of these proteins have been characterized structurally and functionally, once one has identified a posttranslational modification immediate knowledge into how these changes might affect key functions of the respiratory complex is possible. Moreover, using the methods recently developed by Landar and colleagues, the amount of thiol modification in a specific protein can be accurately quantified and directly correlated with protein activity (Hill *et al.*, 2009). It is possible that some of the proteins identified as being modified will be present in lower amounts. This might indicate that posttranslational modifications of thiols interfere with assembly of the OxPhos complexes leading to increased clearance of modified proteins.

7. Conclusion

Using the protocol described herein it will be possible for laboratories to begin to detect and possibly identify specific alterations in mitochondrial protein thiol groups that impact mitochondrial function and contribute to normal physiology and pathophysiology. It has been proposed by many groups that oxidative and nitrosative stress induced by toxicants and disease contributes to mitochondrial dysfunction via oxidation and/or modification of protein cysteinyl groups. Due to the elevated pH in the mitochondrial matrix, the reactivity of protein thiols is increased making them more susceptible to oxidative modification than proteins present in other cellular compartments. Indeed, previous studies from our laboratory and others demonstrate that cysteine resides that are targeted are potentially involved in increased pathobiology from hepatoxicants like ethanol and acetaminophen (Andringa et al., 2008; Moon et al., 2006; Venkatraman et al., 2004b). While the presence of these posttranslational modifications in proteins predicts that the structure and/or function of a protein may be altered, the crucial finding is whether there is a significant decrease or possible increase in activity. Thus, functional assays, if available, must be performed for those proteins identified as containing significant increases in posttranslational modifications. Only then can a direct link between protein modification and function be made.

ACKNOWLEDGMENTS

This work was supported in part by NIH grants AA15172 and DK73775 (S. M. B.).

The authors would like to thank Dr. Michael P. Murphy, MRC-Dunn Human Nutrition Unit, Cambridge, UK, for kindly providing our laboratory with the IBTP reagent and anti-TPP antibody. The authors thank Dr. Steve Barnes and Mr. Landon Wilson of the UAB Mass Spectrometry Shared Facility for mass spectrometry analyses. Mass spectrometers in the Shared Facility came from funds provided by the NCRR grants S10 RR11329, S10 RR13795 plus UAB Health Services Foundation General Endowment Fund. Operational funds came in part from the UAB Comprehensive Cancer Center Core Grant (P30 CA13148), the Purdue-UAB Botanicals Center for Age-Related Disease (P50 AT00477), the UAB Center for Nutrient–Gene Interaction in Cancer Prevention (U54 CA100949), the UAB Skin Disease Research Center (P30 AR050948), and the UAB Polycystic Kidney Disease Center (P30 DK74038). We would also like to thank Drs. Victor Darley-Usmar and Aimee Landar for helpful advice in proteomics studies and analyses and Ms. Adrienne L. King for helpful review and revision of this chapter.

REFERENCES

Andringa, K. K., Bajt, M. L., Jaeschke, H., and Bailey, S. M. (2008). Mitochondrial protein thiol modifications in acetaminophen hepatotoxicity: Effect on HMG-CoA synthase. *Toxicol. Lett.* **177**, 188–197.

Andringa, K., King, A., and Bailey, S. (2009). Blue native-gel electrophoresis proteomics. *Methods Mol. Biol.* **519**, 241–258.

Bailey, S. M., Patel, V. B., Young, T. A., Asayama, K., and Cunningham, C. C. (2001). Chronic ethanol consumption alters the glutathione/glutathione peroxidase-1 system and protein oxidation status in rat liver. *Alcohol. Clin. Exp. Res.* **25**, 726–733.

Bailey, S. M., Landar, A., and Darley-Usmar, V. (2005). Mitochondrial proteomics in free radical research. *Free Radic. Biol. Med.* **38,** 175–188.

Bailey, S. M., Robinson, G., Pinner, A., Chamlee, L., Ulasova, E., Pompilius, M., Page, G. P., Chhieng, D., Jhala, N., Landar, A., Kharbanda, K. K., Ballinger, S., and Darley-Usmar, V. (2006). S-adenosylmethionine prevents chronic alcohol-induced mitochondrial dysfunction in the rat liver. *Am. J. Physiol. Gastrointest. Liver Physiol.* **291,** G857–G867.

Brookes, P. S., Pinner, A., Ramachandran, A., Coward, L., Barnes, S., Kim, H., and Darley-Usmar, V. M. (2002). High throughput two-dimensional blue-native electrophoresis: A tool for functional proteomics of mitochondria and signaling complexes. *Proteomics* **2,** 969–977.

Burwell, L. S., Nadtochiy, S. M., Tompkins, A. J., Young, S., and Brookes, P. S. (2006). Direct evidence for S-nitrosation of mitochondrial complex I. *Biochem. J.* **394,** 627–634.

Costa, N. J., Dahm, C. C., Hurrell, F., Taylor, E. R., and Murphy, M. P. (2003). Interactions of mitochondrial thiols with nitric oxide. *Antioxid. Redox Signal.* **5,** 291–305.

Dahm, C. C., Moore, K., and Murphy, M. P. (2006). Persistent S-nitrosation of complex I and other mitochondrial membrane proteins by S-nitrosothiols but not nitric oxide or peroxynitrite: Implications for the interaction of nitric oxide with mitochondria. *J. Biol. Chem.* **281,** 10056–10065.

Doorn, J. A., Hurley, T. D., and Petersen, D. R. (2006). Inhibition of human mitochondrial aldehyde dehydrogenase by 4-hydroxynon-2-enal and 4-oxonon-2-enal. *Chem. Res. Toxicol.* **19,** 102–110.

Hill, B. G., Reily, C., Oh, J. Y., Johnson, M. S., and Landar, A. (2009). Methods for the determination and quantification of the reactive thiol proteome. *Free Radic. Biol. Med.* **47,** 675–683.

Keeney, P. M., Xie, J., Capaldi, R. A., and Bennett, J. P., Jr. (2006). Parkinson's disease brain mitochondrial complex I has oxidatively damaged subunits and is functionally impaired and misassembled. *J. Neurosci.* **26,** 5256–5264.

Kettenhofen, N. J., Wang, X., Gladwin, M. T., and Hogg, N. (2008). In-gel detection of S-nitrosated proteins using fluorescence methods. *Methods Enzymol.* **441,** 53–71.

Lemasters, J. J., Qian, T., He, L., Kim, J. S., Elmore, S. P., Cascio, W. E., and Brenner, D. A. (2002). Role of mitochondrial inner membrane permeabilization in necrotic cell death, apoptosis, and autophagy. *Antioxid. Redox Signal.* **4,** 769–781.

Lin, T. K., Hughes, G., Muratovska, A., Blaikie, F. H., Brookes, P. S., Darley-Usmar, V., Smith, R. A., and Murphy, M. P. (2002). Specific modification of mitochondrial protein thiols in response to oxidative stress: A proteomics approach. *J. Biol. Chem.* **277,** 17048–17056.

Moon, K. H., Hood, B. L., Kim, B. J., Hardwick, J. P., Conrads, T. P., Veenstra, T. D., and Song, B. J. (2006). Inactivation of oxidized and S-nitrosylated mitochondrial proteins in alcoholic fatty liver of rats. *Hepatology* **44,** 1218–1230.

Schagger, H., and von Jagow, G. (1991). Blue native electrophoresis for isolation of membrane protein complexes in enzymatically active form. *Anal. Biochem.* **199,** 223–231.

Schagger, H., Cramer, W. A., and von Jagow, G. (1994). Analysis of molecular masses and oligomeric states of protein complexes by blue native electrophoresis and isolation of membrane protein complexes by two-dimensional native electrophoresis. *Anal. Biochem.* **217,** 220–230.

Schilling, B., Bharath, M. M. S., Row, R. H., Murray, J., Cusack, M. P., Capaldi, R. A., Freed, C. R., Prasad, K. N., Andersen, J. K., and Gibson, B. W. (2005). Rapid purification and mass spectrometric characterization of mitochondrial NADH dehydrogenase (Complex I) from rodent brain and a dopaminergic neuronal cell line. *Mol. Cell Proteomics* **4,** 84–96.

Sethuraman, M., McComb, M. E., Huang, H., Huang, S., Heibeck, T., Costello, C. E., and Cohen, R. A. (2004). Isotope-coded affinity tag (ICAT) approach to redox proteomics: Identification and quantitation of oxidant-sensitive cysteine thiols in complex protein mixtures. *J. Proteome Res.* **3,** 1228–1233.

Tompkins, A. J., Burwell, L. S., Digerness, S. B., Zaragoza, C., Holman, W. L., and Brookes, P. S. (2006). Mitochondrial dysfunction in cardiac ischemia-reperfusion injury: ROS from complex I, without inhibition. *Biochim. Biophys. Acta* **1762,** 223–231.

Torta, F., Usuelli, V., Malgaroli, A., and Bachi, A. (2008). Proteomic analysis of protein S-nitrosylation. *Proteomics* **8,** 4484–4494.

Venkatraman, A., Landar, A., Davis, A. J., Chamlee, L., Sanderson, T., Kim, H., Page, G., Pompilius, M., Ballinger, S., Darley-Usmar, V., and Bailey, S. M. (2004a). Modification of the mitochondrial proteome in response to the stress of ethanol-dependent hepatotoxicity. *J. Biol. Chem.* **279,** 22092–22101.

Venkatraman, A., Landar, A., Davis, A. J., Ulasova, E., Page, G., Murphy, M. P., Darley-Usmar, V., and Bailey, S. M. (2004b). Oxidative modification of hepatic mitochondria protein thiols: Effect of chronic alcohol consumption. *Am. J. Physiol. Gastrointest. Liver Physiol.* **286,** G521–G527.

Witze, E. S., Old, W. M., Resing, K. A., and Ahn, N. G. (2007). Mapping protein post-translational modifications with mass spectrometry. *Nat. Methods* **4,** 798–806.

Ying, J., Clavreul, N., Sethuraman, M., Adachi, T., and Cohen, R. A. (2007). Thiol oxidation in signaling and response to stress: Detection and quantification of physiological and pathophysiological thiol modifications. *Free Radic. Biol. Med.* **43,** 1099–1108.

Zhang, J., Li, X., Mueller, M., Wang, Y., Zong, C., Deng, N., Vondriska, T. M., Liem, D. A., Yang, J. I., Korge, P., Honda, H., Weiss, J. N., Apweiler, R., and Ping, P. (2008). Systematic characterization of the murine mitochondrial proteome using functionally validated cardiac mitochondria. *Proteomics* **8,** 1564–1575.

CHAPTER SEVEN

Mitochondrial Thioredoxin Reductase: Purification, Inhibitor Studies, and Role in Cell Signaling

Maria Pia Rigobello[*] and Alberto Bindoli[†]

Contents

1. Introduction	110
2. Purification of Thioredoxin Reductase from Isolated Mitochondria, Cultured Cells, and Whole Organs	111
2.1. Preparation and purification of mitochondria	111
2.2. Freeze/thaw cycles and disruption of mitochondria	112
2.3. Heat treatment	113
2.4. Ammonium sulfate fractionation	113
2.5. DEAE-Sephacel chromatography	113
2.6. 2′,5′-ADP-Sepharose 4B affinity chromatography	114
2.7. ω-Aminohexyl-Sepharose 4B	114
2.8. Rechromatography on 2′,5′-ADP-Sepharose 4B	114
2.9. Purification of TrxR2 from whole organs or cultured cells	116
3. Estimation of Thioredoxin Reductase Activity	117
4. Inhibitor Studies of Thioredoxin Reductase	118
5. Role in Cell Signaling	118
References	120

Abstract

Mitochondrial thioredoxin reductase (TrxR2) maintains thioredoxin (Trx2) in a reduced state and plays a critical role in mitochondrial and cellular functions. TrxR2 has been identified in many different tissues and can be purified to homogeneity from whole organs and isolated mitochondria.

Here we describe the detailed steps required to purify this enzyme. A different initial procedure is needed, according to whether purification starts from whole organs or from isolated and purified mitochondria. In the first case, acid precipitation is a critical preliminary step to separate mitochondrial

---

[*] Department of Biological Chemistry, University of Padova, Padova, Italy
[†] Institute of Neuroscience (CNR), Section of Padova, c/o Department of Biological Chemistry, Padova, Italy

thioredoxin reductase from the cytosolic isoform. Preparation involves ammonium sulfate fractionation, heating, and freeze/thaw cycles, followed by chromatographic passages involving DEAE-Sephacel, 2′,5′-ADP-Sepharose 4B affinity, and ω-Aminohexyl-Sepharose 4B columns. The 2′,5′-ADP-Sepharose 4B affinity step can be repeated to remove any contaminating glutathione reductase completely. Although several methods are available to detect the activity of this enzyme, reduction of DTNB is an easy and inexpensive test that can be applied not only to the highly purified enzyme but also to lysed mitochondria, provided non-TrxR2-dependent reaction rates are subtracted. TrxR2, like TrxR1, can be inhibited by several different and chemically unrelated substances, usually acting on the C-terminal containing the cysteine–selenocysteine active site. Many of these inhibitors react preferentially with the reduced form of the C-terminal tail. This condition can be evaluated by estimating enzyme activity after removal of the inhibitor by gel filtration of the enzyme preincubated in oxidizing or reducing conditions. Inhibition of thioredoxin reductase has important consequences for cell viability and can lead to apoptosis. Inhibition of TrxR2 causes large production of hydrogen peroxide, which diffuses from the mitochondrion to the cytosol and is responsible for most of the signaling events observed. Methods to measure hydrogen peroxide in isolated mitochondria or cultured cells are described.

1. INTRODUCTION

Thioredoxin reductases (TrxR; EC 1.8.1.9) belong to the pyridine nucleotide disulfide oxidoreductase family, and their major function is to maintain thioredoxins in a reduced state. Mammalian thioredoxin reductases (high-Mr thioredoxin reductases) are homodimeric enzymes containing a C-terminal selenocysteine involved in catalytic activity (Arnér and Holmgren, 2000; Mustacich and Powis, 2000; Tamura and Stadtman, 1996). In mammals, three major isoforms of thioredoxin reductase have been found, cytosolic (TrxR1), mitochondrial (TrxR2, also called TR3 or TRβ), and testis-specific (TGR) (Arnér, 2009). However, several splice variants of TrxR1 and TrxR2 have been also identified (Arnér, 2009). Interestingly, at variance with the predominant isoform, splice variants of TrxR2 lacking the mitochondrial signaling peptide have been found located in the cytosol (Turanov et al., 2006). Other splice forms of TrxR2 are a protein variant subunit with a shorter interface domain (Miranda-Vizuete and Spyrou, 2002) and another version whose overexpression induces cell apoptosis (Chang et al., 2005). The presence of several different splice variants for thioredoxin reductase makes analysis of its expression and functions rather complex. Disruption of TrxR2 gene is associated with embryonic death, suggesting a crucial role played by this enzyme, particularly in hematopoiesis and heart function (Conrad et al., 2004). The three-dimensional structure of mouse TrxR2 has been solved and found comparable to that of TrxR1 and glutathione reductase (Biterova et al., 2005). In addition to thioredoxin, and like its cytosolic counterpart, TrxR2 is able to reduce several different unrelated substrates such as DTNB (5,5′-dithiobis(2-nitrobenzoic acid)), selenite, and alloxan (Rigobello et al., 1998). It has also been identified in many different tissues (Kawai et al., 2000; Kim et al., 1999; Lescure et al., 1999; Miranda-Vizuete et al., 1999a,b) and isolated and purified to homogeneity from rat liver (Lee et al., 1999), rat liver mitochondria (Rigobello et al., 1998), and mitochondria from bovine adrenal cortex (Watabe et al., 1999).

2. Purification of Thioredoxin Reductase from Isolated Mitochondria, Cultured Cells, and Whole Organs

Mitochondrial thioredoxin reductase can be purified after preliminary preparation of isolated mitochondria or starting directly from whole organs or cultured cells (Fig. 7.1).

2.1. Preparation and purification of mitochondria

Mitochondria can be isolated from tissue homogenates or disrupted cells with conventional procedures involving differential centrifugation (see previous volumes of this series, e.g., vol. 10, describing detailed preparations of mitochondria). Briefly, after centrifugation of nuclei and unbroken cells at $800 \times g$ for 5 min, mitochondria can be obtained from the supernatant after centrifugation at $8000 \times g$ for 10 min. Pellets can be resuspended and recentrifuged at $10,000 \times g$ to wash the mitochondria, which are finally resuspended in a small volume of medium.

Crude rat liver mitochondrial suspensions can be further purified by the silica colloid Percoll which, by centrifugation, results in a density gradient. The method of Hovius et al. (1990) with a few modifications (Rigobello et al., 2001) is described. Mitochondria (5 ml of about 60 mg ml^{-1} suspension) are layered on top of centrifuge tubes containing 45 ml of 30% (v/v) Percoll in 0.225 M mannitol and 1 mM EGTA buffered with 25 mM HEPES (pH 7.4). Samples are centrifuged at $95,000 \times g$ for 30 min. Mitochondria can be collected in the lower fraction at the relative density of 1.070/1.100 g ml^{-1} and washed twice with the desired medium, as previously described.

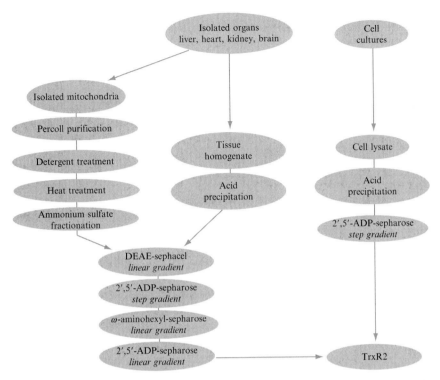

Figure 7.1 Diagram showing major steps for purification of TrxR2 from whole organs, isolated mitochondria or cell cultures.

2.2. Freeze/thaw cycles and disruption of mitochondria

To obtain a sufficient amount of TrxR2 from isolated mitochondria, it is necessary to start from a concentration of 5–6 g of mitochondrial proteins, based on the biuret procedure. We usually start from about 100 ml of a suspension of 50–60 mg protein ml^{-1}. Preparations can be collected over time and stored at $-20\ °C$, even for months. Further processing of mitochondria before the chromatographic steps can be based on sonic irradiation followed by ammonium sulfate and heat treatment, as described in a previous chapter in this series (Bindoli and Rigobello, 2002). Here we describe an alternative procedure based on mitochondria disruption with Triton X-100 (Watabe et al., 1999) which provides higher yields. Frozen stored mitochondria are thawed, pooled, and diluted with distilled water (1:1) containing an antiprotease cocktail ("Complete" Roche, Mannheim, Germany) and 0.2% Triton X-100. They are then subjected to three freeze/thaw cycles (at $-70\ °C$), followed by homogenization in Ultra Turrax (Janke & Kunkel, Staufen, Germany), twice for 30 s each. The resulting suspension is centrifuged at 12,000×g for 60 min at 5 °C. Pellets are discarded, and the supernatant used for further purification of the enzyme.

2.3. Heat treatment

The obtained preparation is heated at 60 °C for 3 min and rapidly cooled in an ice bath at 4 °C. Precipitated proteins are then removed by centrifugation at 105,000×g for 1 h at 5 °C.

2.4. Ammonium sulfate fractionation

The following purification steps are based on modifications of the methods originally described by Luthman and Holmgren (1982) for thioredoxin reductase from rat liver cytosol and Williams *et al.* (1967) for *Escherichia coli*. The clear supernatant obtained after heating is fractionated with ammonium sulfate in two saturation steps. The first stage is 0–50% (w/v) ammonium sulfate fractionation. Precipitated proteins are centrifuged at 37,000×g for 30 min at 5 °C. Pellets are dissolved in 10 mM Tris–HCl (pH 7.5) containing 1 mM EDTA and extensively dialyzed overnight against the same buffer. This enzyme preparation is collected and used for the further chromatographic purification steps. The supernatant obtained after the previous centrifugation can be subjected to a further fractionation step with ammonium sulfate, achieving 85% saturation. Thioredoxin reductase activity is still present in the 50–85% fraction, but is heavily contaminated with glutathione reductase. It is therefore preferable to avoid using it for further purification.

2.5. DEAE-Sephacel chromatography

The dialyzed enzyme preparation is concentrated by ultrafiltration under argon through an Amicon YM/10 membrane and transferred to an anion exchange DEAE-Sephacel column (4 × 16 cm) preequilibrated with 10 mM Tris–HCl (pH 7.5) containing 1 mM EDTA. The column is eluted with 500 ml linear gradient (from 0.0 to 0.3 M) of NaCl in the same buffer. Fractions of 10 ml are collected. Aliquots of these fractions are used to estimate thioredoxin reductase by DTNB method (see below) and protein content by measuring absorbance at 280 nm. Fractions eluted in the 0.13–0.15 M NaCl interval reveal the highest activity of thioredoxin reductase and are pooled and concentrated by ultrafiltration under argon through an Amicon YM/10 membrane. Concentrated samples are dialyzed overnight against 50 mM Tris–HCl (pH 7.5) buffer containing 1 mM EDTA.

2.6. 2′,5′-ADP-Sepharose 4B affinity chromatography

The resulting dialyzed enzyme solution is applied to a 2′,5′-ADP-Sepharose 4B affinity chromatography column (0.8 × 10 cm), preequilibrated with 50 mM Tris–HCl buffer (pH 7.5) containing 1 mM EDTA. The enzyme is then eluted with three discontinuous gradient steps of Na,K-phosphate (0.3 and 0.5 M) and NaCl (0.8 M), in 50 mM Tris–HCl buffer (pH 7.5) containing 1 mM EDTA (Fig. 7.2A). Fractions of 2.5 ml are collected and aliquots used to estimate TrxR2 by DTNB method and protein content at 280 nm. The Na,K-phosphate steps allow the residual TrxR1 to be eluted. Mitochondrial thioredoxin reductase fractions are eluted at 0.8 M NaCl (Fig. 7.2A), pooled, and concentrated, as described previously, in the presence of 0.2% (w/v) octylglucoside (n-octyl-β-D-glucopyranoside) to prevent loss of enzyme activity. Subsequently, the concentrated fraction is dialyzed with 50 mM Tris–HCl buffer (pH 7.5) containing 1 mM EDTA. At this stage, the degree of purity can be assessed by SDS–PAGE.

2.7. ω-Aminohexyl-Sepharose 4B

Enzyme preparation can be further purified in an ω-Aminohexyl-Sepharose 4B column (0.8 × 5 cm) equilibrated with 50 mM Tris–HCl buffer (pH 7.5) and 1 mM EDTA. The enzymatic fraction is eluted with a linear gradient of NaCl (0.0–0.8 M). The flow rate is 0.8 ml min^{-1} and fractions of 2 ml are collected. Pooled fractions are concentrated and dialyzed as described above. Another way of purification of the dialyzed preparation obtained after 2′,5′-ADP-Sepharose 4B affinity is based on fast protein liquid chromatography (FPLC, Pharmacia, Piscataway, NJ). Samples are applied to a Superdex-200 column equilibrated with 20 mM Tris–HCl (pH 7.5), 150 mM NaCl, 10% glycerol, 10 mM mercaptoethanol, and 50 μM PMSF (phenylmethylsulfonyl fluoride). Fractions are collected at a flow rate of 0.4 ml min^{-1}.

2.8. Rechromatography on 2′,5′-ADP-Sepharose 4B

This step is designed to remove contaminating glutathione reductase. If at the end of the ω-Aminohexyl-Sepharose 4B step, the sample still retains glutathione reductase activity, further chromatography on 2′,5′-ADP-Sepharose 4B is required. The dialyzed enzyme is eluted with 50 mM Tris buffer (pH 7.5) containing 1 mM EDTA at a linear gradient from 0.6 to 1.0 M NaCl in the same buffer (Fig. 7.2B). Fractions of 1.2 ml are collected at a flow rate of 0.5 ml min^{-1}.

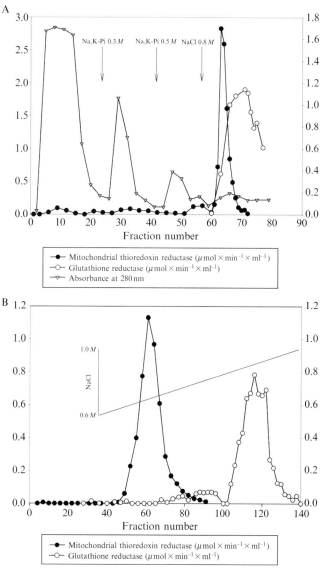

Figure 7.2 (A) Chromatographic purification of mitochondrial thioredoxin reductase by 2′,5′-ADP-Sepharose 4B affinity chromatography. (B) The ω-hexyl column eluate is applied to a 2′,5′-ADP-Sepharose column with a continuous gradient of NaCl. As shown in (B), thioredoxin reductase can be completely separated by glutathione reductase (GR). TrxR2 is estimated by DTNB method, and GR is measured in 0.2 M Tris–HCl buffer (pH 8.1), 1 mM EDTA, 1 mM GSSG and 0.25 mM NADPH. Absorbance is estimated at 340 nm ($\varepsilon_M = 6220\ M^{-1}\ cm^{-1}$).

2.9. Purification of TrxR2 from whole organs or cultured cells

Lee *et al.* (1999) describe a procedure involving acidification to pH 5 of rat liver homogenate, which allows separation and purification of TrxR2 directly from the liver without any preliminary purification of mitochondria. Rat liver is extensively homogenized with Ultra Turrax in 20 mM Tris–HCl buffer (pH 7.8) in the presence of 1 mM EDTA and protease inhibitors ("Complete" Roche), followed by further homogenization in a glass-teflon tissue grinder. After centrifugation at $70,000 \times g$ for 30 min, the resulting supernatant is brought to pH 5.0 by adding 1 M acetic acid dropwise. The resulting cloudy suspension is again centrifuged at $70,000 \times g$ for 30 min and the pellet, resuspended and neutralized, is used as a source of TrxR2. The supernatant contains most of TrxR1. This method is convenient when organs are available in small quantity as may occur with brain, kidney, and heart, and it is consequently difficult to obtain a sufficient amount of mitochondria. The resulting enzyme is slightly less pure, but the yield is satisfactory. Figure 7.3 shows SDS–PAGE and Western blotting separation of TrxR2 obtained from various tissues; by comparison, a sample obtained after preliminary preparation of rat liver mitochondria is also shown (Liver★). This procedure can also be used for cultured cells. Both tissues and cultured cells can be frozen. Cells (at 10^8 density) are lysed with RIPA buffer modified as follows: 150 mM NaCl, 50 mM Tris–HCl (pH 7.4), 1% Triton X-100, 0.1% SDS, 0.5% DOC, 1 mM NaF, 1 mM EDTA, and immediately before use, an antiprotease cocktail ("Complete" Roche) containing PMSF is added. Samples are subjected to acid precipitation as described above for rat liver.

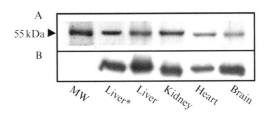

Figure 7.3 SDS–PAGE (A) and Western blotting (B) of mitochondrial thioredoxin reductase prepared from various organs. Purified enzyme obtained from 2′,5′-ADP Sepharose column is separated by polyacrylamide gel electrophoresis, followed by Coomassie brilliant blue staining (A). Polypeptide bands showing molecular weight of about 54 kDa are detected. In (B) proteins are transferred to nitrocellulose membrane and revealed with anti-TrxR2 antibody. Asterisk (★): TrxR2 obtained from previously isolated rat liver mitochondria.

3. ESTIMATION OF THIOREDOXIN REDUCTASE ACTIVITY

In mitochondria prepared from tissues or lysed cells, TrxR2 activity can easily be estimated by DTNB reduction. It has long been known that mitochondria are able to reduce low molecular weight disulfides. In particular, the NADPH-dependent DTNB reductase activity of crude mitochondrial fractions or purified mitochondrial matrix is essentially attributed to thioredoxin reductase activity (Lenartowicz and Wudarczyk, 1995). Either freshly isolated or previously frozen mitochondria can be used for TrxR2 estimation. Mitochondria (5 mg in 200 μl of 0.2 M phosphate buffer (pH 7.4) containing 5 mM EDTA) are treated for 1 h at 0 °C with 75 mM CHAPS (3-[(3-cholamidopropyl)dimethylammonio]-1-propanesulfonate) and subjected to occasional vortexing. Then, 1 mg protein is transferred to both sample and reference cuvettes containing the same medium and added with 1 mM DTNB. The reaction is started by adding 0.25 mM NADPH to the sample cuvette and absorbance is determined for a few minutes at 412 nm (ε_M 13,600 M^{-1} cm^{-1}). Enzyme activity is expressed as nmol min^{-1} mg^{-1} protein and is calculated by taking into account the fact that 1 mol of NADPH yields 2 mol of TNB anion (reduced DTNB). DTNB stock solution can be prepared by bringing the acidic suspension to pH 7.0 by careful addition of 1 M Tris-base, avoiding any local rise above pH 9 to prevent cleavage of the disulfide. Freeze/thaw cycles in the presence of detergent can improve the result of the assay. Mitochondria can also be disrupted by sonic irradiation instead of detergent. Besides, TrxR2 can be measured in the mitochondrial matrix fraction obtained after sonic irradiation followed by centrifugation at 100,000×g for 60 min, to remove mitochondrial membranes. Low concentrations of gold(I) complexes such as auranofin or arsenite are strong inhibitors of thioredoxin reductases (Gromer et al., 1998; Hill et al., 1997; Luthman and Holmgren, 1982; Tamura and Stadtman, 1996) and this property can therefore be exploited to estimate NADPH-dependent DTNB reductase activity other than that of thioredoxin reductase (Hill et al., 1997). Therefore, to a second sample containing all the above reagents, 1–2 μM auranofin (or aurothioglucose, or any other gold(I)complex) is added. The absorbance after addition of NADPH can be subtracted from that of the sample run in the absence of inhibitor. The resulting differential absorbance allows calculation of enzyme activity due solely to TrxR2.

To estimate the activity of the purified enzyme, the same methods employed for cytosolic thioredoxin reductase can be used. They have been thoroughly described in previous issues of this series (Arnér et al., 1999; Gromer et al., 2002; Holmgren and Björnstedt, 1995). The simplest and least expensive method to assess purified TrxR2 is reduction of DTNB, which can be performed in the same conditions as those described above for lysed mitochondria. Other methods rely upon NADPH oxidation or insulin reduction mediated by thioredoxin (Arnér et al., 1999). Spectrophotometric methods can be properly modified to be adapted to microplate readers. This procedure is used for the samples eluted from chromatographic columns.

4. Inhibitor Studies of Thioredoxin Reductase

Due to the highly accessible and reactive C-terminal residue containing a cysteine–selenocysteine group, thioredoxin reductase can easily be inhibited by several chemically unrelated substances. Inhibitors include heavy and transition metals and metal complexes, alkylating agents, dinitrohalobenzenes, quinones, flavonoids, and other polyphenols (Arnér, 2009). In particular, gold complexes are potent inhibitors acting in the nanomolar range of concentration (Gromer et al., 1998). Many of these inhibitors are already established or potential antitumor agents, making this enzyme an interesting molecular target for cancer chemotherapy (Arnér, 2009; Bindoli et al., 2009; Urig and Becker, 2006). The preventive reduction of TrxR2 by NADPH is a critical condition which makes most inhibitors effective. In contrast, oxidized enzyme prevents inhibition. On this basis, it is possible to assess the potential interaction of several different inhibitors with the C-terminal active site of TrxR.

Mitochondrial thioredoxin reductase is incubated in 0.2 M Na,K-phosphate buffer (pH 7.4), 5 mM EDTA in the presence or absence of 0.025 mM NADPH. Preincubation is carried out for 1.5 min and then inhibitors are added at the desired concentrations to a final volume of 60 μl which is applied to a desalting column (Micro Bio-Spin, Bio-Rad Laboratories) and centrifuged at $1000 \times g$ for 5 min. The filtering procedure can be repeated, in order to ensure complete removal of the inhibitor. The eluate is directly used to estimate TrxR2 by DTNB method. As shown in Fig. 7.4, only samples preincubated in reducing conditions show strong inhibition of TrxR2, indicating that preliminary reduction of the catalytic site is a critical condition for inhibitor effectiveness. However, a few inhibitors do not require reducing conditions and presumably interact with a site other than the C-terminal active site. The tightness of the binding of the inhibitor to the enzyme is also an important feature to be considered when this assay is performed.

5. Role in Cell Signaling

Inhibition of mitochondrial thioredoxin reductase leads to oxidation of downstream enzymes such as thioredoxin and peroxiredoxin. This oxidation is essentially due to a concentration increase in hydrogen peroxide

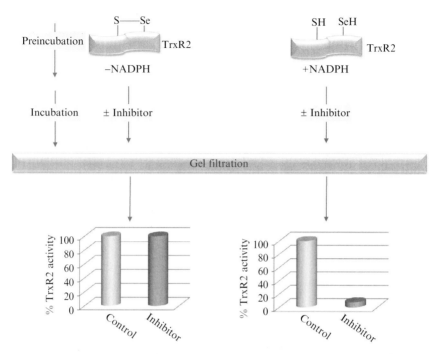

Figure 7.4 Reducing conditions of TrxR2 are critical for occurrence of inhibition. TrxR2 is incubated with an inhibitor (e.g., 1 μM auranofin), in the absence or presence of NADPH. Inhibitor is removed by gel filtration. Thioredoxin reductase is estimated by DTNB procedure. As shown, only prereduction of cysteine–selenocysteine active site causes a strong inhibition by auranofin.

produced by the respiratory complexes and no longer removed by the inhibited thioredoxin system. In addition, both mitochondrial thioredoxin (Trx2) and peroxiredoxin (Prx3) have been shown to be more sensitive to oxidation than the corresponding cytosolic isoforms (Chen et al., 2006; Cox et al., 2008). The increased concentration of mitochondrial hydrogen peroxide, which in turn can be released to the cytosol, has important consequences for cell signaling and apoptosis (Bindoli et al., 2009). Hydrogen peroxide, forming after inhibition of TrxR2, can be estimated in isolated mitochondria with the highly specific fluorescent probe Amplex Red (10-acetyl-3,7-dihydroxyphenoxazine). The assay is based on oxidation of the probe by horseradish peroxidase (HRP), activated by hydrogen peroxide (Mohanty et al., 1997).

Mitochondria (1 mg ml^{-1}), incubated in the presence of inhibitors of TrxR2 in proper medium (e.g., 0.1 M sucrose, 50 mM KCl, 0.5 mM Na, K-phosphate, 20 mM HEPES/Tris buffer (pH 7.4) and respiratory substrates), are supplemented with 10 μM Amplex Red and 0.1 units ml^{-1} of HRP.

The increase in fluorescence is followed spectrofluorometrically at 544 nm (λ_{Ex}) and 620 nm (λ_{Em}). Either a fluorometer or a microplate reader can be used. A H_2O_2 concentration standard curve can be obtained by adding known amounts of hydrogen peroxide to the medium, supplemented with Amplex Red and HRP.

In cultured cells, formation of hydrogen peroxide is assessed with the fluorogenic probes CM-DCFH2-DA (chloromethyl-$2',7'$-dihydrodichlorofluorescein) or DHR-123 (dihydrorhodamine 123) (Molecular Probes, Eugene, OR, USA), according to Royall and Ischiropoulos (1993). Cells (at 2×10^4 density) are washed in phosphate-buffered saline (PBS) containing 10 mM glucose and loaded with 10 μM CM-DCFH2-DA or 15 μM DHR-123 for 20 min in the dark. After washing in the same medium, cells are incubated in the desired conditions with the various inhibitors. Fluorescence increase is estimated in a multiwell fluorescence plate reader at 485 nm (λ_{Ex}) and 527 nm (λ_{Em}). Interestingly, DHR-123 is considered a probe mainly monitoring hydrogen peroxide of mitochondrial origin.

REFERENCES

Arnér, E. S. J. (2009). Focus on mammalian thioredoxin reductases—Important selenoproteins with versatile functions. *Biochim. Biophys. Acta* **1790,** 495–526.

Arnér, E. S. J., and Holmgren, A. (2000). Physiological functions of thioredoxin and thioredoxin reductase. *Eur. J. Biochem.* **267,** 6102–6109.

Arnér, E. S. J., Zhong, L., and Holmgren, A. (1999). Preparation and assay of mammalian thioredoxin and thioredoxin reductase. *Methods Enzymol.* **300,** 226–239.

Bindoli, A., and Rigobello, M. P. (2002). Mitochondrial thioredoxin reductase and thiol status. *Methods Enzymol.* **347,** 307–316.

Bindoli, A., Rigobello, M. P., Scutari, G., Gabbiani, C., Casini, A., and Messori, L. (2009). Thioredoxin reductase: A target for gold compounds acting as potential anticancer drugs. *Coord. Chem. Rev.* **253,** 1692–1707.

Biterova, E. I., Turanov, A. A., Gladyshev, V. N., and Barycki, J. J. (2005). Crystal structures of oxidized and reduced mitochondrial thioredoxin reductase provide molecular details of the reaction mechanism. *Proc. Natl. Acad. Sci. USA* **102,** 15018–15023.

Chang, E. Y., Son, S.-K., Ko, H. S., Baek, S.-H., Kim, J. H., and Kim, J.-R. (2005). Induction of apoptosis by the overexpression of an alternative splicing variant of mitochondrial thioredoxin reductase. *Free Radic. Biol. Med.* **39,** 1666–1675.

Chen, Y., Cai, J., and Jones, D. P. (2006). Mitochondrial thioredoxin in regulation of oxidant-induced cell death. *FEBS Lett.* **580,** 6596–6602.

Conrad, M., Jakupoglu, C., Moreno, S. G., Lippl, S., Banjac, A., Schneider, M., Beck, H., Hatzopoulos, A. K., Just, U., Sinowatz, F., Schmahl, W., Chien, K. R., Wurst, W., Bornkamm, G. W., and Brielmeier, M. (2004). Essential role for mitochondrial thioredoxin reductase in hematopoiesis, heart development, and heart function. *Mol. Cell. Biol.* **24,** 9414–9423.

Cox, A. G., Brown, K. K., Arnér, E. S. J., and Hampton, M. B. (2008). The thioredoxin reductase inhibitor auranofin triggers apoptosis through a Bax/Bak-dependent process that involves peroxiredoxin 3 oxidation. *Biochem. Pharmacol.* **76,** 1097–1109.

Gromer, S., Arscott, L. D., Williams, C. H., Jr., Schirmer, R. H., and Becker, K. (1998). Human placenta thioredoxin reductase. Isolation of the selenoenzyme, steady state kinetics, and inhibition by therapeutic gold compounds. *J. Biol. Chem.* **273**, 20096–20101.

Gromer, S., Merkle, H., Heiner Schirmer, R., and Becker, K. (2002). Human placenta thioredoxin reductase: Preparation and inhibitor studies. *Methods Enzymol.* **347**, 382–394.

Hill, K. E., McCollum, G. W., and Burk, R. F. (1997). Determination of thioredoxin reductase activity in rat liver supernatant. *Anal. Biochem.* **253**, 123–125.

Holmgren, A., and Björnstedt, M. (1995). Thioredoxin and thioredoxin reductase. *Methods Enzymol.* **252**, 199–208.

Hovius, R., Lambrechts, H., Nicolay, K., and de Kruijff, B. (1990). Improved methods to isolate and subfractionate rat liver mitochondria. Lipid composition of the inner and outer membrane. *Biochim. Biophys. Acta* **1021**, 217–226.

Kawai, H., Ota, T., Suzuki, F., and Tatsuka, M. (2000). Molecular cloning of mouse thioredoxin reductases. *Gene* **242**, 321–330.

Kim, K. J., Jang, Y. Y., Han, E. S., and Lee, C. S. (1999). Modulation of brain mitochondrial membrane permeability and synaptosomal Ca^{2+} transport by dopamine oxidation. *Mol. Cell. Biochem.* **201**, 89–98.

Lee, S.-R., Kim, J.-R., Kwon, K.-S., Yoon, H. W., Levine, R. L., Ginsburg, A., and Rhee, S. G. (1999). Molecular cloning and characterization of a mitochondrial selenocysteine-containing thioredoxin reductase from rat liver. *J. Biol. Chem.* **274**, 4722–4734.

Lenartowicz, E., and Wudarczyk, J. (1995). Enzymatic reduction of 5,5′-dithiobis-(2-nitrobenzoic acid) by lysate of rat liver mitochondria. *Int. J. Biochem. Cell Biol.* **27**, 831–837.

Lescure, A., Gautheret, D., Carbon, P., and Krol, A. (1999). Novel selenoproteins identified *in silico* and *in vivo* by using a conserved RNA structural motif. *J. Biol. Chem.* **274**, 38147–38154.

Luthman, M., and Holmgren, A. (1982). Rat liver thioredoxin and thioredoxin reductase: Purification and characterization. *Biochemistry* **21**, 6628–6633.

Miranda-Vizuete, A., and Spyrou, G. (2002). Genomic organization and identification of a novel alternative splicing variant of mouse mitochondrial thioredoxin reductase (TrxR2) gene. *Mol. Cells* **13**, 488–492.

Miranda-Vizuete, A., Damdimopoulos, A. E., Pedrajas, J. R., Gustafsson, J.-Å., and Spyrou, G. (1999a). Human mitochondrial thioredoxin reductase. cDNA cloning, expression and genomic organization. *Eur. J. Biochem.* **261**, 405–412.

Miranda-Vizuete, A., Damdimopoulos, A. E., and Spyrou, G. (1999b). cCDNA cloning, expression and chromosomal localization of the mouse mitochondrial thioredoxin reductase gene. *Biochim. Biophys. Acta* **1447**, 113–118.

Mohanty, J. G., Jaffe, J. S., Schulman, E. S., and Raible, D. G. (1997). A highly sensitive fluorescent micro-assay of H_2O_2 release from activated human leukocytes using a dihydroxyphenoxazine derivative. *J. Immunol. Methods* **202**, 133–141.

Mustacich, D., and Powis, G. (2000). Thioredoxin reductase. *Biochem. J.* **346**, 1–8.

Rigobello, M. P., Callegaro, M. T., Barzon, E., Benetti, M., and Bindoli, A. (1998). Purification of mitochondrial thioredoxin reductase and its involvement in the redox regulation of membrane permeability. *Free Radic. Biol. Med.* **24**, 370–376.

Rigobello, M. P., Donella-Deana, A., Cesaro, L., and Bindoli, A. (2001). Distribution of protein disulphide isomerase in rat liver mitochondria. *Biochem. J.* **356**, 567–570.

Royall, J. A., and Ischiropoulos, H. (1993). Evaluation of 2′,7′-dichlorofluorescein and dihydrorhodamine 123 as fluorescent probes for intracellular H_2O_2 in cultured endothelial cells. *Arch. Biochem. Biophys.* **302**, 348–355.

Tamura, T., and Stadtman, T. C. (1996). A new selenoprotein from human lung adenocarcinoma cells: Purification, properties, and thioredoxin reductase activity. *Proc. Natl. Acad. Sci. USA* **93,** 1006–1011.

Turanov, A. A., Su, D., and Gladyshev, V. N. (2006). Characterization of alternative cytosolic forms and cellular targets of mouse mitochondrial thioredoxin reductase. *J. Biol. Chem.* **281,** 22953–22963.

Urig, S., and Becker, K. (2006). On the potential of thioredoxin reductase inhibitors for cancer therapy. *Semin. Cancer Biol.* **16,** 452–465.

Watabe, S., Makino, Y., Ogawa, K., Hiroi, T., Yamamoto, Y., and Takahashi, S. Y. (1999). Mitochondrial thioredoxin reductase in bovine adrenal cortex. Its purification, properties, nucleotide/aminoacid sequences, and identification of selenocysteine. *Eur. J. Biochem.* **264,** 74–84.

Williams, C. H., Jr., Zanetti, G., Arscott, L. D., and McAllister, J. K. (1967). Lipoamide dehydrogenase, glutathione reductase, thioredoxin reductase, and thioredoxin. *J. Biol. Chem.* **242,** 5226–5231.

CHAPTER EIGHT

MEASURING MITOCHONDRIAL PROTEIN THIOL REDOX STATE

Raquel Requejo,* Edward T. Chouchani,* Thomas R. Hurd,* Katja E. Menger,* Mark B. Hampton,[†] *and* Michael P. Murphy*

Contents

1. Introduction	124
2. Quantification of Mitochondrial Protein Thiols	127
3. Quantification of Glutathionylation of Mitochondrial Proteins	130
3.1. Quantification of protein-bound glutathione	131
3.2. Recycling assay for measurement of mitochondrial GSH, GSSG, and protein-bound GSH	132
3.3. Identification of glutathionylated proteins and cysteine residues	133
4. Assessment of S-Nitrosated Protein Thiols	135
4.1. Quantification of protein S-nitrosothiols	136
4.2. Selective labeling of S-nitrosated mitochondrial protein thiols	137
5. Measurement of the Thioredoxin and Peroxiredoxin Redox States	139
5.1. Western blotting to measure reduced and oxidized peroxiredoxin 3	140
5.2. Measuring thioredoxin redox poise using the PEGylation assay	140
6. Conclusions	143
Acknowledgments	144
References	144

Abstract

Protein thiols are an important component of mammalian intramitochondrial antioxidant defenses owing to their selective interaction with reactive oxygen and nitrogen species (ROS and RNS). Reversible modifications of protein thiols resulting from these interactions are also an important aspect of redox signal transduction. Therefore, to assess how mitochondria respond to oxidative stress and act as nodes in redox signaling pathways, it is important to measure general changes to protein thiol redox states and also to identify the specific

* Medical Research Council Mitochondrial Biology Unit, Wellcome Trust/MRC Building, Cambridge, UK
[†] Free Radical Research Group, Department of Pathology, University of Otago, Christchurch, New Zealand

mitochondrial thiol proteins involved. Here we outline some of the approaches that can be used to accomplish these goals and thereby infer the multiple roles of mammalian mitochondrial protein thiols in antioxidant defense and redox signaling.

1. INTRODUCTION

Mitochondria are central to much of metabolism and are also a major source of reactive oxygen species (ROS) within the cell; consequently, there is considerable interest in mitochondrial interactions with ROS under both normal and pathological conditions (Balaban et al., 2005; Finkel, 2005; Murphy, 2009). The proximal ROS produced in mitochondria is superoxide ($O_2^{\bullet-}$), which is rapidly converted to hydrogen peroxide (H_2O_2) by the action of manganese superoxide dismutase (MnSOD) in the mitochondrial matrix (Murphy, 2009). A major way in which mitochondria deal with and respond to H_2O_2 is through a series of thiol defense systems that exist within the mitochondrial matrix (Fig. 8.1) (Costa et al., 2003; Hurd et al., 2005a,b; Schafer and Buettner, 2001). The mitochondrial glutathione system comprises glutathione peroxidases 1 and 4 (Gpx 1 and 4) which degrade peroxides, converting glutathione (GSH) to glutathione disulfide (GSSG), which is subsequently reduced back to GSH by the action of glutathione reductase (GR), thereby maintaining a high mitochondrial GSH/GSSG ratio. Within mitochondria there are two peroxiredoxins (Prxs), Prx3 and Prx5, which degrade H_2O_2 (Cox et al., 2010a; Rhee et al., 2005). The Prxs are maintained in their active, reduced state by the action of thioredoxin 2 (Trx2), which is in turn reduced by thioredoxin reductase 2 (TrxR2) (Arner and Holmgren, 2000; Lee et al., 1999; Miranda-Vizuete et al., 1999; Spyrou et al., 1997). For both GR and TrxR2 activity the necessary reduction potential is supplied by NADPH. In mitochondria the NADPH/NADP ratio is maintained in a highly reduced state by the action of the transhydrogenase and by NADPH-dependent dehydrogenases such as isocitrate dehydrogenase (Costa et al., 2003). The high abundance and reactivity of Prx3 will ensure it is a major target of mitochondrial H_2O_2 when compared with Gx; however, the importance of Prx3 as an antioxidant will be dependent on the efficiency of turnover by the Trx system (Cox et al., 2009b, 2010a). Thus, the redox poise of the GSH/GSSG, $Prx3_{ox}/Prx3_{red,}$ and $Trx2_{ox}/Trx2_{red}$ redox couples are important representations of the state of mitochondrial antioxidant defenses, and also provide an indication of the exposure of mitochondria to H_2O_2.

The H_2O_2 that evades the peroxidases gives rise to many other ROS—including the hydroxyl radical—and can initiate the formation of a range of oxygen and carbon centered radicals all of which contribute to damage of

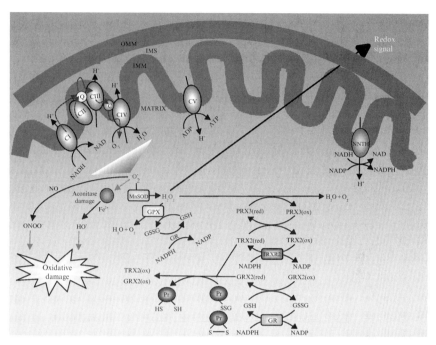

Figure 8.1 The thiol antioxidant defense system of mitochondria. Mitochondria are the primary cellular consumers of oxygen and some components of the electron transport chain are capable of generating superoxide ($O_2^{\bullet-}$), by transferring single electrons to oxygen. Superoxide can damage iron–sulfur proteins ejecting ferrous iron that reacts with hydrogen peroxide to form the very reactive hydroxyl radical ($^{\bullet}OH$). Superoxide also reacts with nitric oxide (NO) to form peroxynitrite ($ONOO^-$). Both peroxynitrite and the hydroxyl radical can cause extensive oxidative damage. To avoid such damage, mitochondria also contain an extensive thiol antioxidant defense system to detoxify the ROS generated by the reactions described above, including nonenzymatic components such as glutathione (GSH) and enzymatic components such as glutathione peroxidase (Gpx), glutathione reductase (GR), peroxiredoxins (Prx3 and Prx5), glutaredoxin (Grx2), thioredoxin (Trx2), and thioredoxin reductase (TrxR2). NADH-NADP-Transhydrogenase (NNTH).

proteins, lipids, and nucleic acids (Beckman and Ames, 1998; Winterbourn, 2008). In addition, in the presence of nitric oxide ($NO^{\bullet}$), $O_2^{\bullet-}$ can form the damaging RNS (reactive nitrogen species) peroxynitrite ($ONOO^-$) (Szabo et al., 2007). This range of ROS and RNS can react with exposed thiols on the surface of proteins leading to a number of thiol modifications, which include sulfenic, sulfinic, and sulfonic acids; S-nitrosothiols (SNOs); and sulfenylamides (Fig. 8.2). Many of these oxidative modifications to protein thiols are rapidly reversed through reactions with other protein thiols, with

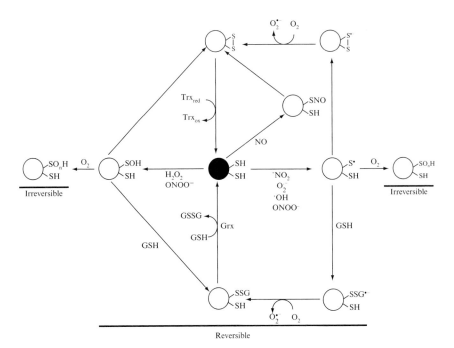

Figure 8.2 Redox modifications to protein thiols and mechanisms of their reversal. A protein model is used to show the oxidative modifications on protein thiols that are produced by ROS. Oxidation by ROS such as $^{\bullet}NO_2$, $O_2^{\bullet-}$, $^{\bullet}OH$, and $ONOO^-$ can generate a thiyl radical which can be irreversibly oxidized to sulfinic or sulfonic acids (RSO_nH), converted to an intramolecular disulfide that can be recycled to the reduced state by Trx/GSH, or form a mixed disulfide with GSH that in this case can be recycled by Grx/GSH. On the other hand, it is possible to have thiol oxidation by H_2O_2 or $ONOO^-$ generating reversible sulfenic acids (RSOH) that can be irreversibly oxidised.

Trx2, or with GSH (Fig. 8.2). In the case of GSH, the reaction is catalyzed by glutaredoxin 2 (Grx2), illustrating the important role played by this protein in antioxidant defenses (Lillig et al., 2004).

In addition to their involvement in antioxidant defense and protein repair, the recycling of oxidatively modified and S-nitrosated protein thiols by GSH and Trx2 may enable the reversible redox regulation of protein activity, whereby the activity of a protein can respond to its local redox environment (Delaunay et al., 2002; Janssen-Heininger et al., 2008). It is possible that these redox modifications to protein thiols—some of which are shown in Fig. 8.2—brought about by exposure to H_2O_2 or $NO^{\bullet}$ metabolites may be an important mode of redox signaling within mitochondria. Similarly, it may be that protein thiols are modified indirectly, through ROS flux affecting the redox state of antioxidant systems such as Prx3, Trx2, or the GSH pool which in turn modify target thiol proteins and thereby allow their activity to respond to the redox environment (Janssen-Heininger et al., 2008). It is likely that only a small subset of protein thiols will be involved in these putative redox signaling pathways, with the local environment of certain protein thiols rendering them either particularly reactive, or stabilizing modified thiols in order to facilitate redox signaling (Winterbourn and Hampton, 2008).

To summarize, reversible oxidative modification to mitochondrial protein thiols plays an important role in defending mitochondria from oxidative damage and may also be important in redox signaling. As both processes contribute to a range of physiological and pathological situations, it is important to be able to measure both the overall protein thiol redox status within mitochondria and assess the thiol redox state of individual proteins. The latter must include methods for the identification of those proteins, which may be crucial in protecting mitochondria from oxidative damage, and also those that may be important nodes in redox signaling. Here we outline some of the methods that can be used to address these issues.

2. QUANTIFICATION OF MITOCHONDRIAL PROTEIN THIOLS

A vital first step in many investigations concerned with mitochondrial protein thiols is to quantify the proportion that exist in a reduced state as well as those that have been modified in the ways shown in Fig. 8.2. Surface protein thiols that are exposed to solvent are those most likely to undergo redox modification and therefore be involved in antioxidant defense and redox signaling. This contrasts with the many other cysteine residues that have structural roles within proteins, such as in iron–sulfur centers. To quantify the total amount of solvent-exposed thiols, we gently lyse mitochondria in mild detergents or by freeze–thawing under conditions that do not lead to protein denaturation. After removing low-molecular-weight thiols by centrifugal gel filtration the total exposed thiols are quantified by the dithionitrobenzoic acid (DTNB) assay (Ellman and Lysko, 1979). In addition, it is also useful to measure the total number of protein thiols, which is accomplished by fully denaturing proteins with sodium dodecyl sulfate (SDS) prior to filtration and quantification with DTNB. Here we describe how we carry out these measurements on isolated mitochondria prepared from rat liver or hearts by homogenization followed by differential centrifugation (Chappell and Hansford, 1972). These methods can be easily applied to a number of other systems including a variety of subcellular fractions.

To measure how exposure to oxidative stress affects the quantity of exposed protein thiols, we typically incubate isolated liver or heart mitochondria (1–2 mg protein/ml) in an appropriate mitochondrial incubation buffer (e.g., 120 mM KCl, 10 mM HEPES, 1 mM EGTA, pH 7.2 in the presence of respiratory substrates, e.g., 10 mM succinate and 4 μg rotenone/ml at 37 °C; alternatively in a sucrose-based buffer) in the presence and absence of the prooxidant of interest. Following incubation, the mitochondria are pelleted by centrifugation (15,000×g for 10 s) and the pellet is then lysed by three freeze/thaw cycles (3 min in a dry ice/ethanol bath followed by 3 min at 30 °C). To obtain the fully reduced protein thiol state as a reference, parallel samples are treated with dithiothreitol (DTT; 1 mM) for 10 min at RT. All preparations are then subjected to centrifugal gel filtration to remove contaminating low-molecular-weight thiols such as GSH, and also to remove DTT from the fully reduced samples. In addition, this procedure excludes low-molecular-weight reductants such as NADPH and NADH that would otherwise lead to extensive artifactual DTNB reduction in the presence of native proteins. Centrifugal gel filtration is done using a spin column (Micro Bio-Spin 6, Bio-Rad) that has been preequilibrated with the incubation buffer containing 1% dodecyl maltoside (DDM), a mild detergent that does not denature samples but solubilizes membrane proteins. Protein thiols are then quantified using the DTNB assay (Ellman and Lysko, 1979). For this, samples (10 μl) are diluted (1:17) with 170 μl DTNB buffer (10 mM DTNB, 0.1 mM NaH$_2$PO$_4$, pH 8), incubated for 30 min at RT in a 96-well plate and the A_{412} measured using a plate reader (SpectraMax Plus 384), relative to a standard curve of 0–250 μM GSH. In parallel the protein concentration is measured to correct for any sample loss during preparation. This is done using the bicinchoninic acid (BCA) assay with bovine serum albumin (BSA) as a standard (Smith et al., 1985). The final data are expressed as nmol protein thiols/mg protein. The overall procedure is illustrated in Fig. 8.3. To measure total protein thiols, rather than only those exposed to the solvent, the protocol is similar, except that SDS (2%, v/v) is added to samples to fully denature the proteins and to expose all protein thiols (if the incubation medium contains high levels of potassium that interfere with SDS solubility, then lithium dodecylsulfate can be used instead). The mechanism of denaturing proteins is not critical as incubation with 8 M urea gave similar total protein thiol content as using 2% SDS (data not shown). A number of controls should be carried out to ensure that the measurements are specific to thiols. For example, pretreatment of mitochondria with 50 mM N-ethylmaleimide (NEM) for 10 min at 37 °C prior to isolation leads to loss of >93% of exposed thiols and >95% of total thiols. Some caveats of this method are that it is difficult to entirely eliminate artifactual oxidation of proteins during sample workup; however, the levels of protein thiols in samples that are not treated with DTT are similar to those that are DTT-treated. A further possible concern is

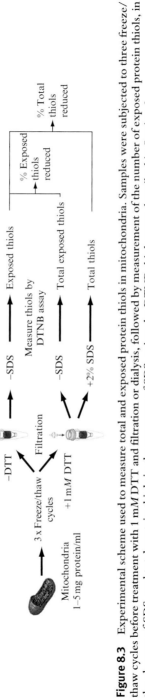

Figure 8.3 Experimental scheme used to measure total and exposed protein thiols in mitochondria. Samples were subjected to three freeze/thaw cycles before treatment with 1 mM DTT and filtration or dialysis, followed by measurement of the number of exposed protein thiols, in the absence of SDS, and total protein thiols in the presence of SDS, using the DTNB thiols assay described in Section 2.

that centrifugal filtration may lead to the selective loss of certain categories or sizes of proteins during the workup. However, alternative measurements where the low-molecular-weight thiols are instead removed by overnight dialysis give similar amounts of protein thiols. Nonetheless, this procedure is significantly more time consuming than the method outlined above, introducing the additional complication that the long incubation may introduce further oxidation.

In our case, these measurements show that fully reduced liver or heart mitochondria contain approximately 50–70 nmol thiols/mg protein in total and that about 70% of these protein thiols are exposed. This indicates that the exposed thiol content of mitochondria is high, and significantly greater than that of GSH, suggesting that within the mitochondrial matrix the quantitatively dominant thiols are those on the surface of proteins, rather than GSH. The proportion of exposed protein thiols that are lost on exposure to oxidative stress can be significant, for example exposure to diamide leads to loss of up to 30% of exposed thiols in isolated mitochondria. This finding implies that surface protein thiols may play an important role as antioxidant defenses within the cell (Requejo et al., 2010). In contrast, exposure to the mitochondria-targeted S-nitrosating agent, MitoSNO leads to negligible loss of total exposed thiols consistent with it S-nitrosating a very small proportion of available cysteine residues (Prime et al., 2009). Measurements of total and exposed protein thiols should be done in parallel with other more sensitive measures of mitochondrial thiol status to help infer whether the modification was specific to a given protein or small group of proteins, or if the modification was part of a general oxidation of protein thiols.

3. Quantification of Glutathionylation of Mitochondrial Proteins

The reversible glutathionylation of proteins is a critical part of the response to oxidative damage. It occurs transiently to protein thiols that have been oxidized to thiyl radicals, sulfenic acids, and S-nitrosothiols, with the consequence that it prevents the formation of higher thiol oxidation states on protein thiols, such as sulfinic and sulfonic acids that cannot be readily reversed. The glutathionylated protein thiol is then deglutathionylated within mitochondria by the action of GSH and Grx2, thereby facilitating the recycling of exposed protein thiols after oxidative stress (Lillig et al., 2004). In addition, it is likely that some protein thiols are more persistently glutathionylated, either by the relatively slow deglutathionylation of particular thiol residues, or by glutathionylation in response to an increase in the GSSG/GSH ratio, catalyzed by Grx2 (Beer et al., 2004). Glutathionylation may be a posttranslational modification akin to phosphorylation affecting enzymes, transcription factors, and transporters by enabling them to respond reversibly to the ambient GSH/GSSG ratio (Hurd et al., 2005a). To assess the importance of protein glutathionylation in redox regulation and antioxidant defense, it is important to be able to quantify the binding of glutathione to proteins. It would also be useful to identify both the proteins affected and the cysteine residues involved. Here we outline some of the methods that can be used to quantify the binding of GSH to mitochondrial proteins and also help identify the proteins and cysteine residues affected.

3.1. Quantification of protein-bound glutathione

One sensitive way to quantify binding of GSH to proteins is to incubate the samples with radiolabeled GSH and measure the amount of radioactivity bound to protein by scintillation counting. This approach is most easily applied to samples in which all proteins are exposed to the incubation medium, such as isolated mitochondrial membranes—mitochondrial fragments that are prepared by disruption of heart mitochondria (Beer et al., 2004; Taylor et al., 2003). This approach can be used to study the binding of GSSG to mitochondrial protein thiols in response to a variable GSH/GSSG ratio. Therefore, it is essential to first ensure that the commercially available [^{3}H]GSH, which is provided as a solution with excess DTT, has the DTT removed by extraction with ethyl acetate (Hurd et al., 2008). Additionally, it is critical that the [^{3}H]GSH is preincubated in the incubation solution so as to come to equilibrium with the GSH/GSSG ratio before mixing with mitochondrial membranes. For example, [^{3}H]GSH (100 μM, 19,246 Bq/mmol, 37 MBq/ml from American Radiolabelled Chemicals Inc.) can be diluted 1:1 with 20 mM unlabeled GSSG to make a 10 mM [^{3}H]GSSG stock solution. This is equilibrated for 30 min and the equilibration confirmed by thin layer chromatography followed by autoradiography. The equilibrated [^{3}H]GSSG can then be incubated with mitochondrial membranes (1 mg protein/ml), in KCl incubation buffer at 30–37 °C in the presence of 2 mM succinate and 4 μg/ml rotenone. To assess binding of [^{3}H]GSSG to membranes at the end of the incubation, the membrane suspension is divided in two with one half incubated with 1 mM DTT and the other with carrier for 2 min. Then the membranes are pelleted by centrifugation (15,000×g for 2 min), washed in incubation buffer, and the protein pellets dissolved in 50 μl 20% Triton X-100 by vortexing and suspended in 3 ml Fluoran-Safe 2 scintillant. The [^{3}H]GSH content is subsequently measured using a Tri-Carb 2 800 TR Perkin Elmer scintillation counter with appropriate quench correction. Samples of the original [^{3}H]GSSG stock solution are measured in parallel to determine the specific activity of the [^{3}H]GSSG and thereby determine the number of nmol of GSH bound to protein. The difference between the samples with and without DTT permits correction for nonspecific binding and occlusion of the incubation medium in the pellet. Parallel measurement of the protein content of the membrane pellet by the BCA assay enables the amount of GSH bound per milligram protein to be determined. These methods have been used to confirm that GSSG leads to glutathionylation of protein thiols in membranes and that Grx2 can rapidly catalyze this reaction (Beer et al., 2004; Taylor et al., 2003). Interestingly, the parallel measurement of the number of exposed protein thiols by the methods described above shows that GSSG leads to loss of far more reduced thiols than are glutathionylated. This could indicate that glutathionylation is a transient intermediate that leads to protein disulfides, suggesting that this mechanism is an important aspect of the interaction of GSSG with protein thiols (Beer et al., 2004).

It is possible to extend the measurements described above to assess the binding of GSH to proteins within isolated intact mitochondria (Hurd et al., 2008). To do this, mitochondria are first incubated with a stock solution of [^{3}H]GSH to allow for its uptake. Mitochondria are washed to remove external [^{3}H]GSH and the [^{3}H]GSH-loaded samples are then incubated under conditions that may lead to oxidative stress. Following incubation, mitochondria are pelleted by centrifugation and the DTT-sensitive radioactivity bound to proteins is assessed as described above for membranes. However, as the [^{3}H]GSH taken up is very significantly diluted by the 1–5 mM GSH pool within the mitochondria, the extent of labeling of proteins is low. In addition, to quantify the amount of GSH bound to protein it is necessary to isolate the internal GSH pool and quantify both its [^{3}H] content and its GSH content in order to determine its specific activity and thereby calculate the amount of GSH bound to the protein. Consequently, this procedure is not done routinely for isolated mitochondria.

3.2. Recycling assay for measurement of mitochondrial GSH, GSSG, and protein-bound GSH

To more effectively quantify the binding of GSH to the mitochondrial proteins under oxidative stress, the glutathione recycling assay is used (Anderson, 1985). For this, mitochondria are incubated under standard incubation conditions as described previously, and exposed to a redox challenge. The mitochondria are then pelleted and the amount of protein-bound GSH and the amount of GSH and GSSG in the matrix are measured using a detailed protocol described below.

Mitochondria are pelleted by centrifugation, the pellets (0.5 mg protein) are resuspended in 100 μl of 5% (w/v) 5-sulfosalicylic acid and 0.2% (w/v) Triton X-100 by vortexing, and the samples are centrifuged (15,000×g, 10 min). To measure total GSH and GSSG the recycling assay is adapted for a 96-well plate reader (Scarlett et al., 1996). To determine the total glutathione equivalents (GSH + 2GSSG), 10 μl of supernatant or 10 μl of standard

(0–70 μM GSH in 5% (w/v) 5-sulfosalicylic acid) is incubated in triplicate in 285 μl of recycling assay buffer (125 mM NaPi (pH 7.5), 5.5 mM EDTA, 183 μM NADPH, and 0.5 mM DTNB) with 0.7 U/ml GR and the slopes are read (A_{405} for 10 min; SpectraMax Plus 384; Molecular Devices). GSSG is measured by incubating 60 μl of supernatant or standard (0–20 μM GSSG in 5% (w/v) 5-sulfosalicylic acid) in 3.4 μl of 2-vinylpyridine and 2.8 μl triethanolamine sealed under argon at 4 °C for 1 h with agitation. Samples in triplicate (10–20 μl) are then added to 285 μl of recycling assay buffer with 3.5 U/ml GR and analyzed as described above.

To measure the amount of GSH bound to protein, the pellet following centrifugation of mitochondria that had been lysed in sulfosalicylic acid is analyzed. To do this the protein pellets are washed in 5% sulfosalicylic acid before being resuspended in 65 μl 8 M urea. Fifteen microliters of Tris–HCl (200 mM, pH 7.4) is then added to adjust the pH of the solution to 7.4 before the addition of 20 μl sodium borohydride (10% (w/v) in Tris–HCl). The sodium borohydride starts to decompose as soon as it is prepared and therefore needs to be made immediately prior to use. Samples are vortexed and incubated at 40 °C for 30 min, with at least one additional vortexing step applied during this time, while the eppendorf lids also are opened intermittently to relieve the pressure from evolving hydrogen gas. To remove borohydride, which interferes with the assay, and to reprecipitate proteins, 25 μl 5% sulfosalicyclic acid is added, the samples are vortexed and left at room temperature with the tube lids off for 15 min, followed by 15 min on ice to facilitate protein precipitation. The tubes are centrifuged (15,000×g for 2 min) and the amount of released GSH is determined using the recycling assay. For this, standards are prepared from a stock 10 mM GSSG solution in water that is diluted to give a 100 μM GSSG stock in 5% sulfosalicyclic acid. These standards are used to make a range of stock solutions of 0, 10, 20, 40, 80, 100, 200, and 400 μM GSSG. These standards are treated with urea and sodium borohydride as described for the sample preparation above. In addition, the concentration of GR in the assay needs is increased to 20 U/ml (instead of 4 U/ml) to account for the presence of urea.

These assays can be used routinely to quantify the extent of protein glutathionylation in isolated mitochondria. As with the methods described in the previous section, these procedures can also be adapted to other systems including intact cells in culture, although in these circumstances it will give an overall average of the various GSH pools.

3.3. Identification of glutathionylated proteins and cysteine residues

In addition to quantifying protein glutathionylation it is often important to identify those proteins that are glutathionylated under conditions of oxidative stress or redox signaling. There are a number of ways to do this in mitochondrial systems. Mitochondrial membranes can be incubated with [^{3}H]GSSG, followed by separation of membrane proteins by nonreducing SDS–PAGE or BN–PAGE and visualization by autoradiography (Beer et al., 2004). Following nonreducing eletrophoresis and transfer to a blotting membrane, it is also possible to identify glutathionylated proteins on western blots with antibodies specific for glutathionylated cysteine residues (Beer et al., 2004; Hurd et al., 2008). However, it should be noted that the antibodies currently available are artifact prone and can lead to both false positives and false negatives. So, we suggest that any candidate proteins identified by these antibodies should be confirmed by labeling with [^{3}H]GSH.

In our studies we have moved toward a candidate protein approach, focusing on complex I. This is done by isolating complex I by BN–PAGE then resolving its component subunits in the second dimension followed by visualization of those glutathionylated by autoradiography or by immunoblotting (Hurd et al., 2008). The identities of the glutathionylated proteins were then confirmed by using specific antibodies or by peptide mass fingerprinting (Hurd et al., 2008). While we have not done so, this procedure can be extended to 2-D gel systems to pick up novel proteins that may be glutathionylated.

The most definitive way of establishing that a protein is glutathionylated is to use mass spectrometry. Usually, this will be done by measuring the mass shift resulting from glutathionylation of a cysteine-containing peptide from a protease digest of a candidate protein. For example, we have found that it is possible to identify the glutathionylation of a particular cysteine residue within complex I by gel electrophoresis followed by mass spectrometry (Hurd et al., 2008). Complex I is first isolated from oxidatively stressed mitochondria by BN–PAGE (Schagger, 1995; Schagger and von Jagow, 1991). To do this, mitochondria (0.5 mg protein in incubation medium) are treated with a prooxidant such as diamide and then with 50 mM NEM for 5 min. The mitochondria are pelleted by centrifugation (15,000×g for 2 min) and the pellet is resuspended in 60 μl extraction buffer (1% DDM, 0.75 M ε-amino-n-caproic acid (ACA), 50 mM Bis-Tris–HCl, pH 7.0 (4 °C)), incubated on ice for 15 min and then clarified by centrifugation in an Airfuge^{TM} (Beckman Coulter) at 17 psi (∼100,000×g) for 15 min. Then, 3.5 μl sample buffer (5% (w/v) Coomassie blue G 250 (Serva, Germany), 500 mM ACA) is added and samples are resolved on a 1-mm-thick 5–12% acrylamide gradient gel containing 0–0.2% (w/v) glycerol, 1.5 M ACA, 150 mM Bis-Tris–HCl pH 7.0 (4 °C), overlaid with a 3.9% acrylamide stacking gel in the same buffer. The anode buffer is 50 mM Bis-Tris pH 7.0 (4 °C) and the cathode buffer is 0.02% (w/v) Coomassie blue G 250, 50 mM tricine, 15 mM Bis-Tris pH 7.0 (4 °C). Gels are run at 4 °C for 1 h at 100 V and then overnight at 40 V in cathode buffer without Coomassie blue. Complex I bands are then excised and incubated for 5 min at room temperature in 125 mM Tris–HCl (pH 7.0), 1% SDS and

50 mM NEM before insertion into the wells of a 3.7% acrylamide stacking gel in 0.13 M Tris–HCl pH 6.8, overlaid on a 1 mm thick SDS–PAGE gel. SDS–PAGE is done using either a 12.5% acrylamide linear gel, or a 5–20% acrylamide gradient gel containing 0–15% (w/v) sucrose and 0.375 M Tris–HCl, pH 8.8. Stacking gel is then polymerized around the excised bands and proteins separated by electrophoresis at 100–120 V in a MiniProtean system (Bio-Rad) using 25 mM Tris, 0.192 M glycine, 0.1% SDS, pH 8.3 as running buffer.

Following SDS–PAGE, protein bands are excised with a razor blade, transferred to a 0.5 ml tube that had been prewashed with 50% methanol and digested by "in-gel" cleavage (Wilm et al., 1996). To do this, the gel slice is washed in HPLC grade water (100 μl) for 30 min; 20 mM Tris–HCl, pH 7.0 (100 μl) for 30 min; 100 μl 50% acetonitrile, 20 mM Tris–HCl, pH 7.0 (100 μl) for 30 min; 100% acetonitrile (20 μl) for 10 min. Gel pieces are dried completely in a Speed Vac at 37 °C for $\sim$20 min, then rehydrated with 3–7 μl 20 mM Tris–HCl, pH 7.0, 5 mM $CaCl_2$ containing trypsin 12.5 ng/μl (Roche Applied Science), or with 25 ng/μl sequencing grade endoproteinase Asp-N (Roche Applied Science) in 20 mM Tris–HCl, pH 7.0, and digested overnight at 37 °C. Subsequently, peptides are extracted from the gel with 4% formic acid (ARISTAR grade, Merck)/60% acetonitrile (Romil). All digests are examined in a MALDI-TOF-TOF mass spectrometer (Model 4700 Proteomics Analyzer, Applied Biosystems) using α-cyano-hydroxy-*trans*-cinnamic acid as the matrix. Peptide sequences are obtained by tandem MS in a MALDI-TOF-TOF mass spectrometer, and peptide masses and fragmentation data are compared against the National Center for Biotechnology Information (NCBI) database using MASCOT (http://www.matrixscience.com; Perkins et al., 1999).

These procedures confirm that a certain subunit of complex I is glutathionylated under conditions of oxidative stress and allows for the identification of the cysteine residues involved (Hurd et al., 2008). While these procedures are time consuming and must be adapted to the particular protein under investigation, they provide definitive evidence that a candidate protein is glutathionylated within mitochondria.

4. Assessment of *S*-Nitrosated Protein Thiols

The *S*-nitrosation of mitochondrial protein thiols to form SNOs may be an important modification in both redox signaling and antioxidant defense (Galkin and Moncada, 2007; Hogg, 2002; Janssen-Heininger et al., 2008; Moncada and Erusalimsky, 2002). This thiol alteration is selective and may be relatively persistent, in that the *S*-nitrosated thiols are retained after their formation (Dahm et al., 2006; Prime et al., 2009).

These modifications may also be transient, with the initial S-nitrosated protein thiol rapidly converted to other redox forms such as sulfenic acids, glutathionylated proteins, sulfenylamides, or protein disulfides (Dahm et al., 2006; Janssen-Heininger et al., 2008). In turn, many of these modifications can be rapidly reduced back to protein thiols, as is outlined in Fig. 8.3. It is important to both quantify the extent of protein S-nitrosation and also to identify the proteins that have been S-nitrosated to best understand the nature and consequences of this unique modification in physiological and pathological scenarios. Some of the ways in which this can be accomplished in mitochondria are outlined below.

4.1. Quantification of protein S-nitrosothiols

In our laboratory we use a chemiluminescence assay to quantify protein S-nitrosothiol formation (Feelisch et al., 2002). Sample protein is first treated with excess NEM to alkylate all free thiols and thus minimize the loss of SNOs during sample workup. Protein-SNOs are then reacted with acidic iodine/iodide to release NO$^{\bullet}$ that then reacts with ozone, and the resulting chemiluminescence is measured relative to nitrite standards. Samples are treated with sufanilimide to derivatize any contaminating nitrite and prevent it from contributing to the chemiluminescent signal. Samples are also treated with or without $HgCl_2$ to selectively degrade any protein-SNOs, and the difference between the samples is calculated to be the signal due to protein-SNOs, thereby eliminating any background due to other modifications such as nitrosoamines. Typical procedures for the assessment of the S-nitrosation of mitochondria and cells are given below.

After incubation with an S-nitrosating agent such as the mitochondria-targeted SNO MitoSNO (Prime et al., 2009) or S-nitroso-N-acetylpenicillamine (SNAP), the mitochondrial suspension is supplemented with 10 mM NEM and 1 min later pelleted by centrifugation (15,000×g for 2 min), resuspended in 1 ml KCl incubation medium supplemented with 10 mM NEM, and pelleted once more. The pellet is then resuspended in 1 ml 25 mM HEPES, pH 7.2, 100 μM DTPA, 10 μM neocuproine, and 10 mM NEM, snap-frozen on dry ice/ethanol, freeze/thawed (3×) and stored at $-20\ °C$ until analysis. For measurements on adherent cell layers, the incubation medium is removed then the cell layer is washed in PBS/10 mM NEM, then scraped into 1 ml PBS/NEM and snap-frozen on dry ice/ethanol, freeze/thawed (3×) and stored at $-20\ °C$ until analysis. Samples (225–450 μl) are thawed rapidly just prior to analysis and made up to a final volume of 250–500 μl containing 0.1 M HCl and 0.5% sulfanilamide $\pm$ 0.2% $HgCl_2$ and incubated in the dark for 30 min. Then, duplicate 100–500 μl samples are injected into 20 ml of an acidic I^-/I_2 solution (33 mM KI, 14 mM I_2 in 38% (v/v) acetic acid) through which bubbled helium carries the released NO through a 1 M NaOH trap to the analyzer (EcoMedics CLD 88 Exhalyzer) where it is mixed with ozone and the chemiluminescence measured. The peak area is compared to a standard curve generated from $NaNO_2$ standards. As $HgCl_2$ selectively degrades SNOs, the SNO content is the difference between the samples with and without $HgCl_2$.

This assay enables the ready quantification of protein SNOs in mitochondria and cells. The parallel measurement of exposed protein thiols indicates that *S*-nitrosating agents such as MitoSNO and SNAP only *S*-nitrosate a small proportion of the available protein thiols (Prime *et al.*, 2009). This suggests that there may be some specificity in the protein thiols that can be selectively *S*-nitrosated. In the next section the selective visualization of these proteins is addressed.

4.2. Selective labeling of *S*-nitrosated mitochondrial protein thiols

As only a small proportion of mitochondrial proteins are persistently *S*-nitrosated and these modifications are likely of physiological and pathological interest (Prime *et al.*, 2009), it would be useful to selectively visualize these proteins for identification purposes. To do this we use a technique developed by Hogg and colleagues as an improvement of the biotin switch method (Wang *et al.*, 2008). Following incubation with an *S*-nitrosating agent, free thiols are blocked by addition of excess NEM. The NEM is then removed by centrifugal gel filtration and the SNOs are selectively reduced by exposure to copper(II) sulfate and ascorbate. This leads to the formation of Cu(I) which selectively reduces the SNO bond, leaving other thiol modifications unaffected. The exposed thiols are then reacted with a cysteine-reactive tag such as maleimide conjugated to a fluorophore, which enables the sensitive detection of *S*-nitrosated protein thiols. These proteins can subsequently be separated by electrophoresis and visualized by fluorescent scanning of the gel. An example of this technique applied to isolated mitochondria is given below.

Isolated rat liver or heart mitochondria (2 mg protein/ml) are suspended in KCl incubation medium supplemented with 10 mM succinate and 4 μg/ml rotenone and are incubated with no additions, 10 μM MitoSNO or 500 μM diamide at 37 °C for 5 min with occasional mixing. Mitochondria are then pelleted by centrifugation and resuspended in a blocking buffer containing 10 mM HEPES, pH 7.7, 1 mM EDTA, 1 mM DTPA, 10 μM neocuproine, 1% SDS, and 50 mM NEM. This blocking reaction is carried out for 5 min at 37 °C. The reaction mixture is then passed three times through MicroBioSpin Columns (6 kDa cutoff, Bio-Rad) to remove NEM. Cy3 maleimide (200 μM; Amersham product number PA 13131), 1 mM ascorbate and 10 μM $CuSO_4$ are added and the mixture gently vortexed and then incubated for 30 min at 37 °C. The protein (10 μg) is then separated on a 12.5% SDS–PAGE gel. After electrophoresis, the fluorescent gel image is acquired with a Typhoon 9410 variable mode imager scanning for Cy3 fluorescence at 532 nm (Fig. 8.4).

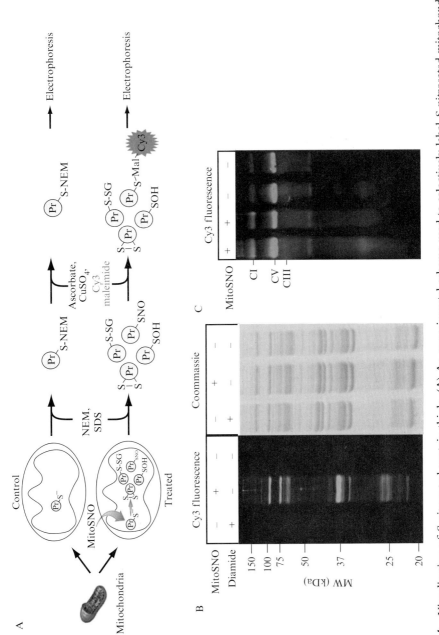

Figure 8.4 Visualization of S-nitrosated protein thiols. (A) An experimental scheme used to selectively label S-nitrosated mitochondrial protein thiols. Isolated mitochondria are separated into untreated and NO-donor treated pools and prepared as described in Section 4.2. Following treatment, sample protein thiols are blocked with NEM and proteins denatured by SDS. Following removal of NEM from the

This technique can also be used to assess S-nitrosation of individual respiratory complexes in bovine heart mitochondrial membranes (Prime et al., 2009). For this, membranes (250 μg protein/ml) are incubated in KCl incubation medium ± 75 μM MitoSNO at 37 °C for 5 min with occasional mixing. Then 10 mM NEM is added and incubated for 5 min at 40 °C. The membranes are then pelleted by centrifugation and washed three times in 1 ml PBS buffer and then resuspended in 100 μl PBS supplemented with 200 μM Cy3 maleimide, 1 mM ascorbate, and 10 μM $CuSO_4$ and incubated for 30 min at 37 °C. The individual respiratory complexes are then separated by BN–PAGE as described in the previous section. Following electrophoresis, the gel image is acquired as described above.

In both the above cases the proteins affected showed up very sensitively due to fluorescent labeling. It is possible to adapt the BN–PAGE approach by cutting out the protien band and running it in the second dimension by conventional SDS–PAGE and then identifying the proteins by cutting out the spot and using mass spectrometry. Additionally, for complex mitochondrial mixtures it is possible to adapt 2-D gel procedures to identify those proteins that are modified by S-nitrosation. Thus, these procedures and possible extensions of them should enable the identification of specific proteins that are S-nitrosated within mitochondria in response to various stimuli.

5. MEASUREMENT OF THE THIOREDOXIN AND PEROXIREDOXIN REDOX STATES

The approaches outlined so far indicate how to quantify general markers of thiol redox changes within mitochondria, including those of the exposed and total protein thiols, the extent of S-nitrosation and the GSH/GSSG, and glutathionylated protein status. In addition, the redox state of critical protein couples can be assessed to indicate further the

---

sample, S-nitrosated protein thiols are selectively degraded by ascorbate and copper in the presence of a fluorescently labeled maleimide dye. When comparing untreated rat heart mitochondria to those treated with the mitochondria-targeted NO-donor MitoSNO that are subsequently resolved by SDS–PAGE and scanned for Cy3 fluorescence (B), significant tagging of protein thiols is selectively observed in the MitoSNO treated samples only. Lack of a significant fluorescent signal in both control and diamide treated samples highlights the selectivity of protein S-nitrosothiol reduction using this method. (C) Resolving the mitochondrial respiratory complexes by BN–PAGE following tagging of S-nitrosated protein thiols shows that complex I is S-nitrosated by MitoSNO. Complex V is labeled in both conditions from this analysis, perhaps due to occluded thiols that were unresponsive to initial NEM treatment becoming exposed later in the workup and reacting with the fluorescent maleimide equally in both conditions.

mitochondrial thiol status. Two protein couples are of most use in this way as they are central to mitochondrial redox chemistry.

5.1. Western blotting to measure reduced and oxidized peroxiredoxin 3

Prx3 is a major regulator of mitochondrial H_2O_2 (Cox et al., 2010a). The enzyme undergoes a cycle where the peroxidatic cysteine within the active site reacts with H_2O_2 to form a sulfenic acid, which subsequently reacts with a resolving thiol on another Prx3 monomer to form an intermolecular dimer. The disulfide is then reduced back to the reduced form by Trx2. However, in some instances of increased oxidative stress the ratio of reduced to disulfide linked Prx3 dimer decreases and is thus a sensitive indication of the extent of H_2O_2 flux and the overall thiol redox state of mitochondria (Brown et al., 2008; Cox et al., 2008a,b, 2009a; Kumar et al., 2009). This ratio can be easily assessed by separating the dimer and monomer by nonreducing SDS–PAGE followed by immunoblotting. The procedure is described in detail in a companion article in this volume (Chapter 4, Cox et al., 2010b).

Here we illustrate the ability of the assay to detect changes in the redox state of Prx3 in isolated mitochondria (Fig. 8.5). Rat liver mitochondria (5 mg protein/ml in 250 mM sucrose, 10 mM HEPES, pH 7.4, 1 mM EGTA, 0.01% BSA) were incubated for 10 min at 37 °C in the presence of selected respiratory substrates and inhibitors. To prevent the oxidation of Prx3 during processing, a 200 μl sample was first quenched with 800 μl buffer containing NEM (250 mM sucrose, 10 mM HEPES, pH 7.8, 1 mM EGTA, 125 mM NEM, 10 μg/ml catalase) for 10 min at 37 °C. Mitochondria were then pelleted by centrifugation and resuspended in sample buffer including NEM (25 mM Tris–HCl, pH 7.0, 5% glycerol, 100 mM NEM, 10 μg/ml catalase). After 5 min, 1% SDS was added to lyse the mitochondria. Samples were then separated by nonreducing SDS–PAGE and Western blotted for Prx3 as described (Chapter 4, Cox et al., 2010b).

5.2. Measuring thioredoxin redox poise using the PEGylation assay

A second protein that provides insight into the thiol redox environment within mitochondria is Trx2 (Arner and Holmgren, 2000; Lee et al., 1999; Miranda-Vizuete et al., 1999; Spyrou et al., 1997). Trx2 plays an important role in the maintenance of the protein redox state within mitochondria. It exists in both an oxidized form characterized by an intramolecular disulfide, and a reduced form in which the dithiol is capable of reducing other protein disulfides. Reduction of other protein thiols by Trx2 regenerates the oxidized form which can then be reduced by TrxR2. Therefore, the ratio of the reduced to oxidized forms of Trx2 will indicate the steady state level of the

A Prx3 redox blot

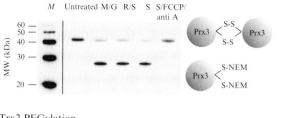

B Trx2 PEGylation

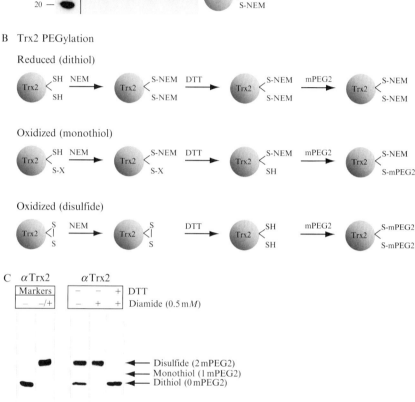

Figure 8.5 Overview of the Prx3 and Trx2 assays. (A) Isolated rat heart mitochondria were treated with the indicated concentrations of glutamate/malate (10 mM), succinate (10 mM), rotenone (4 μg/ml), FCCP (200 nM), antimycin A (10 μM) and then alkylated with NEM as described in the text. Nonreducing SDS–PAGE was combined with Western blotting against Prx3 to enable measurement of the reduced and oxidized forms. (B) Scheme illustrating the PEGylation assay for Trx2. (C) Isolated rat heart mitochondria (1 mg protein/ml) were incubated ± diamide (0.5 mM) in SHE buffer containing glutamate/malate (10 mM) for 5 min at 37 °C. Samples were alkylated with NEM in SHE buffer for 5 min, followed by an additional incubation of 2.5 min in the presence of SDS at 37 °C. NEM was removed from the solution with a size exclusion chromatography column preequilibrated in SHE buffer containing SDS. Oxidized thiols were subsequently reduced with DTT and then reacted with a surplus of 2 kDa PEG maleimide (mPEG2) as in (B). Some samples were also incubated with DTT (10 mM) for 5 min at 37 °C before PEGylation. Markers were prepared by reducing mitochondria with DTT and then alkylating the proteins with mPEG2 and NEM (−/+) or with NEM alone (−).

mitochondrial Trx2 pool. A number of assays for assessing the Trx2 redox state have been devised that function by separating these two redox forms by electrophoresis on native or urea gels and then identifying them by immunoblotting (Bersani et al., 2002; Chen et al., 2002). We have found that selectively tagging the oxidized form of Trx2 with maleimide-modified polyethylene glycol (PEG) polymers of 2 kDa introduces a significant size difference between the reduced and oxidized forms and means that they can be separated on conventional SDS–PAGE. The Trx2 is then transferred to a blotting membrane and visualized using specific antibodies. This procedure has the advantage that, depending on the selectivity and affinity of the antibodies used, the redox poise of the Trx2 pool can be assessed in complex mixtures without any requirement for purification. In this it is vital to ensure that the quenching of the reduced form of Trx2 occurs rapidly and that there is no artifactual oxidation or reduction during workup to ensure that the measured ratio reflects that inside the mitochondria in steady state conditions.

The principle of this assay is described in Fig. 8.5. After exposure to a prooxidant, isolated mitochondria are treated with NEM to block exposed thiols, subsequently lysed, and proteins denatured with SDS. After NEM is removed by centrifugal gel filtration and all oxidized thiols are reduced by the addition of DTT. The reduced thiols are then reacted with maleimide polyethylene glycol (mPEG2). This results in a molecular weight shift of about 4 kDa for the fully oxidized compared to the fully reduced Trx2, and for the monothiol a molecular weight shift of about 2 kDa. Samples are separated by reducing SDS–PAGE, transferred to PVDF and individual Trx2 bands detected with the αTrx2 antibody and suitable secondary antibody. These experiments are carried out using isolated mitochondria or cultured cells, but the method should be easily adaptable to other starting materials such as whole organisms such as flies and nematodes. An example of assessing the Trx2 redox state in isolated mitochondria is given below.

Mitochondria are incubated under standard conditions (e.g.,1 mg protein/ml in SHE buffer, 250 mM sucrose, 10 mM HEPES, 1 mM EGTA, pH 7.4) in the presence and absence of a redox challenge such as diamide. Following incubation, 25–50 mM NEM is added to prevent spontaneous oxidation of thiols. Mitochondria are then isolated by centrifugation and the pellet is resuspended in 50 μl SHE buffer containing 50 mM NEM incubated for 5 min at 37 °C. Then SDS (1% w/v) is added and the suspension is incubated for another 2.5 min at 37 °C. Following alkylation, NEM is removed with a size exclusion chromatography column (Micro Bio-Spin® 6 Chromatography Columns, Bio-Rad) preequilibrated with SHE buffer containing SDS 1% (w/v). Samples are then reduced with 2.5 mM DTT for 10 min at RT, followed by alkylation with 1 vol. of 2 kDa mPEG2 (50 mM mPEG2 in SHE buffer, pH 7.4) for 30 min at 37 °C. To determine protein concentration in the samples a BCA assay is carried out. Then, 50 μg of protein are precipitated using 9 vols. ice-cold acetone and incubated for $\sim$2 h at

−20 °C. The precipitated protein is pelleted by centrifugation (15,000×g, 10 min) and the supernatant discarded. The pellet is washed in 80% (v/v) acetone for 10 min with constant agitation and the samples are centrifuged again (15,000×g, 10 min). The supernatant is discarded and pellets are air dried and then resuspended in conventional SDS–PAGE gel loading buffer with DTT and separated by 12.5% SDS–PAGE under reducing conditions. Following electrophoresis the gel is washed for 15 min at room temperature with constant agitation in transfer buffer (48 mM Tris base, 39 mM glycine, 0.05% (w/v) SDS, 35% (v/v) methanol). The high affinity, low pore size PVDF membrane (0.2 μm Immobilon^{TM}-P^{SQ}, Millipore) is used due to the low-molecular-weight of Trx2. The membrane is first incubated at RT for 15 s in absolute methanol followed by a 2 min wash step in H$_2$O. The transfer is carried out in a Bio-Rad Mini Protean 3 Transfer Cell using transfer buffer for 50 min at 110 V and 4 °C. After transfer the membrane is incubated in blocking buffer (PBST) (137 mM NaCl, 10.2 mM NaHPO$_4$, 2.7 mM KCl, 1.8 mM KH$_2$PO$_4$, pH7.4; and 0.05% (v/v) Tween 20 (NBS Biologicals, UK)) with 5% (w/v) skimmed milk powder at room temperature for 1 h. The blocking buffer is removed and membranes are incubated with the primary antibody αTrx2 antibody (1:5000) (affinity purified goat antihuman/mouse/rat Thioredoxin2/Trx2 antibody, R&D Systems) in blocking buffer at RT for 1 h. Following incubation, the membranes are washed three times for 10 min with PBST (see blocking buffer) and then incubated with the appropriate secondary antibody (1:20,000 in blocking buffer) (rabbit antigoat IgG-peroxidase, Sigma) at RT for 1 h. Secondary antibody incubation is followed by three 10 min washes with PBST. Protein detection is carried out according to manufacturer's instructions (ECL Plus, GE Healthcare). Typical results are shown in Fig 8.5. This procedure can be applied to any mitochondrial incubation and also to cell systems.

6. Conclusions

To better understand oxidative damage and redox signaling within mitochondria, it is essential to both measure global changes to protein thiols and also to try and identify the particular proteins affected, the cysteine residues involved and the nature of the modifications that occur. The methods shown here enable this in isolated mitochondria, but in addition in many cases it may be possible to extend this to mitochondria in cells and intact organisms.

Many of the methods shown here give an overview of the general thiol redox state of mitochondria, which is an important starting point towards understanding redox changes under a variety of conditions. We have also summarized methods for sensitively and selectively detecting protein

S-nitrosothiol formation, which likely have contrasting physiological and pathological effects on protein thiols when compared to other modifications. In addition, we have indicated how it is possible to measure the redox states of Trx2 and Prx3, which play critical roles for many protein thiol modifications. However, it is also important to be able to assess and identify those proteins that are involved in reversible redox modifications both to identify potential nodes of regulation and also those proteins that may be involved in oxidative defenses. To do this, a number of redox proteomic approaches have been developed and have been described in detail (Baty *et al.*, 2005; Fu *et al.*, 2008; Hurd *et al.*, 2009; McDonagh, 2009; Sethuraman *et al.*, 2004), so were not discussed here. However, in all studies the general methods outlined here should be carried out in conjunction with redox proteomic approaches to assist in the identification of the proteins of interest. In the future it is hoped that the approaches outlined here, as well as those based on redox proteomic approaches, can also be applied to intact organisms to enable detection and characterization of thiol changes *in vivo*. Future work should also allow for both the quantification of the extent of the redox modification as well as the identification of the protein thiol involved, and thus give an indication of the important modifications that occur to mitochondrial thiols in response to physiological and pathological events

ACKNOWLEDGMENTS

This work was supported by the Medical Research Council (UK) and by a postgraduate scholarship (to E. T. C.) from the Gates Cambridge Trust.

REFERENCES

Anderson, M. (1985). Determination of glutathione and glutathione disulfide in biological samples. *Methods Enzymol.* **113,** 548–555.

Arner, E. S., and Holmgren, A. (2000). Physiological functions of thioredoxin and thioredoxin reductase. *Eur. J. Biochem.* **267,** 6102–6109.

Balaban, R. S., Nemoto, S., and Finkel, T. (2005). Mitochondria, oxidants, and aging. *Cell* **120,** 483–495.

Baty, J. W., Hampton, M. B., and Winterbourn, C. C. (2005). Proteomic detection of hydrogen peroxide-sensitive thiol proteins in Jurkat cells. *Biochem. J.* **389,** 785–795.

Beckman, K. B., and Ames, B. N. (1998). The free radical theory of aging matures. *Physiol. Rev.* **78,** 547–581.

Beer, S. M., Taylor, E. R., Brown, S. E., Dahm, C. C., Costa, N. J., Runswick, M. J., and Murphy, M. P. (2004). Glutaredoxin 2 catalyzes the reversible oxidation and glutathionylation of mitochondrial membrane thiol proteins: Implications for mitochondrial redox regulation and antioxidant defense. *J. Biol. Chem.* **279,** 47939–47951.

Bersani, N. A., Merwin, J. R., Lopez, N. I., Pearson, G. D., and Merrill, G. F. (2002). Protein electrophoretic mobility shift assay to monitor redox state of thioredoxin in cells. *Methods Enzymol.* **347,** 317–326.

Brown, K. K., Eriksson, S. E., Arner, E. S., and Hampton, M. B. (2008). Mitochondrial peroxiredoxin 3 is rapidly oxidized in cells treated with isothiocyanates. *Free Radic. Biol. Med.* **45,** 494–502.

Chappell, J. B., and Hansford, R. G. (1972). Preparation of mitochondria from animal tissues and yeasts. In "Subcellular Components: Preparation and Fractionation," (G. D. Birnie, ed.), pp. 77–91. Butterworths, London.

Chen, Y., Cai, J., Murphy, T. J., and Jones, D. P. (2002). Overexpressed human mitochondrial thioredoxin confers resistance to oxidant-induced apoptosis in human osteosarcoma cells. *J. Biol. Chem.* **277,** 33242–33248.

Costa, N. J., Dahm, C. C., Hurrell, F., Taylor, E. R., and Murphy, M. P. (2003). Interactions of mitochondrial thiols with nitric oxide. *Antioxid. Redox Signal.* **5,** 291–305.

Cox, A. G., Brown, K. K., Arner, E. S., and Hampton, M. B. (2008a). The thioredoxin reductase inhibitor auranofin triggers apoptosis through a Bax/Bak-dependent process that involves peroxiredoxin 3 oxidation. *Biochem. Pharmacol.* **76,** 1097–1109.

Cox, A. G., Pullar, J. M., Hughes, G., Ledgerwood, E. C., and Hampton, M. B. (2008b). Oxidation of mitochondrial peroxiredoxin 3 during the initiation of receptor-mediated apoptosis. *Free Radic. Biol. Med.* **44,** 1001–1009.

Cox, A. G., Pearson, A. G., Pullar, J. M., Jonsson, T. J., Lowther, W. T., Winterbourn, C. C., and Hampton, M. B. (2009a). Mitochondrial peroxiredoxin 3 is more resilient to hyperoxidation than cytoplasmic peroxiredoxins. *Biochem. J.* **421,** 51–58.

Cox, A. G., Peskin, A. V., Paton, L. N., Winterbourn, C. C., and Hampton, M. B. (2009b). Redox potential and peroxide reactivity of human peroxiredoxin 3. *Biochemistry* **48,** 6495–6501.

Cox, A. G., Winterbourn, C. C., and Hampton, M. B. (2010a). Mitochondrial peroxiredoxin involvement in antioxidant defence and redox signalling. *Biochem. J.* **425,** 313–325.

Cox, A. G., Winterbourn, C. C., and Hampton, M. B. (2010b). Measuring the redox state of cellular peroxiredoxins by immunoblotting. In "Methods in Enzymology," (E. Cadenas and L. Packer, eds.), pp. 51–66. Burlington, US.

Dahm, C. C., Moore, K., and Murphy, M. P. (2006). Persistent S-nitrosation of complex I and other mitochondrial membrane proteins by S-nitrosothiols but not nitric oxide or peroxynitrite: Implications for the interaction of nitric oxide with mitochondria. *J. Biol. Chem.* **281,** 10056–10065.

Delaunay, A., Pflieger, D., Barrault, M. B., Vinh, J., and Toledano, M. B. (2002). A thiol peroxidase is an H2O2 receptor and redox-transducer in gene activation. *Cell* **111,** 471–481.

Ellman, G., and Lysko, H. (1979). A precise method for the determination of whole blood and plasma sulfhydryl groups. *Anal. Biochem.* **93,** 98–102.

Feelisch, M., Rassaf, T., Mnaimneh, S., Singh, N., Bryan, N. S., Jourd'Heuil, D., and Kelm, M. (2002). Concomitant S-, N-, and heme-nitros(yl)ation in biological tissues and fluids: Implications for the fate of NO in vivo. *FASEB J.* **16,** 1775–1785.

Finkel, T. (2005). Opinion: Radical medicine: Treating ageing to cure disease. *Nat. Rev. Mol. Cell Biol.* **6,** 971–976.

Fu, C., Hu, J., Liu, T., Ago, T., Sadoshima, J., and Li, H. (2008). Quantitative analysis of redox-sensitive proteome with DIGE and ICAT. *J. Proteome Res.* **7,** 3789–3802.

Galkin, A., and Moncada, S. (2007). S-nitrosation of mitochondrial complex I depends on its structural conformation. *J. Biol. Chem.* **282,** 37448–37453.

Hogg, N. (2002). The biochemistry and physiology of S-nitrosothiols. *Annu. Rev. Pharmacol. Toxicol.* **42,** 585–600.

Hurd, T. R., Costa, N. J., Dahm, C. C., Beer, S. M., Brown, S. E., Filipovska, A., and Murphy, M. P. (2005a). Glutathionylation of mitochondrial proteins. *Antioxid. Redox Signal.* **7,** 999–1010.

Hurd, T. R., Filipovska, A., Costa, N. J., Dahm, C. C., and Murphy, M. P. (2005b). Disulphide formation on mitochondrial protein thiols. *Biochem. Soc. Trans.* **33,** 1390–1393.

Hurd, T. R., Requejo, R., Filipovska, A., Brown, S., Prime, T. A., Robinson, A. J., Fearnley, I. M., and Murphy, M. P. (2008). Complex I within oxidatively stressed bovine heart mitochondria is glutathionylated on Cys-531 and Cys-704 of the 75-kDa subunit: Potential role of CYS residues in decreasing oxidative damage. *J. Biol. Chem.* **283,** 24801–24815.

Hurd, T. R., James, A. M., Lilley, K. S., and Murphy, M. P. (2009). Chapter 19 Measuring redox changes to mitochondrial protein thiols with redox difference gel electrophoresis (redox-DIGE). *Methods Enzymol.* **456,** 343–361.

Janssen-Heininger, Y. M., Mossman, B. T., Heintz, N. H., Forman, H. J., Kalyanaraman, B., Finkel, T., Stamler, J. S., Rhee, S. G., and van der Vliet, A. (2008). Redox-based regulation of signal transduction: Principles, pitfalls, and promises. *Free Radic. Biol. Med.* **45,** 1–17.

Kumar, V., Kitaeff, N., Hampton, M. B., Cannell, M. B., and Winterbourn, C. C. (2009). Reversible oxidation of mitochondrial peroxiredoxin 3 in mouse heart subjected to ischemia and reperfusion. *FEBS Lett.* **583,** 997–1000.

Lee, S. R., Kim, J. R., Kwon, K. S., Yoon, H. W., Levine, R. L., Ginsburg, A., and Rhee, S. G. (1999). Molecular cloning and characterization of a mitochondrial selenocysteine-containing thioredoxin reductase from rat liver. *J. Biol. Chem.* **274,** 4722–4734.

Lillig, C. H., Lonn, M. E., Enoksson, M., Fernandes, A. P., and Holmgren, A. (2004). Short interfering RNA-mediated silencing of glutaredoxin 2 increases the sensitivity of HeLa cells toward doxorubicin and phenylarsine oxide. *Proc. Natl. Acad. Sci. USA* **101,** 13227–13232.

McDonagh, B. (2009). Diagonal electrophoresis for detection of protein disulphide bridges. *Methods Mol. Biol.* **519,** 305–310.

Miranda-Vizuete, A., Damdimopoulos, A. E., Pedrajas, J. R., Gustafsson, J. A., and Spyrou, G. (1999). Human mitochondrial thioredoxin reductase cDNA cloning, expression and genomic organization. *Eur. J. Biochem.* **261,** 405–412.

Moncada, S., and Erusalimsky, J. D. (2002). Does nitric oxide modulate mitochondrial energy generation and apoptosis? *Nat. Rev. Mol. Cell Biol.* **3,** 214–220.

Murphy, M. P. (2009). How mitochondria produce reactive oxygen species. *Biochem. J.* **417,** 1–13.

Perkins, D. N., Pappin, D. J., Creasy, D. M., and Cottrell, J. S. (1999). Probability-based protein identification by searching sequence databases using mass spectrometry data. *Electrophoresis* **20,** 3551–3567.

Prime, T. A., Blaikie, F. H., Evans, C., Nadtochiy, S. M., James, A. M., Dahm, C. C., Vitturi, D. A., Patel, R. P., Hiley, C. R., Abakumova, I., Requejo, R., Chouchani, E. T., *et al.* (2009). A mitochondria-targeted S-nitrosothiol modulates respiration, nitrosates thiols, and protects against ischemia-reperfusion injury. *Proc. Natl. Acad. Sci. USA* **106,** 10764–10769.

Requejo, R., Hurd, T. R., Costa, N. J., and Murphy, M. P. (2010). Cysteine residues exposed on protein surfaces are the dominant intramitochondrial thiol and may protect against oxidative damage. *FEBS J.* **277,** 1465–1480.

Rhee, S. G., Kang, S. W., Jeong, W., Chang, T. S., Yang, K. S., and Woo, H. A. (2005). Intracellular messenger function of hydrogen peroxide and its regulation by peroxiredoxins. *Curr. Opin. Cell Biol.* **17,** 183–189.

Scarlett, J. L., Packer, M. A., Porteous, C. M., and Murphy, M. P. (1996). Alterations to glutathione and nicotinamide nucleotides during the mitochondrial permeability transition induced by peroxynitrite. *Biochem. Pharmacol.* **52,** 1047–1055.

Schafer, F. Q., and Buettner, G. R. (2001). Redox environment of the cell as viewed through the redox state of the glutathione disulfide/glutathione couple. *Free Radic. Biol. Med.* **30,** 1191–1212.

Schagger, H. (1995). Native electrophoresis for isolation of mitochondrial oxidative phosphorylation complexes. *Methods Enzymol.* **260,** 190–202.

Schagger, H., and von Jagow, G. (1991). Blue native electrophoresis for isolation of membrane protein complexes in enzymatically active form. *Anal. Biochem.* **199,** 223–231.

Sethuraman, M., McComb, M. E., Huang, H., Huang, S., Heibeck, T., Costello, C. E., and Cohen, R. A. (2004). Isotope-coded affinity tag (ICAT) approach to redox proteomics: Identification and quantitation of oxidant-sensitive cysteine thiols in complex protein mixtures. *J. Proteome Res.* **3,** 1228–1233.

Smith, P. K., Krohn, R. I., Hermanson, G. T., Mallia, A. K., Gartner, F. H., Provenzano, M. D., Fujimoto, E. K., Goeke, N. M., Olson, B. J., and Klenk, D. C. (1985). Measurement of protein using bicinchoninic acid. *Anal. Biochem.* **150,** 76–85.

Spyrou, G., Enmark, E., Miranda-Vizuete, A., and Gustafsson, J. (1997). Cloning and expression of a novel mammalian thioredoxin. *J. Biol. Chem.* **272,** 2936–2941.

Szabo, C., Ischiropoulos, H., and Radi, R. (2007). Peroxynitrite: Biochemistry, pathophysiology and development of therapeutics. *Nat. Rev. Drug Discov.* **6,** 662–680.

Taylor, E. R., Hurrell, F., Shannon, R. J., Lin, T. K., Hirst, J., and Murphy, M. P. (2003). Reversible glutathionylation of complex I increases mitochondrial superoxide formation. *J. Biol. Chem.* **278,** 19603–19610.

Wang, X., Kettenhofen, N. J., Shiva, S., Hogg, N., and Gladwin, M. T. (2008). Copper dependence of the biotin switch assay: Modified assay for measuring cellular and blood nitrosated proteins. *Free Radic. Biol. Med.* **44,** 1362–1372.

Wilm, M., Shevchenko, A., Houthaeve, T., Breit, S., Schweigerer, L., Fotsis, T., and Mann, M. (1996). Femtomole sequencing of proteins from polyacrylamide gels by nano-electrospray mass spectrometry. *Nature* **379,** 466–469.

Winterbourn, C. C. (2008). Reconciling the chemistry and biology of reactive oxygen species. *Nat. Chem. Biol.* **4,** 278–287.

Winterbourn, C. C., and Hampton, M. B. (2008). Thiol chemistry and specificity in redox signaling. *Free Radic. Biol. Med.* **45,** 549–561.

CHAPTER NINE

MEASUREMENT OF EXTRACELLULAR (EXOFACIAL) VERSUS INTRACELLULAR PROTEIN THIOLS

Jolanta Skalska,* Steven Bernstein,* *and* Paul Brookes[†]

Contents

1. Measurement of Mitochondrial Thiol Status	150
1.1. Isolation of mitochondria	151
1.2. Mitochondrial purity, yield, and special considerations for thiol status	152
1.3. Measuring mitochondrial thiols with common thiol reagents	153
1.4. Measurement of mitochondrial thiol status *in situ* using IBTP	155
2. Measurement of Cytosolic Thiol Status	156
2.1. Global intracellular thiol measurement	156
3. Measurement of Exofacial Thiols	157
3.1. Measurement of exofacial thiol status *in situ*	157
3.2. Measurement of exofacial thiol status by fractionation	158
3.3. Miscellaneous other methods to measurement exofacial thiols	159
4. Exofacial Thiol Status and Cancer	160
References	162

Abstract

In recent years, the importance of compartmentalization in redox signaling has been realized. A number of specific thiol pools exist both inside and outside the cell, and these thiols are regulated via unique mechanisms and serve specific roles in cell signaling. This chapter covers some of the methodologies available for the interrogation of thiol status in various cellular compartments, with a focus on mitochondrial, cytosolic, and exofacial thiols. Finally, the relevance of these thiols to pathological disease states, in particular cancer, will be discussed.

The chapters in the remainder of this volume more than adequately cover the diversity of thiol modifications, describing the specific biochemical nature of these reactions, ranging from *S*-nitrosation through glutathionylation, to oxidation and beyond. Therefore, this topic will not be further addressed here.

---

* James P. Wilmot Cancer Center, University of Rochester Medical Center, Rochester, New York, USA
[†] Department of Anesthesiology, University of Rochester Medical Center, Rochester, New York, USA

Similarly, general methodological considerations are considered to have been dealt with in the remainder of this volume, including requirements for subdued lighting, avoidance of reducing agents and transition metals in media, and rapid sample preparation with adequate control over temperature and pH.

1. Measurement of Mitochondrial Thiol Status

There are several factors which render the mitochondrion interesting from the perspective of examining its thiol status. First, mitochondria are the powerhouse of the cell, responsible for the bulk of ATP synthesis (Brookes et al., 2004). This synthesis is achieved largely by a set of enzymes which rely on tight control of redox status for their function. Thus, mitochondrial thiol status has emerged in recent years as a reliable indicator of overall mitochondrial functional status (Circu et al., 2008).

Secondly, mitochondria contain a large pool of reduced glutathione (GSH), amounting to a local concentration in the organelle of 5–10 mM (Kimura et al., 2010). This GSH pool is essential for the maintenance of redox state of mitochondrial protein thiols. While traditional GSH-depleting agents such as L-buthionine-sulfoximine (and inhibitor of GSH synthesis) are very effective at depleting whole cell GSH levels, the mitochondrial GSH pool is somewhat refractory to such treatments (Seyfried et al., 1999). In this regard, a reagent which has received very little attention is 3-hydroxy-4-pentenoate (3HP; Shan et al., 1993). This molecule is not reactive to thiols in its native state, but upon delivery to the mitochondrion it undergoes β-oxidation yield 3-oxo-4-pentenoate, a Michael acceptor that reacts directly with GSH and thereby selectively depletes the mitochondrial pool of this important antioxidant. Although 3HP is not commercially available, its synthesis is simple. Wider use of such a reagent may allow a better understanding of the role of the mitochondrial GSH pool in various pathological disease states.

Third, a gradient of 0.25–0.50 pH units exists across the mitochondrial membrane, which renders the mitochondrial matrix more alkaline than the cytosol (Addanki et al., 1967). Since the pK_a of the thiol group on protein cysteine residues is ~ 8, the result will be that cysteines in the mitochondrial matrix are relatively more unprotonated than those in the cytosol. Furthermore, many cysteine thiol modifications are heavily influenced by pH (i.e., they preferentially occur on either the −SH or the thiolate anion state). This has led to speculation that the mitochondrion may be a "hot-spot" for certain types of thiol modification (e.g., S-nitrosation) which are favored by thiolates (Burwell and Brookes, 2008).

There exist several methods for the assessment of mitochondrial thiol status, with the key experimental decision being whether to isolate mitochondria first, or whether to assess mitochondria in situ, within a cell or tissue.

1.1. Isolation of mitochondria

Isolation of mitochondria from tissues, such as liver, heart, or brain, is simple and is covered by several excellent chapters in previous volumes (most recently #456) of this book series (Burgess and Deutscher, 2009), so will not be covered here. Rather, a method is presented for isolation of mitochondria from cells in culture, which is becoming increasingly popular.

Starting material is suggested to be at least $3\times$ T75 flasks of confluent cells in culture. Cell size should be taken into consideration, with smaller cells (e.g., lymphoid cells) requiring far higher numbers than larger ones (e.g., myocytes). Some common cell lines from which mitochondria can easily be isolated include H9C2, Clone 9, HEK293, PC12, and HepG2. It is to be noted that many such cell lines are immortalized and thus favor a glycolytic phenotype, with concomitant downregulation of mitochondrial number. Thus, one possible intervention (provided it does not interfere with the experimental question being asked) is to grow the cells in medium containing only a nonfermentable carbon source (e.g., galactose) to force the phenotype back to mitochondrial oxidative phosphorylation (Rossignol *et al.*, 2004).

Isolation medium (IM) for mitochondria is typically composed of some osmotic support (sucrose, KCl, or mannitol), a Ca^{2+} chelator such as EGTA, and a pH buffer. The use of free acids is recommended, rather than cheaper sodium salts, since these can interfere with mitochondrial Ca^{2+} homeostasis due to the mitochondrial Na^+/Ca^{2+} exchanger (Cox and Matlib, 1993). In addition, the purest deionized water available (ideally $>$ 18 mΩ resistivity) should be used. A typical mitochondrial IM is listed as follows:

Sucrose: 250 mM
EGTA: 1 mM
HEPES: 10 mM
pH to 7.4 with KOH/HCl, store at 4 °C for no more than 1 week.

N.B. Several commercially available kits for mitochondrial isolation will provide isolation media containing preservatives, and many such reagents will affect thiol status. Avoid any kit if the ingredients contain DTT, GSH, or other reductants. Similarly, avoid kits which use an osmotic swelling step to break open cells, as this can also disrupt mitochondrial membranes.

The most important step in any mitochondrial isolation (after the medium) is the method of homogenization. In brief, small plastic microtube pestle-type homogenizers (motorized pellet disruptors) do not generate sufficient shearing forces to break open the cell membrane and release mitochondria. Better methods include a glass Dounce homogenizer, or nitrogen cavitation. Dounce homogenizers are available in a variety of sizes, down to 0.2 ml volume. Ideally, the cell pellet will be resuspended in 5 ml of IM, and homogenized by at least 50 passes with the Dounce pestle. Nitrogen cavitation bombs (cell disruption bombs) are available from Parr Inc., with a typical protocol involving stirring the cells in 5 ml of IM at 2500 psi N_2 for 15 min at 4 °C, followed by rapid release of the pressure resulting in plasma membrane rupture.

Following cell disruption, the homogenate should be diluted at least $3\times$ in IM, and centrifuged (in a refrigerated centrifuged) at $700–1000 \times g$ for 5–10 min to pellet nuclear debris and unbroken cells. The supernatant from this spin is then centrifuged at $8000–10{,}000 \times g$ for 10 min to yield a crude mitochondrial pellet. Depending on the cell type, this initial pellet should be light brown in color. The supernatant is discarded, and the pellet carefully resuspended in the original volume of IM, and subjected to another high-speed centrifugation step. Finally, a third washing and high-speed centrifugation is performed, yielding a relatively pure mitochondrial preparation. The final pellet should be suspended in the smallest volume possible (50–200 μl). At each pellet resuspension step, it is important to work quickly, and to hold the tube in such a way that manual heat does not reach the pellet. If the pellet is covered with a fluffy white layer, this should be removed at each stage, favoring the darker brown material which is enriched in mitochondria. The later stages of the preparation may be transferred to a smaller centrifuge (e.g., microcentrifuge using 1.5 ml plastic tubes), still maintaining tight temperature control.

1.2. Mitochondrial purity, yield, and special considerations for thiol status

Mitochondrial *purity* can only be measured by determining the degree of contamination of the mitochondrial preparation by other membrane fractions. Western blotting for the surface and endoplasmic reticular Ca^{2+} ATPase (SERCA) is typically used to monitor ER/SR contamination. (*Note*: the terms purity and enrichment are frequently misexchanged in the literature. Monitoring the level of a mitochondrial protein informs nothing about the purity of a mitochondrial preparation.) In contrast, mitochondrial *enrichment* can be measured by following the level of a mitochondrial protein through the isolation procedure, and looking for an increasing level associated with the mitochondrial fraction at each stage of the preparation. Typically, Western blotting for Voltage Dependent Anion Channel (VDAC) is used, or the activity of the mitochondrial matrix marker enzyme citrate synthase is measured (Shepherd and Garland, 1969). Mitochondrial yield is typically assessed by measuring the protein concentration of the final sample, by any common method.

Depending on the exact thiol modification that is being examined, it is often necessary to add reagents to the mitochondrial IM to prevent degradation of such PTMs. Common examples include addition of *N*-ethylmaleimide (NEM; 200–500 μM) to block free –SH groups and prevent modifications from transferring between thiols and diethylenetriamine pentaacetic acid (DTPA; 100 μM) to chelate-free transition metals which can contribute to S-nitrosothiol degradation (Mani *et al.*, 2006).

1.3. Measuring mitochondrial thiols with common thiol reagents

Once mitochondria have been isolated, they can be stored on ice for 3–4 h for use in experiments. For treatments of mitochondria with various thiol-reactive reagents, a typical incubation medium consists of:

KCl: 120 mM
Sucrose: 25 mM
MgCl$_2$: 5 mM
KH$_2$PO$_4$: 5 mM
EGTA: 0.5 mM
HEPES: 10 mM, pH 7.4 at 37 °C
Sterile filter, store frozen in 50 ml aliquots for up to 1 year.

Mitochondria are incubated in the above medium at 1 mg protein/ml (typically in 1.5 ml plastic microcentrifuge tubes). Respiratory substrates (5 mM succinate, 5 mM glutamate, 2.5 mM malate) and/or ADP (500 μM) can also be added from stock solutions prepared in the same medium, in order to modulate the respiratory state of the mitochondria during treatment.

Typically, mitochondria are incubated in the above medium with the thiol-reactive reagent present in the medium before mitochondria are added. Incubation times are typically 5–10 min. Reagents available for assaying thiol function are discussed widely throughout this volume, and include:

Biotin-maleimide
Biotin-PEO-maleimide (linker-arm facilitates thiol interactions)
Biotin-HPDP
Biotin-BMCC (maleimide based)
Biotin-IAA or IAM
Biotinylated-glutathione-ethyl ester ("BioGEE").

All can be used at concentrations of 0.5–1 mM. Due to the pH optima of iodo-reagents (they preferentially react with thiols at alkaline pH), their use is encouraged for assaying intramitochondrial thiols, because of the slightly alkaline pH of the mitochondrial matrix. Note that the choice of thiol reagent determines the downstream processing of the samples. Maleimide-based reagents form a covalent –C–S–C– (thioether) bond, which cannot be reduced by common reducing conditions (DTT, β-ME, etc.), and so is well suited to subsequent analysis by reducing SDS–PAGE. In contrast, reagents such as HPDP form a dithiol bond (–C–S–S–C–) which can be broken under reducing conditions, and so samples can only be analyzed by nonreducing SDS–PAGE. Another approach is to break the mitochondrial membrane with detergents prior to addition of the thiol-labeling reagent, thus affording greater access to the entire mitochondrial thiol pool. Addition of 0.1% Triton X-100 into the incubation media is typically sufficient to accomplish this.

Detection of modified thiols is typically accomplished by SDS–PAGE, Western blotting, and probing with antibiotin antibodies or streptavidin-conjugated peroxidase. A "loss-of-signal" mode is typically used, that is, the presence of a biotinylated band on the blot indicates a protein with a thiol (–SH) that is free to react with the biotin reagent, and the loss of a band indicates that particular thiol is no longer in the free (–SH) state and has therefore been modified to something else. The method does not inform about what the exact modification is. A typical example of this approach is shown in Fig. 9.1, in which isolated rat heart mitochondria were treated with a 200 μM bolus of $ONOO^-$, followed by analysis using Biotin-PEO-maleimide. The $ONOO^-$ treatment led to loss of reactivity in a selected number of proteins, particularly those in the 35 kDa range.

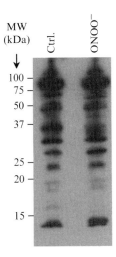

Figure 9.1 Biotin-PEO-maleimide labeling of rat heart mitochondria. Isolated rat heart mitochondria were incubated at 1 mg protein/ml in standard incubation medium at 37 °C. Where indicated, samples were treated with a single 200 μM bolus of $ONOO^-$, with rapid mixing. Five minutes later, 200 μM Biotin-PEO-maleimide and 0.1% Triton X-100 were added and after a further 5 min incubation samples were processed for SDS–PAGE and Western blotting. Blots were probes with antibiotin antibodies (Pierce).

BioGEE is a unique case, since it is used to probe the thiol modification known as glutathionylation (also sometimes called glutathiolation), that is, the direct addition of a GSH molecule to form a mixed disulfide with a protein thiol. There has been much interest in this modification in recent years, since it is hypothesized that other thiol modifications such as *S*-nitrosation may serve to "prime" protein thiols for glutathionylation (Martinez-Ruiz and Lamas, 2007).

BioGEE labeling is accomplished by incubating isolated mitochondria with the BioGEE reagent (typically 250–500 μM) prior to the treatment which induces stress. The BioGEE mixes with the regular GSH pool, and it is anticipated that if glutathionylation of a protein occurs, at least some of the resulting biotinylated GSH will participate, resulting in biotin labeling of the protein. Naturally, the disulfide nature of the bond renders this modification labile to reducing agents, so downstream processing must be on nonreducing gels. Furthermore, unlike the other biotin labeling strategies discussed above, in which a thiol modification results in a loss of signal, BioGEE labeling is a gain-of-signal assay, that is, the presence of a biotin label indicates the protein was modified by glutathionylation.

1.4. Measurement of mitochondrial thiol status *in situ* using IBTP

In addition to the study of mitochondrial thiols by using isolated mitochondria, a unique method has been developed to monitor mitochondrial thiols in intact cells, namely the reagent iodobutyl triphenylphosphonium (IBTP).

The triphenylphosphonium (TPP) moiety of IBTP is both positively charged and hydrophobic, and this functional group has been successfully applied to the targeting of a number of molecules to mitochondria within cells, intact organs, and whole animals. The uptake of TPP-conjugated drugs is driven by the high internal negative membrane potential ($\Delta\Psi$) of the mitochondria, typically resulting in a 500–1000-fold accumulation of the drug over the external concentration (Smith *et al.*, 2008). This strategy is the basis of a number of experimental therapeutics including mitochondrially targeted antioxidants and nitric oxide donors (Murphy, 2008; Smith *et al.*, 2008).

The other end of the IBTP molecule is an iodoakyl-based thiol alkylating group, similar to that found in iodoacetic acid (IAA) or iodoacetamide (IAM), and it therefore reacts with free –SH groups. The elegant method devised to detect such modifications is the development of an antibody which detects the IBTP–thiol conjugate (Lin *et al.*, 2002). Thus, complex biological systems such as cells or organisms can be treated with IBTP, which will accumulate in mitochondria and selectively label mitochondrial thiols. The resulting analysis does not require cellular subfractionation to isolate mitochondria; the whole cell or tissue sample can simply be separated by SDS–PAGE and Western blotted with the IBTP antibodies, with the result being that each band on the blot corresponds to a mitochondrial protein with a free –SH group that reacted with IBTP.

Despite the noncommercial nature of this reagent, the generosity of its developer (Michael Murphy of Cambridge University, UK) has resulted in its application to a number of interesting situations, including the study of mitochondrial thiols in cardiac ischemia, hepatic ethanol toxicity, and lipid oxidation conditions. More recently, the reagent has been utilized in a similar vein to 3HP, to selectively inactivate/alkylate the mitochondrial thiol pool, in order to probe the role of this specific subset of intracellular thiols in various pathophysiological processes.

2. Measurement of Cytosolic Thiol Status

2.1. Global intracellular thiol measurement

Most of the available cytosol thiol-labeling methods are based on utilizing thiol-reactive probes, like DTNB (5,5′-dithiobis(2-nitrobenzoic acid)), and the product of this reaction can be measured spectrophotometrically (absorption measured at 412 nm). This compound reacts quantitatively with protein thiols and GSH and hence can be used to determine the number of free thiols on purified proteins, in cell lysates and in cell fractions (Tassi *et al.*, 2009). While the method does not afford information on specific proteins, it is a good first step in determining overall cellular thiol state in any pathophysiological condition.

To determine the level of GSH in intact cells, a fluorescence probe, monochlorobimane (MCB; λ_{ex} 380 nm, λ_{em} 461 nm) is widely used as described elsewhere (Salazar *et al.*, 2006), although the specificity of this probe for GSH versus other thiols is not clear. One observation that has remained unexplained for several years is the origin of the large vesicular bodies which appear in MCB stained cells (e.g., Fig. 2A in Sebastia *et al.*, 2003). Such patchiness in staining is not observed with a similar probe, 7-amino-4-chloromethylcoumarin (CMAC; Sebastia *et al.*, 2003).

Another method for the assessment of cytosolic thiols is the dye mercury orange [1-(4-chloromercuryphenyl-azo-2-naphthol)], which forms an insoluble fluorescent reaction product with nonprotein –SH groups. In this case the staining can be performed on frozen sections obtained from various biopsies, and the tissue distribution of reduced thiols can be measured by a quantitative image analysis system (Ceccarelli *et al.*, 2008; Hedley *et al.*, 2005).

Proteomic tools, for investigating reversible oxidative modifications of thiol proteins, have been also developed, and is known as redox difference gel electrophoresis (DIGE; Hurd *et al.*, 2009). This technique involves labeling thiols with two different fluorescent dyes containing maleimide groups (and hence reacting with thiols) as described elsewhere (Hurd et al., 2009). This method has been adopted from the DIGE technique, which has been used for quantification of changes in protein expression level (Lilley and Friedman, 2004; Unlu, Morgan et al., 1997).

3. Measurement of Exofacial Thiols

Extracellular redox state and thiol environment have a tremendous influence on cell behavior. Cell surface protein thiols (i.e., exofacial thiols) are targets of redox regulation and their redox status influences critical cellular functions. For example, the redox status of exofacial thiols on T-cells regulates their activation and proliferation (Gelderman et al., 2006). The redox status of critical thiols on the immunoglobulin superfamily member CD4 regulates both T-cell binding to antigen presenting cells as well as HIV-1 entry into $CD4^+$ T-cells (Ou and Silver, 2006). In fact, the levels of free exofacial thiols on normal donor lymphocytes differ from those seen on cells from patients with HIV (Sahaf et al., 2003, 2005).

It has been reported that thiol redox status of integrin α-4 affects integrin-mediated cell adhesion (Laragione et al., 2003). Similarly, the thiol redox status of the tumor necrosis factor receptor superfamily member 8 (TNFRSF8/CD30) influences whether this receptor can engage its cognate ligand and transduce its downstream signaling in lymphocytes (Schwertassek et al., 2007). Furthermore, various membrane receptors, ion channels (Beech and Sukumar, 2007), and extracellular thiol–disulfide oxidoreductases themselves are redox sensitive (Janssen-Heininger et al., 2008) and hence the necessity of method development to evaluate the state of –SH groups on the cell surface.

3.1. Measurement of exofacial thiol status *in situ*

Similar to the staining of free SH groups in cytosol, methods for detection of free SH groups on the cell surface compromise the ability of protein thiol groups to react covalently with maleimide. Hence maleimide is an active component of many commercially available thiol-reactive probes, which can be successfully used for detection of plasma membrane free SH groups.

The most common fluorescent dyes tagged to maleimide are Alexa fluor® dyes, which give both a broad spectrum of detection and quantification of changes in free exofacial thiols due to the fact that maleimide coupled with the charged Alexa fluor® dye is cell impermeable (Sahaf et al., 2003).

Staining the free exofacial thiols on cells cultured in suspension can be achieved by using a simple protocol. As an example, Granta cells (mantle

B-cell lymphoma line) at a concentration of 10^6 ml^{-1} should be washed in 1× PBS, resuspended in RPMI, plated onto 35 mm Petri dishes and treated with desired drugs which modify free SH groups. At the end of the experiment, cells should be collected (e.g., scraping), washed twice with ice-cold 1× PBS, suspended in 1 ml of 1× PBS containing 1 μM Alexa Fluor 633 C$_5$ maleimide, and incubated on ice for 15 min in the dark. Incubation at low temperature assures that the dye does not penetrate the cell plasma membrane. After washing twice with ice-cold 1× PBS, suspended in 1× PBS (1 ml) cells can be analyzed using fluorescence-assisted cell sorting (e.g., Becton Dickinson FACScan Instrument) and histograms generated showing the changes of free SH groups on cell surface as shown in Fig. 9.2.

3.2. Measurement of exofacial thiol status by fractionation

Another strategy to selectively examine the surface thiols of cells is to label with a nonspecific thiol reagent, and then fractionate the cells to isolate the plasma membrane fraction. Typical reagents used for this approach are alkylating sulfhydryl reactants (maleimides, iodoalkylators, see Section 1). The detection of cell surface protein with biotinylated iodoacetamide (BIAM)

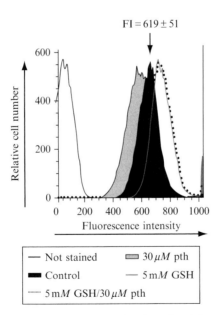

Figure 9.2 Parthenolide attenuates the levels of exofacial free thiols as assessed by flow cytometry. The change in Granta cell free exofacial thiol groups was assessed by staining cells with 1 μM Alexa 633 maleimide with analysis by flow cytometry after exposure to 30 μM parthenolide alone for 3 h, 5 mM GSHee alone for 2 h, 5 mM GSH for 2 h followed by 30 μM parthenolide for 3 h, or no treatment (control).

and examination of thiol modification via Western blotting can be achieved by a simple protocol. As an example of the staining, 3×10^7 Granta cells at a concentration of 10^6 ml^{-1} are washed in 1× PBS, resuspended in RPMI, treated with desired drugs which modify cell surface SH groups, and at the end of experiments centrifuged, washed twice with prewarmed 1× PBS, and then suspended in RPMI supplemented with 100 μM N-biotinoyl-N'-(iodoacetyl)ethylendiamine (BIAM; Invitrogen) and incubated for 15 min at 37 °C as previously described (Laragione *et al.* 2003). Then the cells are collected, washed twice with 1× PBS, and the pellets kept in -80 °C for 12 h.

Following thawing of the sample, cells are homogenized in ice-cold 1× PBS (for 3×10^7 Granta cells, the homogenization volume should be around 50 μl) with protease inhibitor cocktail (Calbiochem). For separation of the plasma membrane fraction proteins, the homogenate is centrifuged at $700 \times g$ for 10 min to pellet nuclear debris, and the resulting supernatant fraction centrifuged again at $100,000 \times g$ for 35 min (all centrifugation steps at 4 °C). The remaining pellet contains the plasma membrane proteins, and can be resuspended in PBS with protease inhibitor cocktail. A minimum of 20 μg of proteins can be then resolved by 15% SDS–PAGE, and transferred to a nitrocellulose membrane (Bio-Rad) and probed with 1:2000 streptavidin–peroxidase (Fig. 9.3 A).

Two-dimensional gel separation of plasma membrane proteins labeled with BIAM is also possible, but before the procedure the plasma membrane protein fraction (as described above) needs to be extracted with mild detergent such as lauryl-maltoside or CHAPS. The proteins should be suspended in buffer containing 0.5% lauryl-maltoside, 2% CHAPS, 8 M urea, and 0.5% carrier ampholytes, and run-on immobilized pH gradient of pH range 3–10 (IPG strips, Bio-Rad). The IPG strips should be then embedded on 15% SDS–PAGE gel for the second dimension, and can be blotted on nitrocellulose and probed with 1:2000 streptavidin–peroxidase (Fig. 9.3 B).

3.3. Miscellaneous other methods to measurement exofacial thiols

The methods described above stain almost all free SH groups on the cell surface. For selective staining of SH groups with low pK_a (reactive thiols), the staining with BIAM should be performed with low pH as well (pH = 6.5). However, this staining is difficult to perform on living cells in buffered media, and is rather used for protein *in vitro* modification, where pH can be well controlled (Fang *et al.*, 2005; Rhee *et al.*, 2000).

Correlation between the redox state of extracellular nonprotein thiols (NPSH) in different tissues is usually validated with the use of mercury orange staining (see Section 2.1 above) (Ceccarelli *et al.*, 2008; Hedley *et al.*, 2005).

In the presence of oxidants, protein thiols can form mixed disulfides with GSH, where GSH binds covalently to reactive cysteines within

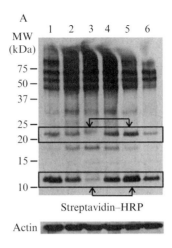

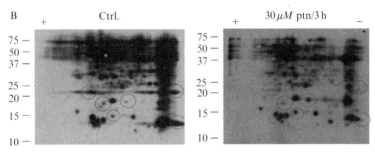

Figure 9.3 Parthenolide attenuates the levels of exofacial free thiols as assessed by Western blot analysis. (A) Plasma membrane enriched cell lysates were obtained from Granta cells treated as follows: (lane 1) untreated control cells; (lane 2) 10 μM parthenolide for 3 h; (lane 3) 30 μM parthenolide for 3 h; (lane 4) 5 mM GSH alone for 2 h; (lane 5) 5 mM GSH for 2 h followed by 30 μM parthenolide for 3 h; (lane 6) 100 μM H_2O_2 for 3 h. The lysates were stained with the thiol probe BIAM and blotted with streptavidin–HRP. As a protein loading control, the blots were also probed for actin (bottom panel). (B) Two-dimensional protein separation—BIAM labeling pattern. Samples treated with parthenolide for 3 h probed against streptavidin–HRP.

proteins, forming mixed disulfides. The detailed protocol of measuring this modification on cell surface of various tissues has been described elsewhere (Aesif et al., 2009).

4. Exofacial Thiol Status and Cancer

In vitro data indicate the extracellular redox state of the cells has tremendous impact on critical physiological functions such as proliferation, differentiation, and apoptosis. On the other hand, under physiological conditions the extracellular space is considered to have a more oxidized redox state compared to interior of the cell (Moriarty-Craige and Jones, 2004). Hence, the reduced state might be referred to as an *extracellular microenvironment*. Also, the existence of differences in redox state of thiol/disulfide pools in human plasma between individuals, cultured cells, and perfused tissues indicate that there must be mechanisms that control this state precisely and efficiently.

The extracellular redox state is determined by several factors. One of them is the presence (on plasma membrane) of the redox-modulating proteins, such as NADPH oxidase, extracellular superoxide dismutase (EC-SOD), thioredoxin 1 (TRX1), peroxiredoxin-IV, and extracellular glutathione peroxidase 3 (GPx3; Nakamura *et al.*, 2006; Yu *et al.*, 2007). The ratio of extracellular thiol/disulfide couples, such as GSH/GSSG, reduced/oxidized thioredoxin, and cysteine/cystine—also have a significant impact on extracellular redox potential. The reductive microenvironment is in majority dependent on cysteine/cystine pool outside the cell, and the Cys/CySS system is mainly present in disulfide form of cystine (Moriarty-Craige and Jones, 2004). However, it has been demonstrated that the reduced microenvironment is dependent on cystine uptake through the x_C^- (anionic amino acid transport system), intracellular reduction by NADP-dependent redox protein systems, and secretion of cysteine to the extracellular environment. Interestingly, overexpression of the x_C^- gen did not influence significantly the level of intracellular GSH, but did caused increases in intracellular, and also mostly extracellular, cysteine levels. The study conducted on the human colon carcinoma cell line, Caco2, confirmed that the redox state of the extracellular cysteine/cystine pool can affect cell proliferation at the external surface of the cells (Banjac *et al.*, 2008). This result is in agreement with another observation that proliferation of many human and murine lymphoma and leukemia cell lines is dependent on free thiols and the extracellular redox potential (defined by the ratio of extracellular cysteine and cystine). This ratio seems to be critical for cell division and is independent of cellular GSH level (Falk *et al.*, 1993).

Other studies conducted on cancer cell lines (i.e., human prostate carcinoma and lung cancer cell lines), which differ in terms of their aggressiveness, show that more invasive cell lines display different extracellular redox states and contain more reduced SH groups on plasma membrane proteins (Ceccarelli *et al.*, 2008; Chaiswing *et al.*, 2007, 2008; Olm *et al.*, 2009). This observation suggests that extracellular microenvironment, particularly created by reduced thiols, might play very significant role in cancer formation, malignancy, and metastasis.

One interesting consequence of such a link between extracellular redox and proliferation is the particularly aggressive nature of cancers of the glioma type. It has long been known that in the brain, trafficking of GSH occurs from astrocytes to neurons (SJ Heales), since neurons do not have a very active GSH synthesis apparatus. Astrocytes also secrete large quantities of EC-SOD to protect the trafficked GSH from oxidative damage in the extracellular space. Thus, it may be speculated that this highly reduced microenvironment in the peri-astrocytic space is perfect for glioma development.

From a therapeutic standpoint, the notion of specifically targeting the exofacial thiols of cancer cells and tumors has received very little attention, and is an area ripe for development.

REFERENCES

Addanki, S., Cahill, F. D., and Sotos, J. F. (1967). Intramitochondrial pH and intra-extramitochondrial pH gradient of beef heart mitochondria in various functional states. *Nature* **214,** 400–402.

Aesif, S. W., Anathy, V., Havermans, M., Guala, A. S., Ckless, K., Taatjes, D. J., and Janssen-Heininger, Y. M. (2009). In situ analysis of protein S-glutathionylation in lung tissue using glutaredoxin-1-catalyzed cysteine derivatization. *Am. J. Pathol.* **175,** 36–45.

Banjac, A., Perisic, T., Sato, H., Seiler, A., Bannai, S., Weiss, N., Kolle, P., Tschoep, K., Issels, R. D., Daniel, P. T., Conrad, M., and Bornkamm, G. W. (2008). The cystine/cysteine cycle: A redox cycle regulating susceptibility versus resistance to cell death. *Oncogene* **27,** 1618–1628.

Beech, D. J., and Sukumar, P. (2007). Channel regulation by extracellular redox protein. *Channels (Austin)* **1,** 400–403.

Brookes, P. S., Yoon, Y., Robotham, J. L., Anders, M. W., and Sheu, S. S. (2004). Calcium, ATP, and ROS: A mitochondrial love-hate triangle. *Am. J. Physiol. Cell. Physiol.* **287,** C817–C833.

Burgess, R. R., and Deutscher, M. P. (2009). Methods in enzymology. Preface. *Methods Enzymol.* **463,** xxv–xxxvi.

Burwell, L. S., and Brookes, P. S. (2008). Mitochondria as a target for the cardioprotective effects of nitric oxide in ischemia-reperfusion injury. *Antioxid. Redox Signal.* **10,** 579–599.

Ceccarelli, J., Delfino, L., Zappia, E., Castellani, P., Borghi, M., Ferrini, S., Tosetti, F., and Rubartelli, A. (2008). The redox state of the lung cancer microenvironment depends on the levels of thioredoxin expressed by tumor cells and affects tumor progression and response to prooxidants. *Int. J. Cancer* **123,** 1770–1778.

Chaiswing, L., Bourdeau-Heller, J. M., Zhong, W., and Oberley, T. D. (2007). Characterization of redox state of two human prostate carcinoma cell lines with different degrees of aggressiveness. *Free Radic. Biol. Med.* **43,** 202–215.

Chaiswing, L., Zhong, W., Cullen, J. J., Oberley, L. W., and Oberley, T. D. (2008). Extracellular redox state regulates features associated with prostate cancer cell invasion. *Cancer Res.* **68,** 5820–5826.

Circu, M. L., Rodriguez, C., Maloney, R., Moyer, M. P., and Aw, T. Y. (2008). Contribution of mitochondrial GSH transport to matrix GSH status and colonic epithelial cell apoptosis. *Free Radic. Biol. Med.* **44,** 768–778.

Cox, D. A., and Matlib, M. A. (1993). A role for the mitochondrial Na(+)-Ca2+ exchanger in the regulation of oxidative phosphorylation in isolated heart mitochondria. *J. Biol. Chem.* **268,** 938–947.

Falk, M. H., Hultner, L., Milner, A., Gregory, C. D., and Bornkamm, G. W. (1993). Irradiated fibroblasts protect Burkitt lymphoma cells from apoptosis by a mechanism independent of bcl-2. *Int. J. Cancer* **55,** 485–491.

Fang, J., Lu, J., and Holmgren, A. (2005). Thioredoxin reductase is irreversibly modified by curcumin: A novel molecular mechanism for its anticancer activity. *J. Biol. Chem.* **280,** 25284–25290.

Gelderman, K. A., Hultqvist, M., Holmberg, J., Olofsson, P., and Holmdahl, R. (2006). T cell surface redox levels determine T cell reactivity and arthritis susceptibility. *Proc. Natl. Acad. Sci. USA* **103,** 12831–12836.

Hedley, D. W., Nicklee, T., Moreno-Merlo, F., Pintilie, M., Fyles, A., Milosevic, M., and Hill, R. P. (2005). Relations between non-protein sulfydryl levels in the nucleus and cytoplasm, tumor oxygenation, and clinical outcome of patients with uterine cervical carcinoma. *Int. J. Radiat. Oncol. Biol. Phys.* **61,** 137–144.

Hurd, T. R., James, A. M., Lilley, K. S., and Murphy, M. P. (2009). Chapter 19 Measuring redox changes to mitochondrial protein thiols with redox difference gel electrophoresis (redox-DIGE). *Methods Enzymol.* **456,** 343–361.

Janssen-Heininger, Y. M., Mossman, B. T., Heintz, N. H., Forman, H. J., Kalyanaraman, B., Finkel, T., Stamler, J. S., Rhee, S. G., and van der Vliet, A. (2008). Redox-based regulation of signal transduction: Principles, pitfalls, and promises. *Free Radic. Biol. Med.* **45,** 1–17.

Kimura, Y., Goto, Y. I., and Kimura, H. (2010). Hydrogen sulfide increases glutathione production and suppresses oxidative stress in mitochondria. *Antioxid. Redox Signal* **12,** 1–13.

Laragione, T., Bonetto, V., Casoni, F., Massignan, T., Bianchi, G., Gianazza, E., and Ghezzi, P. (2003). Redox regulation of surface protein thiols: Identification of integrin alpha-4 as a molecular target by using redox proteomics. *Proc. Natl. Acad. Sci. USA* **100,** 14737–14741.

Lilley, K. S., and Friedman, D. B. (2004). All about DIGE: quantification technology for differential-display 2D-gel proteomics. *Expert Rev. Proteomics* **1,** 401–409.

Lin, T. K., Hughes, G., Muratovska, A., Blaikie, F. H., Brookes, P. S., Darley-Usmar, V., Smith, R. A., and Murphy, M. P. (2002). Specific modification of mitochondrial protein thiols in response to oxidative stress: A proteomics approach. *J. Biol. Chem.* **277,** 17048–17056.

Mani, A. R., Ebrahimkhani, M. R., Ippolito, S., Ollosson, R., and Moore, K. P. (2006). Metalloprotein-dependent decomposition of S-nitrosothiols: Studies on the stabilization and measurement of S-nitrosothiols in tissues. *Free Radic. Biol. Med.* **40,** 1654–1663.

Martinez-Ruiz, A., and Lamas, S. (2007). Signalling by NO-induced protein S-nitrosylation and S-glutathionylation: Convergences and divergences. *Cardiovasc. Res.* **75,** 220–228.

Moriarty-Craige, S. E., and Jones, D. P. (2004). Extracellular thiols and thiol/disulfide redox in metabolism. *Annu. Rev. Nutr.* **24,** 481–509.

Murphy, M. P. (2008). Targeting lipophilic cations to mitochondria. *Biochim. Biophys. Acta* **1777,** 1028–1031.

Nakamura, H., Masutani, H., and Yodoi, J. (2006). Extracellular thioredoxin and thioredoxin-binding protein 2 in control of cancer. *Semin. Cancer Biol.* **16,** 444–451.

Olm, E., Fernandes, A. P., Hebert, C., Rundlof, A. K., Larsen, E. H., Danielsson, O., and Bjornstedt, M. (2009). Extracellular thiol-assisted selenium uptake dependent on the x (c)-cystine transporter explains the cancer-specific cytotoxicity of selenite. *Proc. Natl. Acad. Sci. USA* **106,** 11400–11405.

Ou, W., and Silver, J. (2006). Role of protein disulfide isomerase and other thiol-reactive proteins in HIV-1 envelope protein-mediated fusion. *Virology* **350,** 406–417.

Rhee, S. G., Bae, Y. S., Lee, S. R., and Kwon, J. (2000). Hydrogen peroxide: A key messenger that modulates protein phosphorylation through cysteine oxidation. *Sci. STKE* **2000,** ★★p.pe.1.

Rossignol, R., Gilkerson, R., Aggeler, R., Yamagata, K., Remington, S. J., and Capaldi, R. A. (2004). Energy substrate modulates mitochondrial structure and oxidative capacity in cancer cells. *Cancer Res.* **64,** 985–993.

Sahaf, B., Heydari, K., and Herzenberg, L. A. (2003). Lymphocyte surface thiol levels. *Proc. Natl. Acad. Sci. USA* **100,** 4001–4005.

Sahaf, B., Heydari, K., and Herzenberg, L. A. (2005). The extracellular microenvironment plays a key role in regulating the redox status of cell surface proteins in HIV-infected subjects. *Arch. Biochem. Biophys.* **434,** 26–32.

Salazar, M., Rojo, A. I., Velasco, D., de Sagarra, R. M., and Cuadrado, A. (2006). Glycogen synthase kinase-3beta inhibits the xenobiotic and antioxidant cell response by direct phosphorylation and nuclear exclusion of the transcription factor Nrf2. *J. Biol. Chem.* **281,** 14841–14851.

Schwertassek, U., Balmer, Y., Gutscher, M., Weingarten, L., Preuss, M., Engelhard, J., Winkler, M., and Dick, T. P. (2007). Selective redox regulation of cytokine receptor signaling by extracellular thioredoxin-1. *EMBO J.* **26,** 3086–3097.

Sebastia, J., Cristofol, R., Martin, M., Rodriguez-Farre, E., and Sanfeliu, C. (2003). Evaluation of fluorescent dyes for measuring intracellular glutathione content in primary cultures of human neurons and neuroblastoma SH-SY5Y. *Cytometry A* **51,** 16–25.

Seyfried, J., Soldner, F., Schulz, J. B., Klockgether, T., Kovar, K. A., and Wullner, U. (1999). Differential effects of L-buthionine sulfoximine and ethacrynic acid on glutathione levels and mitochondrial function in PC12 cells. *Neurosci. Lett.* **264,** 1–4.

Shan, X., Jones, D. P., Hashmi, M., and Anders, M. W. (1993). Selective depletion of mitochondrial glutathione concentrations by (R, S)-3-hydroxy-4-pentenoate potentiates oxidative cell death. *Chem. Res. Toxicol.* **6,** 75–81.

Shepherd, D., and Garland, P. B. (1969). The kinetic properties of citrate synthase from rat liver mitochondria. *Biochem. J.* **114,** 597–610.

Smith, R. A., Adlam, V. J., Blaikie, F. H., Manas, A. R., Porteous, C. M., James, A. M., Ross, M. F., Logan, A., Cocheme, H. M., Trnka, J., Prime, T. A., Abakumova, I., *et al.* (2008). Mitochondria-targeted antioxidants in the treatment of disease. *Ann. N. Y. Acad. Sci.* **1147,** 105–111.

Tassi, S., Carta, S., Vene, R., Delfino, L., Ciriolo, M. R., and Rubartelli, A. (2009). Pathogen-induced interleukin-1beta processing and secretion is regulated by a biphasic redox response. *J. Immunol.* **183,** 1456–1462.

Unlü, M., Morgan, M. E., and Minden, J. S. (1997). Difference gel electrophoresis: a single gel method for detecting changes in protein extracts. *Electrophoresis* **18,** 2071–2077.

Yu, Y. P., Yu, G., Tseng, G., Cieply, K., Nelson, J., Defrances, M., Zarnegar, R., Michalopoulos, G., and Luo, J. H. (2007). Glutathione peroxidase 3, deleted or methylated in prostate cancer, suppresses prostate cancer growth and metastasis. *Cancer Res.* **67,** 8043–8050.

CHAPTER TEN

Redox Clamp Model for Study of Extracellular Thiols and Disulfides in Redox Signaling

Young-Mi Go *and* Dean P. Jones

Contents

1. Introduction	166
2. Key Concepts for Use	166
3. Principles for Experimental Design	167
4. Summary of Available Redox Clamp Studies	171
5. Perspectives and Conclusion	177
Acknowledgments	178
References	178

Abstract

Extracellular thiol/disulfide redox environments are highly regulated in healthy individuals and become oxidized in disease. This oxidation affects the function of cell surface receptors, ion channels, and structural proteins. Downstream signaling due to changes in extracellular redox potential can be studied using a redox clamp in which thiol and disulfide concentrations are varied to obtain a series of controlled redox potentials. Previous applications of this approach show that cell proliferation, apoptosis, and proinflammatory signaling respond to extracellular redox potential. Furthermore, gene expression and proteomic studies reveal the global nature of redox effects, and different cell types, for example, endothelial cells, fibroblasts, monocytes, and epithelial cells, show cell-specific redox responses. Application of the redox clamp to studies of different signaling pathways could enhance the understanding of redox transitions in many aspects of normal physiology and disease.

Department of Medicine, Division of Pulmonary, Allergy and Critical Care Medicine, Emory University, Atlanta, Georgia, USA

1. Introduction

Recognition of the highly regulated nature of extracellular thiol/disulfide couples, measured as GSH/GSSG and cysteine/cystine (Cys/CySS) redox potentials (Jones et al., 2000), prompted studies to determine whether variation in extracellular redox state changed with cell differentiation (Nkabyo et al., 2002) or affected cell growth and proliferation (Jonas et al., 200, 2003). The latter studies revealed that human cells in culture regulate the Cys/CySS redox potential (E_hCySS) of the culture medium to the same value as found in plasma of young health adults (Fig. 10.1). Subsequent human research has revealed that E_hCySS values are more oxidized in association with disease risk factors and specific diseases (Jones and Liang, 2009). In the present chapter, we describe the redox clamp approach which is useful to study mechanisms whereby variation in extracellular E_h value contributes to disease.

2. Key Concepts for Use

The GSH/GSSG and Cys/CySS couples represent distinct redox signaling nodes (Jones et al., 2004) which have not been fully delineated. The initial observations showed that the GSH/GSSG redox potential (E_hGSSG) in human plasma is maintained at a more reducing steady-state

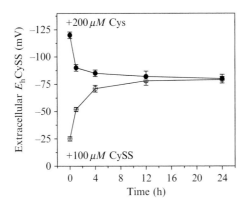

Figure 10.1 Changes in cysteine/cystine redox potential in culture medium of HT29 cells. Cells were grown to 90% confluence, culture media was removed, and cells were washed once with PBS. Cys and CySS-free DMEM with 10% neonatal calf serum was added with Cys and CySS concentrations calculated as in Table 10.1 to give either −150 or 0 mV. Samples were collected over a 24-h time course and Cys and CySS concentrations were measured by HPLC. E_h was calculated using the Nernst equation.

value (approximately -140 mV) compared to the E_hCySS value (approximately -80 mV). Importantly, the Cys/CySS pool size (>90 μM in Cys equivalents) in human plasma is at least 10-fold greater than the GSH/GSSG pool (<9 μM in GSH equivalents). Cell culture media commonly contain CySS, and only some specialized media contain Cys, GSH, or GSSG. Consequently, the most commonly used redox clamp has been created by varying Cys and CySS concentrations.

Under cell culture conditions, cells slowly release GSH into the culture media, and GSH reacts with CySS to produce the disulfide of Cys and GSH, CySSG, and a small amount of GSSG (Reed and Beatty, 1978). The rates of these processes are relatively slow in tissue culture so that unlike the *in vivo* situation for plasma, E_hCySS and E_hGSSG are equilibrated in the culture medium. Thus, studies with a Cys/CySS redox clamp do not discriminate between effects of E_hCySS and E_hGSSG. One can anticipate that systematic variation in E_hGSSG could yield cellular responses which are distinct from those of E_hCySS due to differences in accessibility and reactivity of the couples with specific protein thiols but this has not been explicitly shown. Differences are expected because GSH and GSSG are negatively charged at physiologic pH while Cys and CySS are neutral. Moreover, in lung lining fluid and bile, the GSH/GSSG pools are in higher abundance than the Cys/CySS pools. Consequently, in this description of the redox clamp, we include conditions for systematic variation of both Cys/CySS and GSH/GSSG pools to facilitate studies to discriminate effects of each couple (see Tables 10.1, 10.2, 10.3). Similar studies with a homocysteine/homocystine clamp could be useful to understand possible mechanisms of total homocysteine in cardiovascular disease (CVD).

Human studies show that some disease risks associate with plasma E_hCySS while others associate with plasma E_hGSSG (Jones and Liang, 2009). As summarized below, several studies using a Cys/CySS redox clamp show cell responses to variation in E_hCySS. In addition, Essex and Li (2003) found that platelet activation was maximal with a mixture of GSH and GSSG. The optimal value for E_hGSSG calculated from their experimental conditions corresponded to the physiologic E_hGSSG found in human plasma. These results highlight the need to consider cellular responses to E_hGSSG separately from E_hCySS.

3. Principles for Experimental Design

- *Desired E_h values are obtained by addition of selected concentrations of thiol and disulfide.*

A redox clamp is obtained by using a relatively large pool of Cys plus CySS in which the concentrations of Cys and CySS are set to values which give desired E_h values according to the Nernst equation:

$$E_h \text{CySS} = E_o + (RT/nF)\ln([\text{CySS}]/[\text{Cys}]^2).$$

In this equation, E_o is the standard half-cell potential (-250 mV for pH 7.4) for the Cys/CySS couple relative to a standard hydrogen electrode, R is the gas constant, T is absolute temperature, and F is Faraday's constant. The value for n is 2 for a two-electron transfer, so that an alternative form of the equation, with combined constants is

$$E_h \text{CySS} = -250 + 29.5\log([\text{CySS}]/[\text{Cys}]^2),$$

where concentrations are expressed in molar values and $E_h\text{CySS}$ is expressed in millivolts.

- *A total cysteine pool of 200 μM provides a reasonable compromise condition for most experiments.*

A redox clamp can be obtained with different pool sizes. We typically use a pool size of 200 μM Cys equivalents, that is, CySS concentration is multiplied by 2 to express in Cys equivalents, and this is added to Cys concentration (Table 10.1). This provides an initial pool which is at the upper limit of physiologic concentrations but is useful because the concentrations are too high for most cell lines to normalize the E_h within 12 h (Fig. 10.2A). Consequently, with this condition, culture media is changed every 12 h to maintain the clamp (Fig. 10.2B) or more frequently if necessary. It is important to note that rapidly proliferating cells tend to adjust the E_h of the culture medium more rapidly than slowly proliferating cells, and that addition of growth factors increases this rate (Jonas et al., 2003). Thus, the redox clamp as described does not provide precisely

Table 10.1 Cys and CySS concentrations and respective redox potentials in pool size of 200 μM Cys equivalents

Cys (μM)	CySS (μM)	E_hCySS (mV)
0.6	99.7	0.9
4.0	98.0	−49.8
14.0	93.0	−82.5
40.0	80.0	−111.3
80.0	60.0	−132.8
130.0	35.0	−152.1

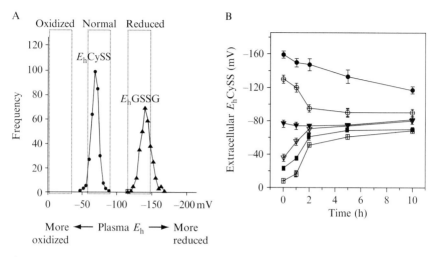

Figure 10.2 Human plasma redox states of thiol/disulfide couples, GSH/GSSG, and Cys/CySS, and control of extracellular E_hCySS by cultured cells. (A) Histogram showing frequency of E_hCySS and E_hGSSG in human plasma. Data are fasting values for young, healthy, and physically fit individuals. More reduced, normal and more oxidized ranges of plasma redox potential are indicated by the labels at the top. (B) Changes in extracellular E_hCySS with time were measured in cultured bovine aortic endothelial cells (Go and Jones, 2005), human retinal pigment epithelial cells (Jiang et al., 2005), and human colonic cancer cells, HT-29 (Anderson et al., 2007). Cells were incubated with different initial E_hCySS (-159, -130, -77, -35, -23, -8 mV) and concentrations of Cys and CySS were quantified as function of time (0, 1, 2, 4, 10 h). E_hCySS values calculated from the Nernst equation are shown as mean $\pm$ S.E.M., $n = 3$.

clamped values but rather E_h ranges (see Fig. 10.2A). More precise control can be obtained with a biofermenter which provides continuous control of E_h, along with control of pH and pO$_2$ (Hwang and Sinske, 1991).

- *Verification of E_h is obtained by analysis of respective thiol and disulfide concentrations or measurement with a potentiometric electrode.*

Cys is relatively unstable in solution but CySS is relatively stable. Most cell culture media contains CySS in the range of 200–400 μM without Cys, GSH, or GSSG. Most conditioned medium contains Cys in the low micromolar range so that upon change of culture media, E_hCySS values are in the range of 0 to $+10$ mV. To perform redox clamp studies involving more negative values which are characteristic of healthy *in vivo* conditions, cell culture medium without CySS, Cys, GSH, or GSSG is used to allow appropriate additions of thiol and disulfide to create desired E_h values. Verification of values is obtained by measurement of respective concentrations of Cys and CySS and calculation of the E_h value (Jones and Liang, 2009) or by potentiometric measurements (Ramirez et al., 2007). Values obtained using potentiometric electrodes are not precisely the same as those calculated from the Nernst equation because of electrode response characteristics; however, the systematic variation in E_h with altered concentrations of Cys and CySS can be readily determined and are in agreement.

- *Serum has small effect on initial E_h value.*

Many cells are grown in the presence of serum, which contains a substantial amount of CySS with lesser amounts of Cys, GSH, and GSSG. However, by avoiding use of more than 10% serum, redox clamp studies can be performed with only small adjustments for the contributions of serum (0.5% fetal bovine serum for the studies of endothelial cells). On the other hand, serum contains growth factors and affects cell growth rate. Consequently, serum should be maintained constant in experimental designs.

- *Controls are needed for unambiguous interpretation of redox clamp results.*

For any given redox clamp experiment, variations in E_hCySS result in systematic variation in three other parameters, Cys concentration, CySS concentration, and [Cys]/[CySS] ratio. Thus, although it is convenient to discuss effects of redox clamp experiments in terms of association with E_hCySS, the redox clamp results typically do not allow an unambiguous interpretation. For instance, Cys is required for protein synthesis so that variation in E_h could have effects because the conditions alter the Cys concentration and this affects protein synthesis. This may or may not have any relationship to the prevailing E_h value. Similarly, CySS concentration could determine the rate of cysteinylation of a signaling protein, and this could be kinetically limited by the CySS concentration and not determined by the equilibration with the E_hCySS. Alternatively, CySS could be used for cysteinylation but in a rapid and reversible manner. In this case, the equilibrium reaction,

$$CySS + PrSH \leftrightarrow CySSPr + Cys,$$

would determine the amount of cysteinylation so that the [Cys]/[CySS] would be the most relevant expression. Cysteinylation involving monothiols results in a 10-fold change in reduced:oxidized ratio with a 60 mV change and does not have as much of a gain-of-function per millivolt as the corresponding dithiol interaction,

$$CySS + Pr(-SH)_2 \leftrightarrow 2Cys + Pr(-SS-),$$

where a 60 mV change corresponds to a 100-fold change in reduced:oxidized protein ratio (Gilbert, 1990). As is apparent from these multiple possibilities, that is, dependence upon Cys concentration, CySS concentration, Cys/CySS ratio, or E_hCySS, careful experimental designs are needed to discriminate possible mechanisms.

- *Variation in pool size to discriminate effects of E_h from respective thiol and disulfide variations.*

A relatively straightforward test to distinguish possibilities is to use different pool sizes with adjustments of both Cys and CySS concentrations to obtain the same E_h values (Table 10.2). With this approach, mechanistic links to Cys or CySS concentration will be separated from those due to E_h. Such an approach indicated that E_h, rather than Cys or CySS concentration, was the important parameter in redox signaling of monocyte adhesion to endothelial cells (Go and Jones, 2005).

- *Semilog plots allow discrimination of functions related to E_h and to thiol/disulfide ratio.*

To distinguish effects related to E_hCySS from those due to [Cys]/[CySS], one can plot cell response as a function of the log of [Cys]/[CySS] (Gilbert, 1990). For functions dependent upon the ratio, the slope will be 1. In contrast, for processes dependent upon E_hCySS, the slope will be 2. Such an approach was used as the basis to conclude that nuclear erythroid 2-related factor-2 (Nrf2) signaling in response to cellular E_hGSSG operates through a monothiol mechanism (Hansen *et al.*, 2004).

4. SUMMARY OF AVAILABLE REDOX CLAMP STUDIES

Hwang and Sinskey (1991) provided an important contribution to the understanding of redox potential and mammalian cell growth in a study designed to optimize *in vitro* conditions for production of biologic products using cultured mammalian cells. They noted that three parameters were

Table 10.2 Cys and CySS concentrations and respective redox potentials in different pool sizes

Total pool size of Cys equivalents (μM)	Cys (μM)	CySS (μM)	E_hCySS (mV)
100	0.4	49.8	0.5
	10.0	45.0	−83.2
	78.0	11.0	−153.9
200	0.6	99.7	−0.9
	14.0	93.0	−82.5
	130.0	35.0	−152.1
400	0.8	199.6	0.6
	20.0	190.0	−82.5
	220.0	90.0	−153.6

Table 10.3 GSH and GSSG concentrations and respective redox potentials in different pool sizes

Total pool size of GSH equivalents (μM)	GSH (μM)	GSSG (μM)	E_hGSSG (mV)
100	0.2	49.9	4.3
	5	47.5	−78.8
	55	22.5	−149.8
	99.4	0.3	−220.3
200	0.4	99.8	−4.5
	8	96	−81.8
	92	54	−151.8
	198	1	−222.5
400	0.5	199.8	−1.4
	10	195	−78.4
	140	130	−151.3
	391	4.5	−220.7

critical to rapidly obtain maximum cell density, i.e., pO$_2$, pH, and redox potential. They found that with pO$_2$ and pH controlled, the redox potential (measured with a potentiometric electrode) could be maintained by controlled supply of cysteine. Although they did not show that the measured potential was related to the E_hCySS and did not examine underlying mechanisms, they showed that maximum cell density could be obtained with a measured potential about −60 mV and that this value was similar for over 20 different mammalian cell lines. The studies of Hwang and Sinskey were performed with a biofermenter which allowed continuous control of pO$_2$, pH, and E_h. Studies summarized below using standard cell culture conditions (95% air, 5% CO$_2$, pH buffered at 7.4) confirm that with nominally constant pO$_2$ and pH conditions, cell density varies as a function of E_h. Importantly, these studies indicate that the effects are specifically linked to E_hCySS and are mediated through effects on both cell proliferation and cell death.

A study of a human colon carcinoma cell line (Caco-2) showed that cell proliferation was altered by systematic variation in extracellular E_hCySS over a range (0 to −150 mV) that occurs in human plasma (Jonas et al., 2002). Caco-2 cells grow slowly in the absence of serum and respond to growth factors with increased rate of cell division. Incorporation of 5-bromo-2-deoxyuridine (BrdU) showed that DNA synthesis was lowest at the most oxidized extracellular E_h (0 mV). Incorporation increased as a function of redox state, attaining a 100% higher value at the most reduced condition (−150 mV). Addition of insulin-like growth factor-1 (IGF-1) or epidermal growth factor (EGF) increased the rate of BrdU incorporation at more oxidizing redox conditions (0 to −80 mV) but had no effect at −150 mV. Cellular GSH was not significantly affected by extracellular E_h. In the absence of growth factors, extracellular E_h values were largely maintained for 24 h. However, IGF-1 or EGF stimulated a change in extracellular redox to values similar to that for E_hCySS in plasma of young, healthy individuals. The results showed that extracellular E_hCySS modulates cell proliferation rate and that this control interacts with growth factor signaling apparently independent of cellular GSH.

In a follow-up study with Caco-2 cells (Jonas et al., 2003), the effects of extracellular E_hCySS on glutamine (Gln) and keratinocyte growth factor (KGF)-stimulated cell proliferation were studied. Gln (10 mM) or KGF (10 μg/l) did not alter BrdU incorporation at reducing E_h (−131 to −150 mV), but significantly increased incorporation at more oxidizing E_h values (Gln at 0 to −109 mV; KGF at −46 to −80 mV). Cellular E_hGSSG was unaffected by Gln, KGF, or variations in extracellular E_hCySS. Control cells largely maintained extracellular E_h at initial values after 24 h (−36 to −136 mV). However, extracellular E_hCySS shifted toward a narrow physiological range with Gln and KGF treatment (Gln, −56 to −88 mV; KGF, −76 to −92 mV). The results showed that E_hCySS is an important determinant of Caco-2 cell proliferation induced by Gln and KGF, that this control is independent of intracellular GSH redox status, and that both Gln and KGF enhance the capability of Caco-2 cells to modulate extremes of extracellular redox.

A more reduced extracellular E_hCySS activates cell proliferation in Caco-2 cells through the mitogenic p44/p42 mitogen-activated protein kinase (MAPK) pathway (Nkabyo et al., 2005). In the absence of added growth factors, −150 mV induced an 80% increase in EGFR phosphorylation, and this was followed by a marked increase in phosphorylation of p44/p42 MAPK. Inhibitors of EGFR (AG1478) and p44/p42 MAPK (U0126) phosphorylation blocked redox-dependent p44/p42 phosphorylation, indicating that signaling occurred by EGFR. These effects were inhibited by pretreatment with a nonpermeant alkylating agent, showing that signaling involved thiols accessible to the extracellular space. The EGFR ligand, transforming growth factor-α (TGF-α), was increased in culture medium at more reduced redox states. Redox-dependent phosphorylation of EGFR was completely prevented by a metalloproteinase inhibitor (GM6001), and an antibody to TGF-α partially inhibited the phosphorylation of p44/p42 MAPK. Together, the data showed that an E_hCySS-dependent activation of a metalloproteinase stimulates the mitogenic p44/p42 MAPK pathway by a TGF-α-dependent mechanism.

Redox clamp studies were used with cultured aortic endothelial cells and monocytes to determine whether oxidized values of E_hCySS could contribute in a causal way to atherosclerosis development (Go and Jones, 2005). Endothelial cells were exposed to initial E_hCySS from −150 to 0 mV.

Compared with the more reduced E_h, oxidized E_hCySS stimulated cellular H_2O_2 but not nitric oxide production, activated nuclear factor-κB, increased expression of adhesion molecules (intercellular adhesion molecule-1, platelet endothelial cell adhesion molecule-1, P-selectin), and stimulated monocyte binding to endothelial cells. Measurement of protein thiols in the extracellular membrane proteins showed that extracellular E_hCySS regulated redox states of membrane proteins, indicating that variation in extracellular E_hCySS is detected and signaled at the cell surface.

A subsequent study (Go et al., 2009a) showed that mitochondria are a major source of oxidant production in the NF-κB signaling response to a more oxidized extracellular E_hCySS. Analyses with mitochondrial ROS-sensitive reagents, MitoSox and MitoTracker showed that oxidant production in response to E_hCySS in mouse aortic endothelial cells occurred in mitochondria and that this production was blocked in cells from thioredoxin-2 (Trx2) transgenic mice. Mass spectrometry-based redox proteomics showed that several classes of plasma membrane and cytoskeletal proteins involved in inflammation responded to oxidized E_hCySS, including vascular cell adhesion molecule, integrins, actin, and several ras family GTPases. Together, the data show that the proinflammatory effects of oxidized plasma E_hCySS in endothelial cells are due to a mitochondrial signaling pathway which is mediated through redox control of downstream effector proteins. Because E_hCySS is oxidized in association with risk factors for CVD, including age, smoking, type II diabetes, obesity, and alcohol abuse, the data support a cause–effect relationship of extracellular E_hCySS in cell signaling pathways associated with CVD, including those which control monocyte adhesion to endothelial cells.

Microarray analysis and mass spectrometry-based proteomics were used to evaluate global changes in protein redox state, gene expression, and protein abundance in THP1 monocytes in response to E_hCySS (Go et al., 2009b). The percentage oxidized protein thiols for −150 and 0 mV were determined using nanoLC-MS/MS with a redox ICAT method (Go et al., 2009c). Oxidized E_hCySS in THP1 monocytes resulted in approximately 10% of the peptides detected being more oxidized. Despite the global nature of these proteomic changes, oxidation of the major cytoplasmic thiol systems, GSH and Trx1, was not detected. The relative abundance of proteins for −150 and 0 mV were determined using nanoLC-MS/MS with a quantitative ICAT method. Protein abundance results showed that oxidized extracellular E_hCySS stimulated toxicologic and canonical pathways mediated by Nrf2. Abundance of six proteins in this pathway were found to be higher at 0 mV compared with −150 mV, including ACTB (actin-beta, P60709), CAT (catalase, P04040), CCT7 (chaperonin containing TCP1, Q99832), GSTM3 (glutathione S-transferase M3, P21266), HSP90AB1 (heat shock protein 90kD, P08238), and VCP (valosin-containing protein, P55072). Thus, the quantitative proteomics analyses showed that oxidized E_hCySS increased abundance of components linked to oxidative stress and glutathione metabolism. Microarray and pathway analysis also showed activation of stress/detoxification pathways (e.g., NADPH:quinone oxidoreductase-1, ferritin, maf) at 0 mV and revealed that interleukin (IL)-1β-related pathways and cell death pathways were also increased by the oxidized E_hCySS.

In contrast, components of cell growth and proliferation pathways were increased by −150 mV. The genes expressed highly by the reduced redox state included transcription factors and regulators [early growth response protein (egr1, 2, 3), cyclin-l-1 (ccnl1), nuclear receptor subfamily-4 (nr4a2), polo-like kinase-2 (plk2), and prostaglandin-endoperoxide synthase2 (ptgs2)] supporting positive regulation of cell growth and proliferation. Phenotypic studies confirmed that a cell stress response occurred with oxidized E_hCySS and that cell proliferation was stimulated with reduced E_hCySS. The results from this study Go et al. (2009b) with THP1 cells support the conclusion that plasma E_hCySS provides a control over monocyte phenotype which could contribute to CVD risk and provide a novel therapeutic target for disease prevention.

The effect of oxidized E_hCySS on proinflammatory signaling was studied in more detail in U937 monocytes, specifically examining whether oxidized E_hCySS is a determinant of IL-1β levels (Iyer et al., 2009). Results showed a 1.7-fold increase in secreted pro-IL-1β levels in U937 monocytes exposed to −46 mV compared to controls exposed to a physiological E_h of −80 mV. This response was directly confirmed in LPS-challenged mice, where preservation of plasma E_hCySS from oxidation by dietary sulfur amino acid (SAA) supplementation was associated with a 1.6-fold decrease in plasma IL-1β compared to control mice fed an isonitrogenous SAA-adequate diet. Similarly, analysis of E_hCySS and IL-1β in human plasma revealed a significant positive association between oxidized E_hCySS and IL-1β after controlling for age, gender, and BMI. Together, the data substantiate the value of the redox clamp approach in showing that oxidized extracellular E_hCySS is a determinant of IL-1β levels.

Additional support for stimulation of cell death pathways by an oxidized E_hCySS was obtained using the redox clamp in study of oxidant-induced apoptosis in human retinal pigment epithelial (hRPE) cells (Jiang et al., 2005). hRPE cells were incubated in culture medium with E_hCySS varied from +16 mV (most oxidized) to −158 mV (most reduced). The hRPE were sensitized to tert-butylhydroperoxide (tBH)-induced apoptosis in the more oxidized extracellular conditions (E_h > −55 mV) compared with the reduced conditions (E_h < −89 mV). Loss of mitochondrial membrane potential ($\Delta\mu_m$), release of cytochrome c, and activation of caspase 3 after tBH treatments all increased under the more oxidized conditions. In contrast, E_hCySS did not affect expression of Fas or FasL in hRPE cells.

The results suggest that the oxidized E_hCySS, which has been found in patients with age-related macular degeneration (Moriarty-Craige et al., 2005), could contribute to susceptibility of hRPE to oxidant-induced apoptosis through the intrinsic mitochondrial pathway, and thereby contribute to an age-related decline in cell populations. The studies do not exclude a contribution of the Fas pathway to macular degeneration, but suggest that the increased sFasL, which has been found to increase in human plasma with age (Jiang et al., 2008), is not likely to occur directly in the hRPE in response to an oxidized E_hCySS.

Additional support for Nrf2-dependent activation of cell protective systems in response to oxidized E_hCySS was obtained by applying the redox clamp to NIH 3T3 fibroblasts (Imhoff and Hansen, 2009). An oxidized extracellular Cys/CySS redox potential was found to activate nuclear factor-erythroid 2-related factor 2 (Nrf2) and induce an antioxidant response. Cellular and mitochondrial oxidant production increased in cells at 0 and -46 mV compared to -80 mV, and mitochondrial thioredoxin (Trx2) became oxidized. The study confirmed the findings described above that oxidized extracellular E_hCySS stimulates mitochondrial oxidant generation and activates Nrf2-regulated gene expression.

Studies with the redox clamp also show that oxidized E_hCySS activates profibrotic signaling pathways which could contribute to pulmonary fibrosis by stimulating lung fibroblast proliferation and matrix deposition (Ramirez et al., 2007). Such studies are important because mechanisms that link oxidant stress to fibrogenesis remain only partially elucidated. Unlike cells described above, which proliferated more rapidly with more reduced E_hCySS (<-80 mV), primary murine lung fibroblasts were stimulated to proliferate when exposed to an oxidized E_hCySS (>-46 mV). The oxidized condition also stimulated expression of fibronectin, a matrix glycoprotein highly expressed in fibrotic lung diseases and implicated in lung injury. This stimulatory effect was dependent on protein kinase C activation. Oxidized E_hCySS increased the phosphorylation of cAMP response element binding protein, a transcription factor known for its ability to stimulate fibronectin expression, and increased the expression of mRNAs and proteins coding for the transcription factors, nuclear factor, NF-κB and mothers against decapentaplegic homolog 3 (SMAD3). Fibroblasts cultured in normal (-80 mV) or reduced (-131 mV) E_hCySS showed less induction. Fibronectin expression in response to an oxidized E_hCySS value was associated with expression of TGF-β1 and was inhibited by an anti-TGF-β1 antibody and SB-431542, a TGF-β1 receptor inhibitor. Thus, the redox clamp approach provided important evidence that an oxidized E_hCySS could contribute to pulmonary fibrosis by stimulating lung fibroblast proliferation and matrix expression through upregulation of TGF-β1.

5. Perspectives and Conclusion

The redox clamp approach as described provides a straightforward and convenient means to test for redox signaling processes which are dependent upon extracellular E_hCySS or E_hGSSG. The cumulative evidence indicates that multiple extracellular and cell surface proteins are responsive to thiols and disulfides in the plasma and other extracellular fluids. Although information on E_h values for thiol/disulfide couples in biologic fluids is limited, the available data has been recently reviewed (Go and Jones, 2008) and provides sufficient information to design studies for many cell types. Importantly, the available data show that cell types differ in their responses to extracellular redox potential so that there is a considerable opportunity to apply the redox clamp approach to different biologic research questions.

Presently, there is little information concerning the polarity of redox responses. Given that epithelial surfaces are highly polarized, different effects are likely to occur with oxidative or reductive changes on the poles of the cell. Such effects could be very important under conditions where barrier integrity fails. Indeed, irrigation solutions designed to provide correct E_h values for biologic surfaces may be useful for a range of surgical and other interventional procedures.

A subject of particular interest for cancer prevention is the possibility that the stimulated cell proliferation due to a more negative E_hCySS could contribute to tumor growth. Cells with poor vascularization are likely to be more reduced. The studies with Caco2 cells show that more reduced E_hCySS stimulates cell proliferation independent of growth factors and to an extent equivalent to growth factor stimulation. Thus, E_hCySS-stimulated growth could be relevant to uncontrolled growth in solid tumors.

Finally, the mechanisms which normally control extracellular redox potentials are largely unknown. *In vitro* and *in situ* perfused organ studies show that cells have a considerable activity in redox control. Studies in hepatocytes suggested that regulation of extracellular E_h could involve GSH release and thiol–disulfide exchange reactions (Reed and Beatty, 1978). However, subsequent research in colon carcinoma (HT29) cells showed that depletion of GSH by pretreatment with buthionine sulfoximine did not impair the cells capacity to regulate extracellular E_hCySS (Anderson *et al.*, 2007). The best available data supports the function of a cysteine–cystine shuttle mechanism (Dahm and Jones, 2000) in which CySS is transported into cells, reduced to Cys, and Cys is released to the extracellular fluid.

In summary, the presently described method for use of a thiol–disulfide redox clamp in cell culture provides a means to study effects of thiol/disulfide redox potential on cell functions. The method is limited by the requirement that cell culture media be periodically changed and the lack of continuous, precise control of E_h values. However, the approach has been used in a number of studies with different cell types and reveals fundamental cell properties, including cell proliferation and cell death, are sensitive to E_h. Moreover, pathways relevant to cardiovascular and lung diseases are clearly responsive to E_h when studied using this model. Consequently, the redox clamp should be considered as a general approach to incorporate into *in vitro* studies of redox signaling.

ACKNOWLEDGMENTS

This work was supported by NIH grants ES011195 and ES009047.

REFERENCES

Anderson, C. L., Iyer, S. S., Ziegler, T. R., et al. (2007). Control of extracellular cysteine/cystine redox state by HT-29 cells is independent of cellular glutathione. *Am. J. Physiol. Regul. Integr. Comp. Physiol.* **293,** R1069–R1075.

Dahm, L. J., and Jones, D. P. (2000). Rat jejunum controls luminal thiol-disulfide redox. *J. Nutr.* **130,** 2739–2745.

Essex, D. W., and Li, M. (2003). Redox control of platelet aggregation. *Biochemistry* **42,** 129–136.

Gilbert, H. F. (1990). Molecular and cellular aspects of thiol-disulfide exchange. *Adv. Enzymol. Relat. Areas Mol. Biol.* **63,** 69–172.

Go, Y. M., and Jones, D. P. (2005). Intracellular proatherogenic events and cell adhesion modulated by extracellular thiol/disulfide redox state. *Circulation* **111,** 2973–2980.

Go, Y. M., and Jones, D. P. (2008). Redox compartmentalization in eukaryotic cells. *Biochim. Biophys. Acta* **1780,** 1273–1290.

Go, Y. M., Park, H., Koval, M., et al. (2009a). A key role for mitochondria in endothelial signaling by plasma cysteine/cystine redox potential. *Free Radic. Biol. Med.* **48,** 275–283.

Go, Y. M., Craige, S. E., Orr, M., et al. (2009b). Gene and protein responses of human monocytes to extracellular cysteine redox potential. *Toxicol. Sci.* **112,** 354–362.

Go, Y. M., Pohl, J., and Jones, D. P. (2009c). Quantification of redox conditions in the nucleus. *Methods Mol. Biol.* **464,** 303–317.

Hansen, J. M., Watson, W. H., and Jones, D. P. (2004). Compartmentation of Nrf-2 redox control: Regulation of cytoplasmic activation by glutathione and DNA binding by thioredoxin-1. *Toxicol. Sci.* **82,** 308–317.

Hwang, C., and Sinskey, A. J. (1991). The role of oxidation-reduction potential in monitoring growth of cultured mammalian cells. In "Production of biologicals from animal cells in culture," (R. E. Spier, J. B. Griffiths, and B. Meignier, eds.) **991,** pp. 548–567. Halley Court, Oxford.

Imhoff, B. R., and Hansen, J. M. (2009). Extracellular redox status regulates Nrf2 activation through mitochondrial reactive oxygen species. *Biochem. J.* **424,** 491–500.

Iyer, S. S., Accardi, C. J., Ziegler, T. R., et al. (2009). Cysteine redox potential determines pro-inflammatory IL-1beta levels. *PloS One* **4,** e5017.

Jiang, S., Moriarty-Craige, S. E., Orr, M., et al. (2005). Oxidant-induced apoptosis in human retinal pigment epithelial cells: Dependence on extracellular redox state. *Invest. Ophthalmol. Vis. Sci.* **46,** 1054–1061.

Jiang, S., Moriarty-Craige, S. E., Li, C., et al. (2008). Associations of plasma-soluble fas ligand with aging and age-related macular degeneration. *Invest. Ophthalmol. Vis. Sci.* **49**, 1345–1349.

Jonas, C. R., Ziegler, T. R., Gu, L. H., et al. (2002). Extracellular thiol/disulfide redox state affects proliferation rate in a human colon carcinoma (Caco2) cell line. *Free Radic. Biol. Med.* **33**, 1499–1506.

Jonas, C. R., Gu, L. H., Nkabyo, Y. S., et al. (2003). Glutamine and KGF each regulate extracellular thiol/disulfide redox and enhance proliferation in Caco-2 cells. *Am. J. Physiol. Regul. Integr. Comp. Physiol.* **285**, R1421–R1429.

Jones, D. P., and Liang, Y. (2009). Measuring the poise of thiol/disulfide couples in vivo. *Free Radic. Biol. Med.* **47**, 1329–1338.

Jones, D. P., Carlson, J. L., Mody, V. C., et al. (2000). Redox state of glutathione in human plasma. *Free Radic. Biol. Med.* **28**, 625–635.

Jones, D. P., Go, Y.-M., Anderson, C. L., et al. (2004). Cysteine/cystine couple is a newly recognized node in the circuitry for biologic redox signaling and control. *FASEB J.* **18**, 1246–1248.

Moriarty-Craige, S. E., Adkison, J., Lynn, M., et al. (2005). Antioxidant supplements prevent oxidation of cysteine/cystine redox in patients with age-related macular degeneration. *Am. J. Ophthalmol.* **140**, 1020–1026.

Nkabyo, Y. S., Ziegler, T. R., Gu, L. H., et al. (2002). Glutathione and thioredoxin redox during differentiation in human colon epithelial (Caco-2) cells. *Am. J. Physiol. Gastrointest. Liver Physiol.* **283**, G1352–G1359.

Nkabyo, Y. S., Go, Y. M., Ziegler, T. R., et al. (2005). Extracellular cysteine/cystine redox regulates the p44/p42 MAPK pathway by metalloproteinase-dependent epidermal growth factor receptor signaling. *Am. J. Physiol. Gastrointest. Liver Physiol.* **289**, G70–G78.

Ramirez, A., Ramadan, B., Ritzenthaler, J. D., et al. (2007). Extracellular cysteine/cystine redox potential controls lung fibroblast proliferation and matrix expression through upregulation of transforming growth factor-beta. *Am. J. Physiol.* **293**(4), L972–L981.

Reed, D. J., and Beatty, P. (1978). In "Functions of Glutathione in Liver and Kidney," (H. Sies and A. Wendel, eds.), pp. 13–21. Springer-Verlag, Berlin.

CHAPTER ELEVEN

REDOX STATE OF HUMAN SERUM ALBUMIN IN TERMS OF CYSTEINE-34 IN HEALTH AND DISEASE

Karl Oettl* and Gunther Marsche[†]

Contents

1. Background	182
2. HPLC Analysis	182
2.1. Sample preparation	185
2.2. Sample storage	186
2.3. Other species	186
3. Albumin Thiol State and Exercise	187
4. Influence of Supplementation	188
5. Albumin Oxidation in Disease	188
5.1. Nonserum albumin	189
5.2. Diabetes	189
5.3. Liver disease	189
5.4. Hemodialysis	191
6. Albumin Thiol State During Aging	191
7. Summary	193
References	193

Abstract

Albumin, the main protein in plasma, is prone to different mechanisms of oxidative modification since extracellular fluids contain only small amounts of antioxidant enzymes. The redox state of cysteine-34 of human albumin defines three fractions which allow to monitor albumin oxidation: mercaptalbumin with a free thiol group, nonmercaptalbumin1 containing a disulfide, and nonmercaptalbumin2 with a sulfinic or sulfonic acid group. These fractions can be separated by HPLC and detected with UV or fluorescence detection. The method is very rugged and only simple sample preparation is needed. It has been used to demonstrate albumin oxidation during exercise, aging, and pathologies like

---

* Institute of Physiological Chemistry, Medical University of Graz, Harrachgasse, Graz, Austria
[†] Institute of Experimental and Clinical Pharmacology, Medical University of Graz, Universitätsplatz, Graz, Austria

diabetes, liver disease, or renal disease. Problems may arise when high endogenous concentrations of bilirubin or certain drugs are present. The redox state of albumin shows high variability and is hence a valuable tool for the investigation of reversible and irreversible modification of plasma protein.

1. Background

The redox state of thiols in general and the extracellular thiol/disulfide redox state have gained broad attention in recent years. The situation in plasma is often described in terms of cysteine, cysteinylglycine, homocysteine, and glutathione which occur as thiols homo- or mixed disulfides. However, by far the largest thiol pool in plasma is comprised by albumin which contains a single cysteine (cys-34) not involved in one of the 17 intramolecular disulfide bonds (Peters, 1996).

As the main plasma protein albumin is one of the major targets of oxidative modification (Himmelfarb and McMonagle, 2001; Shacter et al., 1994) and several authors describe oxidized albumin or albumin redox state (for review see Oettl and Stauber, 2007). However, there are a number of possibilities to oxidize albumin, therefore the terms need further explanation. Cysteine-34 is a particular redox sensitive site of albumin and therefore it is legitimate to speak of the redox state of albumin in terms of cys-34.

Chromatographic separation of albumin gives three fractions according to cys-34: (i) the free thiol form called human mercaptalbumin (HMA); (ii) the disulfide with a small thiol compound, mainly cysteine or cysteinylglycine but also homocysteine and glutathione, called human nonmercaptalbumin1 (HNA1); and (iii) the higher oxidized form with cysteine as sulfinic or sulfonic acid (Fig. 11.1) (Peters, 1996). The sulfenic acid form has been described to be unstable and occur as an intermediate in the formation of the disulfide (Carballal et al., 2006; Turell et al., 2009). In a healthy young person HMA accounts for 70–80%, HNA1 for 20–30%, and HNA2 for about 2–5% of total albumin. It has been described that during numerous diseases and aging the fractions of the oxidized forms are increased.

2. HPLC Analysis

Chromatographic methods for the separation of albumin fractions according to the sulfhydryl content have been described in the 1960s (Anderson, 1965; Janatova et al., 1968). In the 1980s a method for the separation of HMA and HNA has been published (Era et al., 1988; Sogami

Albumin Redox State

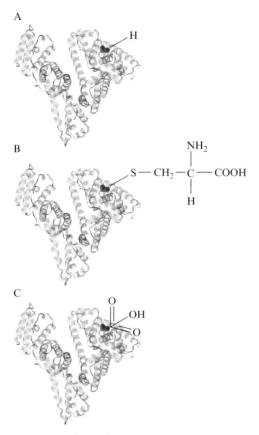

Figure 11.1 Structure of albumin: (A) HMA with a free thiol group, (B) HNA1 as disulfide with cysteine, and (C) HNA2 with a sulfonic acid group. Cysteine-34 is shown in space-fill representation. The graph was created using a PDB-file of Curry *et al.* (1998) (PDB-ID 1BJ5).

et al., 1985) and developed further to separate HMA, HNA1, and HNA2 (Hayashi *et al.*, 2000). The application of this method will be described here.

Albumin was separated on a Shodex-Asahipak ES-502N DEAE anion exchange column (7.5 × 100 mm) at 35 °C. The elution was carried out with a gradient using 50 mM sodium acetate, 400 mM sodium sulfate, pH 4.85 and up to 6% ethanol at a flow rate of 1.0 ml/min. For the first 5 min elution was carried out without ethanol, from 5 to 25 min ethanol was raised linearly to 6% which was kept for 5 min. From 30 to 35 min ethanol was reduced to 0% again. After further 5 min without ethanol the next run was started. Detection can be carried out by UV absorption at 280 nm or fluorescence at 280/340 nm wavelength for excitation and emission, respectively. As no baseline separation of albumin fractions can be achieved by this method, quantification was described by fitting Gaussian functions to the peak shapes by a simulation software (Hayashi *et al.*, 2000). The use of peak heights for quantification is also possible.

The parameters of interest concerning the redox state of albumin in terms of cys-34 are the relative fractions. The determination of the redox potential as it is defined by the Nernst equation and applied to glutathione or cysteine would afford the knowledge of the absolute concentrations of albumin fractions and the small thiol compounds of interest. In any case, the absolute quantification of HMA, HNA1, and HNA2 are problematic as no standards are available. HMA and HNA1 are reversibly convertible. HMA may be generated from HNA1 by incubation with glutathione and HNA1 can be synthesized from HMA by incubation with cystine. Also HNA2 can be made from HMA by oxidation with hydrogen peroxide (Fig. 11.2) (Hayashi *et al.*, 2002b). However, commercially available albumin preparations

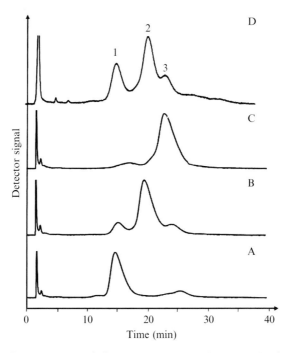

Figure 11.2 Chromatograms of albumin preparations: (A) HMA (peak 1), (B) HNA1 (peak 2), (C) HNA2 (peak 3), and (D) commercial human serum albumin. HMA was prepared from albumin shown in (D) (20% albumin solution from Behring GmbH) by incubation of 0.1 mM albumin in phosphate-buffered saline, pH 7.4 with 5 mM glutathione for 1 h at room temperature. HNA1 was prepared from the same commercial albumin by incubation with 2 mM cystine at room temperature for 2 days. HNA2 was prepared from HMA by incubating 0.1 mM HMA with 5 mM hydrogen peroxide at room temperature for 3 h.

contain different amounts of HNA2 which cannot be reduced and anyone who is working with commercial albumin should bear that in mind. Consequently, preparations of HMA and HNA1 are always contaminated by HNA2. A typical albumin preparation (20% solution from Behring GmbH, Vienna, Austria) consists of about 50% HMA, 41% HNA1, and 9% HNA2 (Fig. 11.1A). As HNA2 is not completely separated from HNA1 and comprises only a minor fraction of albumin, the accurate quantification of HNA2 is not always possible. If the accurate amount of HNA2 is of interest, an improvement of HNA2 quantification would be the reduction of HNA1 to HMA by incubation of the sample with glutathione. The result would be the appearance of HNA2 as a fully separated peak which then could be quantified reliably (Fig. 11.2).

2.1. Sample preparation

Plasma, serum, aqueous humor, synovial fluid, cell culture media, or commercial albumin solutions have been used as samples. Plasma samples were diluted 1:100 with 0.3 M NaCl, 100 mM sodium phosphate (pH 6.87) and filtered through a Whatman 0.45 μm nylon filter. Comparison of samples from 11 donors showed no significant differences between serum and plasma (Table 11.1).

As the sample volume needed is quite low due to dilution (we normally take 10 μl of samples and dilute with 990 μl buffer), plasma or serum prepared from capillary blood taken from the finger tip may be used. However, in this case we found slightly but significantly lower values of HMA and accordingly higher values of HNA1. HNA2 did not differ significantly (Table 11.1).

It was reported that albumin from human aqueous humor or synovial fluid may also be analyzed by this method when the gradient was modified (up to 10% ethanol) (Hayashi et al., 2000; Tomida et al., 2003).

It is important that plasma or serum preparations are not hemolytic. Addition of erythrocyte lysate to plasma to mimic hemolysis showed that upon hemolysis albumin fractions are shifted from HNA1 to HMA (Fig. 11.3). This is presumably due to the release of glutathione from red blood cells.

Table 11.1 Albumin in plasma and serum

	HMA (%)	HNA1 (%)	HNA2 (%)
Serum, venous	57.5 ± 7.0	38.5 ± 6.6	4.0 ± 0.6
Plasma, venous	56.8 ± 6.7	38.3 ± 6.2	4.4 ± 0.8
Serum, capillary	53.7 ± 8.0 (0.007)	42.0 ± 7.7 (0.011)	4.3 ± 0.6

The percentages of albumin fractions in different kinds of samples are given (mean values ± S.D.). Numbers in parentheses give the p-values for capillary versus venous serum (paired Student's t-test). All other parameters did not differ significantly.

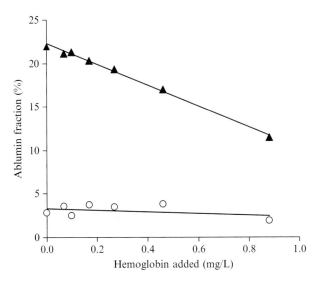

Figure 11.3 Influence of hemolysis on albumin fractions: Erythrocytes were lysed by ultrasonication for 60 s and added to plasma. Albumin fractions were analyzed by HPLC after 15 min incubation at room temperature. Hemoglobin was analyzed by a routine clinical chemical method (▲, HNA1; ○, HNA2).

Once the plasma sample is diluted it is stable at 4 °C for at least overnight runs. Performing 26 consecutive HPLC runs of the same diluted sample (which takes almost 17 h), we observed relative standard deviations of 0.4%, 1.3%, and 5.1% for HMA, HNA1, and HNA2, respectively.

2.2. Sample storage

The storage of the samples is another important factor. We stored plasma samples at 4, −20, and −70 °C, respectively. Frozen samples were thawed and frozen again up to nine times. The influence of the storage temperature is shown in Fig. 11.4. If the samples are stored at 4 °C it is necessary that analysis is performed at the same day. Otherwise there will be a shift from HMA to HNA1. At −20 °C a marked shift to HNA1 was observed from the second freeze/thaw cycle after 2 weeks. At −70 °C there is only a slight shift to HNA1 which is, however, held constant over five freeze/thaw cycles within 8 weeks. HNA2 did not change in our experiments. It has been reported that storage at −80 °C for 170 days without thawing results in only minimal increases of HNA1 (Hayashi *et al.*, 2002a).

2.3. Other species

Albumin from rat serum could be separated by this method in mercaptalbumin and nonmercaptalbumin (Hayashi *et al.*, 2002a). Analyzing commercial bovine serum albumin (A6003 albumin from bovine serum from Sigma,

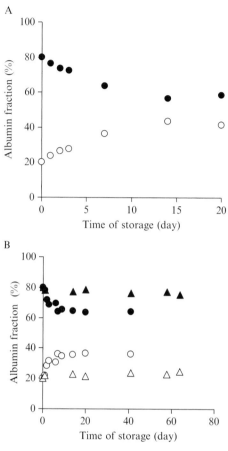

Figure 11.4 Influence of storage temperature on albumin fractions: EDTA plasma was stored at different temperatures for the time indicated and then analyzed by HPLC: (A) 4 °C: ●, HMA; ○, HNA1; (B) −20 °C: ●, HMA; ○, HNA1; and −70 °C: ▲, HMA; △, HNA1. In (B) the same samples were thawed and frozen again.

Vienna, Austria), we found one major peak which was identified as mercaptalbumin as the peak declined after incubation with cystine. The disulfide formed could be only partially separated from the initial peak (data not shown). Using mouse plasma and applying different modified gradients, we found three peaks which could not be separated sufficiently enough for reliable quantification (data not shown).

3. Albumin Thiol State and Exercise

It was reported that during strenuous exercise albumin fractions are shifted to more oxidized forms. After a 4-day Kendo training HMA was decreased, while HNA1 and HNA2 were increased (Imai *et al.*, 2002,

2005). In the case of exercise the study protocol has to pay regard to intensity of exercise and the time point of sample collection. We found an increasing intensity-depending shift from HMA to HNA1 after 40 min exercise at 70%, 75%, and 80% of VO_{2max} in a group of healthy trained men (69.5 ± 2.4%, 67.9 ± 1.9%, 67.2 ± 2.3%, and 65.9 ± 2.0% HMA at baseline and increasing intensities, respectively). The fraction of HNA2 was not changed significantly in this setting (Lamprecht et al., 2008). Immediately and 30 min after exercise the described shifts could be found. Regeneration of albumin redox state after exercise occurred at least in part overnight (Imai et al., 2005) and complete regeneration was found after 30 h (Lamprecht et al., 2009). Even overcompensation was reported a day after the 4-day Kendo training by an increase of the mean HMA fraction from initially 72.6 to 83.2% (Imai et al., 2005). However, data concerning the physiological kinetics of albumin oxidation and regeneration in the course of exercise are scarce.

The changes of HMA and HNA1 levels due to exercise may seem only slight. However, one has to bear in mind that albumin occurs in a concentration range around 600 μM and that a decrease of only 1% HMA means a drop of the albumin thiol pool of 6 μM!

4. Influence of Supplementation

Only two reports exist describing the effect of nutritional supplementation on albumin redox state. A high-dose supplementation of propolis tablets during a short time period suppressed albumin oxidation caused by strenuous exercise (Imai et al., 2005). On the other hand, a low-dose supplementation with a fruit and vegetable extract for 28 weeks did neither change the baseline levels of albumin fractions nor did it inhibit exercise induced shift from HMA to HNA1 (Lamprecht et al., 2009). In contrast, supplementation of thiol compounds like N-acetylcysteine may release cysteine from HNA1 (Harada et al., 2004).

5. Albumin Oxidation in Disease

A number of diseases are accompanied by oxidative stress (Halliwell and Gutteridge, 1999). Therefore, it could be expected that albumin redox state is altered under pathophysiological conditions. The method of albumin analytics described here has been used to investigate albumin from patients suffering different diseases like noninsulin-dependent diabetes mellitus (NIDDM; Oettl et al., 2010; Suzuki, et al. 1992), cataract (Hayashi et al., 2000; Oettl et al., 2010), joint disorders (Tomida et al.,

2003), chronic renal disease (Marsche et al., 2009; Matsuyama et al., 2009; Terawaki et al., 2004), and liver disease (Oettl et al., 2008, 2009; Sogami et al. 1985).

5.1. Nonserum albumin

Albumin from aqueous humor of cataract patients was shown to be significantly different from the corresponding serum albumin. While serum albumin consisted of 66.4 ± 4.5% HMA, 31.0 ± 4.4% HNA1, and 2.6 ± 0.8% HNA2, aqueous humor albumin was distinctly shifted to oxidized fractions (3.5 ± 5.5% HMA, 84.5 ± 8.1% HNA1, and 12.0 ± 4.4% HNA2) (Hayashi et al., 2000). However, the separation of aqueous humor albumin into fractions was by far less good compared to serum albumin. Investigating cataract patients we found no difference in serum albumin fractions compared to age-matched controls (Oettl et al., 2010).

Analogous to aqueous humor in synovial fluid, HNA1 was increased in both controls and patients with temporomandibular joint disorders, while HNA2 was increased only in the patient group (Tomida et al., 2003). For the analysis of aqueous humor and synovial fluid a modified gradient was applied with no ethanol from 0 to 1 min, a linear increase to 10% ethanol from 1 to 50 min, and a decrease back to 0% ethanol from 50 to 55 min.

5.2. Diabetes

Suzuki et al. reported clearly lower levels of HMA in NIDDM patients compared to controls (63 ± 7% vs. 75 ± 3%). However, in this study the control subjects were more than 30 years younger compared to the patients (Suzuki et al., 1992). Investigating albumin of age-matched persons we found only slightly but significantly decreased fractions of HMA (53.5 ± 1.0% vs. 56.8 ± 0.8%) and increased fractions of HNA1 (42.8 ± 0.9% vs. 39.7 ± 0.9%) and HNA2 (3.7 ± 0.9% vs. 3.5 ± 0.1%) in NIDDM patients compared to persons without NIDDM (Oettl et al., 2010). While in the earlier report complications like retinopathy, nephropathy, and neuropathy did not affect the albumin redox state, we found further shift from HMA to HNA1 in patients with proliferating diabetic retinopathy (50.8 ± 2.3% HMA and 45.4 ± 2.3% HNA1) (Oettl et al., 2010).

5.3. Liver disease

Severe oxidative modifications of serum albumin have been reported in patients with liver disease (Sogami et al., 1985). Liver cirrhosis was found to be accompanied by a shift of albumin from HMA to HNA1. On the other hand, in acute-on-chronic liver failure (ACLF), an acute deterioration of liver function over a period of 2–4 weeks, very high levels of HNA2 have been detected (Table 11.2, Fig. 11.5) (Oettl et al., 2008, 2009). In the case of ACLF patients an important methodological problem emerged. ACLF is accompanied by enormously high levels of bilirubin (up to 400 mg/l

Table 11.2 Oxidized albumin in healthy persons and patients

	n	Age (year)	HNA1 (%)	HNA2 (%)
Control (children)	13	13.9 ± 0.3	21.2 ± 2.5	3.0 ± 0.6
Controls (adults)	15	55.3 ± 5.1	29.1 ± 4.0	3.0 ± 0.5
LC	10	57.2 ± 6.0	39.6 ± 6.1	7.5 ± 2.4
ACLF	8	58.4 ± 4.0	41.4 ± 7.4	15.4 ± 3.4
HD	11	69.4 ± 14.1	59.2 ± 14.7	n.d.
NIDDM	53	73.1 ± 1.3	42.8 ± 0.9	3.7 ± 0.9

Mean values ± S.D. are given. LC, liver cirrhosis; ACLF, acute-on-chronic liver failure; HD, hemodialysis; NIDDM, noninsulin-dependent diabetes mellitus; n.d., not determined.

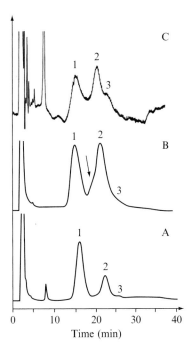

Figure 11.5 Representative HPLC chromatograms of plasma samples: (A) plasma of a young healthy control, (B) HD patient, and (C) ACLF patient. In (A) and (B) fluorescence detection and in (C) UV detection was used (Peak 1, HMA; Peak 2, HNA1; Peak 3, HNA2). The arrow shows an additional peak in the chromatogram of HD plasma.

compared to < 10 mg/l in controls). The detection method we apply routinely in albumin analysis utilizes fluorescence. Bilirubin, however, is a quencher of fluorescence and in albumin samples from ACLF patients it turned out that fluorescence is predominantly quenched in HMA and much less in HNA1 and HNA2. Hence fluorescence detection and UV detection gave different results concerning the relative amount of albumin fractions in ACLF samples. Therefore, in the case of ACLF patients, UV detection is superior compared to fluorescence detection. On the other hand, we used the observed selective fluorescence quenching to conclude a preferred binding of bilirubin to HMA (Oettl et al., 2008).

5.4. Hemodialysis

It has been shown that in predialysis patients with chronic renal disease oxidized albumin fractions increased with decreasing kidney function (Matsuyama et al., 2009; Terawaki et al., 2004). In patients undergoing maintenance hemodialysis (HD), increased levels of HNA have been reported (Soejima et al., 2004), albumin was shown to be a major target of oxidative modification (Himmelfarb and McMonagle, 2001) and we found the highest levels of HNA1 of all patient groups in HD patients (Table 11.2). Because of the high level of HNA1, quantification of HNA2 was not possible reliably (Fig. 11.5 B). In an earlier study essentially the same HNA2 levels were found in HD patients and controls (Soejima et al., 2004). In 6 out of 11 chromatograms of HD patients a characteristic fronting of the HNA1 peak appeared. However, the nature of this shoulder has not been characterized up to now. Recently, we could show that albumin from HD patients obtained new binding properties and displaces high-density lipoprotein from its receptor scavenger receptor class B, type I (Marsche et al., 2009). There is strong evidence that phagocytes are activated during the dialysis procedure generating the myeloperoxidase catalyzed oxidant hypochlorous acid (Capeillere-Blandin et al., 2006). We could recently show that hypochlorous acid oxidized albumin accumulates in human atherosclerotic lesions (Marsche et al., 2007).

6. ALBUMIN THIOL STATE DURING AGING

Different parameters of oxidative stress have been shown to be increased in aged persons (Halliwell and Gutteridge, 1999). Plasma thiol compounds were described to be shifted to disulfides with increasing age (Dröge, 2002; Era et al., 1988, 1995; Jones et al., 2002). We found that the percentage of HNA1 significantly increases with age ($r = 0.649, p < 0.001$, Fig. 11.6). The picture is similar to the course of GSSG or cystine during

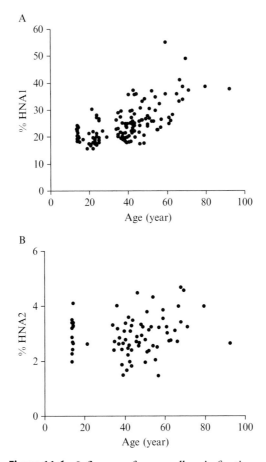

Figure 11.6 Influence of age on albumin fractions.

aging (Jones et al., 2002). From our data we cannot conclude whether the albumin redox state remains constant during the first decades and then increases or it increases continuously from childhood on. With increasing age it is also getting more and more difficult to rule out that pathophysiological changes have developed in the subjects investigated. A principal problem of the parameter albumin redox state is evident in Fig. 11.6 A. Even in healthy children the variation of HNA1 is very high and the percentage ranges from about 17% to 27% which reflects an albumin thiol concentration of about 100–160 μM, assuming a total albumin concentration of 600 μM. No data are available to conclude possible reasons or consequences of these high variations.

While there is a highly significant correlation of HNA1 with age, no such correlation was found for HNA2 (Fig. 11.6 B). The variations are even higher as found for HNA1. However, one has to bear in mind that the quantification of the small HNA2 peak (which is often only a shoulder of the HNA1 peak) is less accurate.

7. Summary

It has been well documented that albumin is quite vulnerable to reactive oxidant species *in vivo*. In this chapter we described the separation of albumin according to the redox state of cys-34. Changes in the redox state of cys-34 involve a reversible oxidation of HMA to the disulfide HNA1 and the irreversible oxidation to HNA2. HPLC analyses of these albumin fractions provides much more information compared to the measurement of plasma total or protein thiol groups. It should be noted that *in vivo* multiple mechanisms may contribute to oxidize albumin. This is reflected by different patterns of albumin oxidation in different diseases. However, aged subject groups with distinct pathologies often show considerable overlaps. Therefore, one has to take care in selecting subjects for studies and statistical methods for analyzing data (Oettl *et al.*, 2010). The variations of albumin fractions in patients and healthy persons are quite high. Hence, it is difficult to draw conclusions from the momentary redox state of albumin. On the other hand, especially in longitudinal and comparative studies it is a very useful tool for the evaluation of the systemic redox state in health, disease, and aging.

REFERENCES

Anderson, L. O. (1965). The heterogeneity of bovine serum albumin. *Biochim. Biophys. Acta* **117,** 115–133.

Capeillere-Blandin, C., Gausson, V., Nguyen, A. T., Descamps-Latscha, B., Drueke, T., and Witko-Sarsat, V. (2006). Respective role of uremic toxins and myeloperoxidase in the uremic state. *Nephrol. Dial. Transplant.* **21,** 1555–1563.

Carballal, S., Alvarez, B., Turell, L., Botti, H., Freeman, B. A., and Radi, R. (2006). Sulfenic acid in human serum albumin. *Amino Acids* **32,** 543–551.

Curry, S., Mandelkow, H., Brick, P., and Franks, N. (1998). Crystal structure of human serum albumin complexed with fatty acid reveals an asymmetric distribution of binding sites. *Nat. Struct. Biol.* **5,** 827–835.

Dröge, W. (2002). Aging-related changes in the thiol/disulfide redox state: Implication for the use of thiol antioxidants. *Exp. Gerontol.* **37,** 1331–1343.

Era, S., Hamaguchi, T., Sogami, M., Kuwata, K., Suzuki, E., Miura, K., Kawai, K., Kitazawa, Y., Okabe, H., Noma, A., and Miyata, S. (1988). Further studies on the resolution of human mercapt- and nonmercaptalbumin and on human serum albumin in the elderly by high-performance liquid chromatography. *Int. J. Pept. Protein Res.* **31,** 435–442.

Era, S., Kuwata, K., Imai, H., Nakamura, K., Hayashi, T., and Sogami, M. (1995). Age-related change in redox state of human serum albumin. *Biochim. Biophys. Acta* **1247,** 12–16.

Halliwell, B., and Gutteridge, J. M. C. (1999). Free Radicals in Biology and Medicine. Oxford University Press, New York.

Harada, D., Anraku, M., Fukuda, H., Naito, S., Harada, K., Suenaga, A., and Otagiri, M. (2004). Kinetic studies of covalent binding between N-acetyl-L-cysteine and human serum albumin through a mixed-disulfide using an N-methylpyridinium polymer-based column. *Drug Metab. Pharmacokin.* **19,** 297–302.

Hayashi, T., Era, S., Kawai, K., Imai, H., Nakamura, K., Onda, E., and Yoh, M. (2000). Observation for redox state of human serum and aqueous humor albumin from patients with senile cataract. *Pathophysiology* **6,** 237–243.

Hayashi, T., Imai, H., Kuwata, K., Sogami, M., and Era, S. (2002a). The importance of sample preservation temperature for analysis of the redox state of human serum albumin. *Clin. Chim. Acta* **316,** 175–178.

Hayashi, T., Suda, K., Imai, H., and Era, S. (2002b). Simple and sensitive high-performance liquid chromatographic method for the investigation of dynamic chances in the redox state of rat serum albumin. *J. Chromatogr. B* **772,** 139–146.

Himmelfarb, J., and McMonagle, E. (2001). Albumin is the major plasma protein target of oxidant stress in uremia. *Kidney Int.* **60,** 358–363.

Imai, H., Hayashi, T., Negawa, T., Nakamura, K., Tomida, M., Koda, K., Tajima, T., Koda, Y., Suda, K., and Era, S. (2002). Strenuous exercise-induced change in redox state of human serum albumin during intensive *Kendo* training. *Jpn. J. Physiol.* **52,** 135–140.

Imai, H., Era, S., Hayashi, T., Negawa, T., Matsuyama, Y., Okihara, K., Nakatsuma, A., and Yamada, H. (2005). Effect of propolis supplementation of human serum albumin during high-intensity *Kendo* training. *Adv. Exerc. Sports Physiol.* **11,** 109–113.

Janatova, J., Fuller, J. K., and Hunter, M. J. (1968). The heterogeneity of bovine albumin with respect to sulfhydryl and dimer content. *J. Biol. Chem.* **243,** 3612–3622.

Jones, D. P., Mody, V. C., Carlson, J. L., Lynn, M. J., and Sternberg, P. (2002). Redox analysis of human plasma allows separation of pro-oxidant events of aging from decline in antioxidant defenses. *Free Radic. Biol. Med.* **33,** 1290–1300.

Lamprecht, M., Greilberger, J. F., Schwaberger, G., Hofmann, P., and Oettl, K. (2008). Single bouts of exercise affect albumin redox state and carbonyl groups on plasma protein of trained men in a workload-dependent manner. *J. Appl. Physiol.* **104,** 1611–1617.

Lamprecht, M., Oettl, K., Schwaberger, G., Hofmann, P., and Greilberger, J. F. (2009). Protein modification responds to exercise intensity and antioxidant supplementation. *Med. Sci. Sports Exerc.* **41,** 55–163.

Marsche, G., Semlitsch, M., Hammer, A., Frank, S., Weigle, B., Demling, N., Schmidt, K., Windischhofer, W., Waeg, G., Sattler, W., and Malle, E. (2007). Hypochlorite-modified albumin colocalizes with RAGE in the artery wall and promotes MCP-1 expression via the RAGE-Erk1/2 MAP-kinase pathway. *FASEB J.* **21,** 1145–1152.

Marsche, G., Frank, S., Hrzenjak, A., Holzer, M., Dirnberger, S., Wadsack, C., Scharnagl, H., Stojakovic, T., Heinemann, A., and Oettl, K. (2009). Plasma-advanced oxidation protein products are potent high-density lipoprotein receptor antagonists in vivo. *Circ. Res.* **104,** 750–757.

Matsuyama, Y., Terawaki, H., Terada, T., and Era, S. (2009). Albumin thiol oxidation and serum protein carbonyl formation are progressively enhanced with advancing stages of chronic kidney disease. *Clin. Exp. Nephrol.* **13,** 308–315.

Oettl, K., and Stauber, R. E. (2007). Physiological and pathological changes in the redox state of human serum albumin critically influence its binding properties. *Br. J. Pharmacol.* **151,** 580–590.

Oettl, K., Stadlbauer, V., Petter, F., Greilberger, J., Putz-Bankuti, C., Hallstrom, S., Lackner, C., and Stauber, R. E. (2008). Oxidative damage of albumin in advanced liver disease. *Biochim. Biophys. Acta* **1782,** 469–473.

Oettl, K., Stadlbauer, V., Krisper, P., and Stauber, R. E. (2009). Effect of extracorporeal liver support by molecular adsorbents recirculating system and prometheus on redox state of albumin in acute-on-chronic liver failure. *Ther. Apher. Dial.* **13,** 431–436.

Oettl, K., Reibnegger, G., and Schmut, O. (2010). The redox state of human serum albumin in eye diseases with and without complications. *Acta Ophthalmol.* doi: 10.1111/j.1755-3768.2009.01824.x.

Peters, T. (1996). All About Albumin; Biochemistry, Genetics and Medical Applications. Academic Press, San Diego, CA.

Shacter, E., Williams, J. A., Lim, M., and Levine, R. L. (1994). Differential susceptibility of plasma proteins to oxidative modification: examination by western blot immunoassay. *Free Radic. Biol. Med.* **17,** 429–437.

Soejima, A., Matsuzawa, N., Hayashi, T., Kimura, R., Ootsuka, T., Fukuoka, K., Yamada, A., Nagasawa, T., and Era, S. (2004). Alteration of redox state of human serum albumin before and after hemodialysis. *Blood Purif.* **22,** 525–529.

Sogami, M., Era, S., Nagaoka, S., Kuwata, K., Kida, K., Shigemi, J., Miura, K., Suzuki, E., Muto, Y., Tomita, E., Hayano, S., Sawada, S., *et al.* (1985). High-performance liquid chromatographic studies on non-mercapt in equilibrium with mercapt conversion of human serum albumin. II. *J. Chromatogr.* **332,** 19–27.

Suzuki, E., Yasuda, K., Takeda, N., Sakata, S., Era, S., Kuwata, K., Sogami, M., and Miura, K. (1992). Increased oxidized form of human serum albumin in patients with diabetes mellitus. *Diabetes Res. Clin. Pract.* **18,** 153–158.

Terawaki, H., Yoshimura, K., Hasegawa, T., Matsuyama, Y., Negawa, T., Yamada, K., Matsushima, M., Nakayama, M., Hosoya, T., and Era, S. (2004). Oxidative stress is enhanced in correlation with renal dysfunction: Examination with the redox state of albumin. *Kidney Int.* **66,** 1988–1993.

Tomida, M., Ishimaru, J. I., Hayashi, T., Nakamura, K., Murayama, K., and Era, S. (2003). The redox states of serum and synovial fluid of patients with temporomandibular joint disorders. *Jpn. J. Physiol.* **53,** 351–355.

Turell, L., Botti, H., Carballal, S., Radi, R., and Alvarez, B. (2009). Sulfenic acid—A key intermediate in albumin thiol oxidation. *J. Chromatogr. B* **877,** 3384–3392.

CHAPTER TWELVE

Methods for Studying Redox Cycling of Thioredoxin in Mediating Preconditioning-Induced Survival Genes and Proteins

Chuang C. Chiueh

Contents

1. Hormetic Mechanism: Role of Redox Cycling of Thioredoxin	198
2. Redox Functioning of Thioredoxin	199
3. Implications of Preconditioning Protection from Preclinical and Clinical Studies	200
4. Drugs Mimic Thioredoxin-Medicated Preconditioning-Induced Signaling and Protection in Cells	202
5. Methods and Materials	203
5.1. The *in vitro* preconditioning human cell model	203
5.2. Methods for measuring short-lived free radicals	204
5.3. Methods for studying the redox function of endogenous thioredoxin	206
5.4. Molecular biology method for redox-sensitive genes, mRNA, and proteins	206
5.5. Measurement of apoptosis	207
5.6. *Ex vivo* autoradiographic image of nigrostriatal dopamine terminals in the rat brain	207
5.7. Materials	208
5.8. Statistical analysis	209
References	209

Abstract

Recent advances in molecular biology provide methods and tools for studying cell signaling pathways underlying hormetic mechanisms produced by radiation hormesis, ischemic, remote ischemic, and chemical preconditioning as well as withholding of nutrients and/or trophic factors. Most of the proposed key

Division of Clinical Pharmacy, School of Pharmacy and Taipei Medical University—Shuang Ho Hospital, Taipei, Taiwan signaling pathways of hormetic mechanisms remain to be elucidated. For the investigation of possible role of thiol redox signaling systems in hormesis, a serum deprivation preconditioned human cell model, free radical assays, and molecular biological methods are employed for studying whether free radicals, the •NO–cGMP–PKG cell signaling pathway, and the redox protein thioredoxin (Trx) play any roles in the hormetic mechanism against cytotoxicity caused by serum deprivation and also neurotoxin 1-methyl-4-phenyltetrahydropyridinium ion (MPP$^+$). This •NO-dependent cell signaling pathway of the redox protein Trx may play a key role in the cellular protective mechanism of several potential neuroprotective agents such as S-nitrosoglutathione (GSNO), 17β-estradiol, selegiline as well as ebeselen, sildenafil, and rasagiline. Consistently, exogenously administered Trx (<1 μM) provides a concentration-dependent protection for human neuroblasts against MPP$^+$-induced oxidative injury. This newly discovered role of the redox protein of Trx in preconditioning-induced cell signaling and protection could lead to the development of new lead compounds for upregulation of Trx and related thiol redox proteins for cell survival, repair, proliferation, and neuronal plasticity.

1. Hormetic Mechanism: Role of Redox Cycling of Thioredoxin

Several studies indicate that preconditioning cascades are triggered by a brief surge of reactive oxygen species (ROS) and a prolonged release of nitric oxide (•NO) (Andoh et al., 2000, 2002a; Chiueh et al., 2005a; Huang, 2004; Kapinya et al., 2002; Tapuria et al., 2008; Zhao and Zuo, 2004). Pretreatment of antioxidants blocks most cellular protections induced by preconditioning procedures (Andoh et al., 2000, 2002a; Gidday, 2006). Therefore, nonlethal ROS serve as signaling molecules that upregulate host defenses and protect against oxidative stress during adaptive response. Moreover, prior results of Andoh et al. (2000) suggest that a brief nonlethal levels of ROS does not cause oxidative injury while activate stress genes and proteins such as redox effector factor-1 (Ref-1/REF-1) and a sustained expression of neuronal nitric oxide synthase (nNOS/NOS1). This result indicates that nNOS is inducible which do not support an early notion that only iNOS is inducible. The induction of nNOS by preconditioning procedure leads to hormetic protection instead of cytotoxicity because both endogenous •NO and exogenous •NO donors mimic cell signaling and protection induced by preconditioning procedures through the cGMP-dependent cascade (Andoh et al., 2002a, 2003). The activation of the •NO–cGMP–PKG cell signaling pathway eventually leads to a delay induction of thiol proteins such as thioredoxin (Trx) (Andoh et al., 2000, 2002b; Das, 2004; Turoczi et al., 2003; Yamawaki et al., 2003); The redox cycling of Trx in the presence of NADPH, Trx reductase, and selenium-containing enzyme Trx peroxidase results in antioxidation and activation of thiol-containing transcription factors, proteins, and enzymes resulting in new protein synthesis for antiapoptosis, anti-inflammation, cellular repair, cell proliferation, and neuronal plasticity (Andoh et al., 2002a; Berndt et al., 2007; Chiueh et al., 2005a; Hashemy and Holmgren, 2008; Holmgren and Bjornstedt, 1995; Lu et al., 2009). Trx-mediated preconditioning protection and cell signaling are prevented by the reductase inhibitors 5,5′-dithiobis (2-nitrobenzoic acid) (DTNB), supporting the role of redox cycling of Trx in hormesis (Andoh et al., 2000, 2002b, 2003). Moreover, the possible mediator role of Trx in hormetic mechanism is further supported by the evidence that preconditioning-induced signaling and protection can be blocked by the transfection of cells with antisense oligonucleotides but not sense or antisense mutant against human Trx mRNA for blocking Trx induction in Human SHSY-5Y cells (Chiueh et al., 2005a,b).

2. Redox Functioning of Thioredoxin

Reduced Trx and its highly conserved active site of Cys^{32}-Gly-Pro-Cys^{35} that bind to apoptosis signal-regulating kinase-1 (ASK-1) inhibiting the apoptotic chain of events (Saitoh et al., 1998). Trx also upregulates antiapoptotic BCL-2 that decreases the release of cytochrome c and stops the cascade of mitochondria-induced apoptosis and blocks mitochondria-mediated apoptosis (Andoh et al., 2000, 2002a,b). There are two major programmed cell death pathways such as (i) the release of intermembrane cytochrome c after opening permeable transition pore for calcium entry and (ii) the activation of proapoptotic pathway of caspase-3 and -9 (see review of Wang, 2001) all of which can be blocked by Trx ($<1,000$ nM). Released cytochrome c activates apoptotic peptidase activating factor 1 (Apaf-1), which in turn activates a downstream caspase program. Also, Trx directly inhibits caspase-3 and -9 (Andoh et al., 2002b, 2003) and thus minimizing the expression of death substrate poly (ADP-ribose) polymerase (PARP) and associated programmed cell death.

Owing to its thiol redox recycling, Trx (nM concentrations) is far more potent than trolox, S-nitrosoglutathione (GSNO; μM), ascorbate, and glutathione (GSH; mM) in the scavenging of hydroxyl radical (•OH) and the suppression of lipid peroxidation (Andoh et al., 2003; Rauhala et al., 1996, 1998). The upregulation of mitochondrial MnSOD instead of cytosolic Cu/ZnSOD by Trx would protect mitochondria from oxidative stress and injury (Andoh et al., 2000, 2002a,b). Furthermore, recent reports indicate that Trx blocks the adhesion and migration of reactive neutrophils which may inhibit pulmonary infiltration/fibrosis and inflammation in acute respiratory distress syndrome and perhaps sepsis (Hofer et al., 2009; Kondo et al., 2006). Finally, our data suggest that Trx is a key component in the redox cell signaling pathway for hormesis because the preconditioning protection can be blocked by not only the antisense against Trx mRNA but also the Trx reductase inhibitor DTNB, Ellman's reagent (Andoh et al., 2002a,b, 2003; Chiueh et al., 2005a).

Jones (2008) proposed that oxidative stress could occur in proteins containing cysteine (Cys) residues as a consequence of disruption of thiol redox cycling, which normally regulate cell signaling and associated physiological functions. The redox states of protein thiol systems (P–(SH)$_2$ ↔ P–S–S–P) are controlled by Trx as well as GSH, glutaredoxin, and peroxiredoxin to minimize the oxidative stress of protein Cys residues due to redox imbalance. Oxidation of thiol functional groups of proteins could lead to the formation of GSH (P–S–SG), the oxidation to sulfenic (P–SOH), sulfinic (P–SO$_2$H), and sulfonic (P–SO$_3$H) acid and S-nitrosylation (P–S–NO) as well (Rhee et al., 2005). It is postulated that the oxidation of GSH to GSSG might also produce glutathionyl radicals (GS•) which subsequently react with •NO radicals to generate GSNO (Chiueh and Rauhala, 1999); the antioxidative activity of GSNO is at least three orders of magnitude greater than GSH (Rauhala et al., 1996, 1998).

The redox protein Trx contains multiple Cys moieties with the known active site of Cys^{32}-Gly-Pro-Cys^{35} which undergoes reversible disulfite oxidation during redox cycling by Trx reductase and Trx peroxidase in reaction with other Cys-containing protein and/or transcription factors (Holmgren et al., 2005). In addition to S-nitrosylation (Trx–S–NO) by •NO and/or GSNO, Cys^{62} can also form a disulfide bridge with Cys^{69}. Finally, Cys^{73} of Trx is possibly for dimerization of Trx that can also undergo S-glutathionylation (Trx–S–S–G) in the presence of GSH and/or GSNO. Therefore, Cys residues of Trx have different thiol redox sensitivities (Hashemy and Holmgren, 2008; Jones, 2008). In cell nuclei, Trx reduces REF-1, which regulates conserved Cys residues of transcription factors (i.e., NF-κB, Nrf-2, AP-1, P53, HIF, estrogen receptor, and glucocorticoid receptor) in the reduced form that is required for DNA binding (Grippo et al., 1985; Hayashi et al., 1997; Saitoh et al., 1998). Through the activation of redox-sensitive nuclear transcription factors and enzymes, Trx controls virtually all aspects of cellular functions including survival, repair, proliferation, and prevent programmed cell death as well. Trx may also participate in neuronal plasticity and synaptogenesis (Zhong et al., in preparation).

3. Implications of Preconditioning Protection from Preclinical and Clinical Studies

A surprising hormetic finding of lowering cancer incidence by low-dose irradiation in Taiwan confirms a previously proposed notion of irradiation hormesis based on early preclinical reports using cell and animal preparations (Chen et al., 2004). Chemical preconditioning of heart by inhalational anesthetics produces myocardial protection in patients during prolonged cardiosurgical procedures (Berndt et al., 2007; Kapinya et al., 2002; Kitano et al., 2007; Mortier et al., 2000; Sano et al., 1992; Sedlic et al., 2009; Zhao and Zuo, 2004). This has already been successfully performed with anticipated cardioprotection in cardiosurgery patients. Conversely, the expression of high levels of Trx in malignant cancer cells may lead to drug resistance or tolerance against chemotherapy. Chemical preconditioning induced by anticancer agents might occur in cancer cells during the routing treatment sessions of chemotherapy. Sedlic et al. (2009) reported that anesthetics desflurane or sevoflurane protects isolated rat cardiomyocytes from oxidative stress-induced cell death. Desflurane exhibits greater effect on stimulation of ROS production and mitochondrial uncoupling than that of sevoflurane. Scavenging of ROS abolishes the preconditioning effect of both anesthetics and attenuates anesthetic-induced mitochondrial uncoupling. This is consistent with early report of Andoh et al. (2000, 2002a) that •OH scavengers prevent the serum deprivation-induced preconditioning induction of Trx and associated cytoprotection. In fact, the use of intracerebral trapping of •OH by microdialysis of salicylate (Chiueh et al., 1992, 1993) observed that a decrease of body temperature from 37 to 32 °C by a controlled dose of barbiturate significantly blocked the generation of ROS and associated lipid peroxidation in the rat brain (C. C. Chiueh, unpublished observations). Medically induced coma bring on by a controlled dose of a barbiturate drug may be beneficial for preventing spraying depression of brain neurons and a better recovery of brain function after brain trauma and stroke as well (Mortier et al., 2000). These adaptive responses of preconditioning from lowering oxygen and temperature appear in tandem and the induction of Trx that may have clinical implications in myocardial preservation for heart transplantation (Chiueh et al., 2005a; Das, 2004; Karck et al., 1996; Tapuria et al., 2008; Turoczi et al., 2003; Yamawaki et al., 2003). The underlying hormetic mechanisms of above-mentioned preconditioning procedures could be elucidated because a common observation that •OH triggers while •NO maintains preconditioning-induced cell signaling and protection. Both ischemic preconditioning and remote ischemic preconditioning may protect tissue from subsequent severe ischemic insults in animal experiments probably medicate by both •OH and •NO (Shahid et al., 2008). It is likely that the present proposed working hypothesis of •NO/cGMP/PKG and Trx redox cycling for hormesis signaling (Andoh et al., 2002a; Chiueh et al., 2005a,b) could be confirmed in the remote ischemic preconditioning model for myocardial and pulmonary protection. Finally, this upregulation of Trx system by cGMP (Andoh et al., 2000, 2002b, 2003) is consistent with the key role of •NO during the fetal-to-neonatal transition; the observed acute increase of the thiol redox system of Trx in the lung at birth could be an important biochemical event for regulating and maintaining normal pulmonary function (Das et al., 1999). It remains to be investigated whether this cGMP-dependent expression of Trx also plays a key role in calorie restriction-induced cardioprotection and longevity against aging process (Ingram *et al.*, 2006; Maalouf *et al.*, 2009).

4. Drugs Mimic Thioredoxin-Medicated Preconditioning-Induced Signaling and Protection in Cells

The newly discovered hormetic role of the induction of the thiol redox protein Trx through the activation of the •NO–cGMP–PKG signaling pathway in free radical-triggered preconditioning cascade for protection against damages caused by withdrawal of oxygen, trophic factors, and nutrients and neurotoxins is very interesting. This Trx-medicated thiol cell signaling pathway can be used for future research and development of not only new Trx mimetics for cardio- and neuro-protection but also new Trx suppressors as adjunctive medication for managing chemotherapy resistance. Moreover, the upregulation of Trx in animals and cells increases their tolerance against oxidative free radical insults caused by 1-methyl-4-phenyl-1,2,3,6-tetrahydropyridine (MPTP) and its toxic metabolite MPP^+ (1-methyl-4-phenylpyridinium ion) on brain dopamine neurons (Andoh *et al.*, 2002b, 2003; Bai *et al.*, 2007; Chiueh *et al.*, 2000, 2005b). A brief and nonlethal serum deprivation in human SHSY-5Y cell cultures evokes preconditioning protection against not only the lethal oxidative stress caused by a sustained 24-h serum withdrawal but also the severe oxidative stress produced by MPP^+, the active metabolite of the dopamine neurotoxin MPTP. The LD_{50} of MPP^+ shifted to right from 0.1 to 3 mM due to the Trx-mediated hormesis or preconditioning-induced neuroprotection.

Trx is a 12-kDa thiol redox protein and reduced Trx is the active form (Trx–$(SH)_2$) while oxidized Trx (Trx–S–S–Trx) can enter the cell membrane through a yet-to-be-identified mechanism and then reduce back to active form in the presence of Trx reductase and NADPH for regulating normal cell functions and survival (Kondo *et al.*, 2006; Matsuo *et al.*, 2001). There is a significant development of macromolecular drugs since the identification of Trx as the adult T cell leukemia trophic factor. The Trx-based preconditioning molecular mechanism for new drug development aiming cytoprotection in the central nervous system is the one of our research efforts during the past decade. The inducers for Trx and related biologics may have multifaceted indications for enhancing survival of cells, organs, and animals while the inhibitors of the Trx redox cycling may be used for enhancing anticancer agent's potency especially in cancers with drug resistance. For validating this working hypothesis, recent research activities of our and other laboratories are focusing on the evaluation of GSNO, 17β-estradiol, selegiline as well as 8-bromo-cGMP, ebeselen, rasagiline, and sildenafil for their effects on the induction and activation of redox signaling proteins such as Trx for survival, repair, and plasticity against severe oxidative stress in the brain (Andoh et al., 2002a, 2005; Chiueh et al., 2001, 2005a,b; Lee et al., 2003; Takagi et al., 1999). GSNO is a •NO donor and sildenafil decreases the degradation of cGMP by diphosphoesterase 5 all of which lead to the activation of Trx signaling pathway and associated functional improvement in pulmonary system such as acute respiratory distress syndrome, pulmonary hypertension, and high altitude ischemic insults (mountain sickness). Antidepression and mode modification effects of estrogen (i.e., 17β-estradiol) and selegiline may be due to the fact that both compounds increase the expression of Trx resulting in neuroprotection and synaptogenesis in the brain (Andoh et al., 2005; Chen et al., 2009; Lee et al., 2003; Zhong et al., in preparation). Trx antisense prevents the neuroprotective effect of 17β-estradiol and selegiline. Recently, FDA approves a transdermal system of selegiline for the treatment of major depression devoid of typical side effects caused by monoamine oxidase inhibition due mainly to its unconventional neuroprotective property after the induction of the thiol redox system of Trx (Andoh et al., 2005). The selenium compound ebeselen and the more potent selegiline analog rasagiline are developed for clinical indications of stroke and Parkinson's disease, respectively (Chiueh et al., 2005b; Lu et al., 2009; Zhou et al., 2009). Rasagilien is more potent than that of selegiline but selegiline is devoid of the side effects of rasagiline during the clinical trials. Finally, after preclinical trial in rat model of sepsis, Trx is undergoing clinical trials for reducing mortality in patients with sepsis (Hofer et al., 2009). The redox protein family members include not only Trx but also glutaredoxins and peroxiredoxins which may be upregulated for participating in hormetic cascade that requires additional future studies. The other redox protein glutaredoxins are required for cellular iron homeostasis including iron–sulfur cluster assembly and heme biosynthesis in photosynthesis (Rouhier et al., 2008). The relative abundance of peroxiredoxin enzymes appears to protect cellular components by removing the harmful levels of peroxides produced as a result of normal cellular metabolism and the active Cys sites may be overoxidized to cysteinyl sulfinic acid and/or sulfonic acid (Rhee et al., 2005). Interestingly, during the preconditioning stress, $p66^{shc}$ is not induced but rather reduced (Andoh et al., 2000); the reduction of this longevity suppressive gene may help Trx to ensure cell survival during severe oxidative stress.

5. Methods and Materials

5.1. The *in vitro* preconditioning human cell model

The human SHSY-5Y neuroblastoma cells ($\sim 10^6$ cells, kindly provided by Carol Thiele, NCI) are cultured in 1 ml Dulbecco's modified Eagle's medium containing 10% fetal bovine serum; cell cultures are incubated in an incubator (37 °C, 5% CO_2) for 2–3 days before the preconditioning experiments. In these serum deprivation *in vitro* experiments, serum and phenol red are removed from the culture media for either 2 h (nonlethal preconditioning stress) or 24 h (lethal oxidative stress). The nonlethal preconditioning stress is applied 24 h prior to the lethal serum withdrawal stress (Andoh *et al.*, 2000).

5.2. Methods for measuring short-lived free radicals

5.2.1. Trapping hydroxyl radicals by salicylate

A trapping reagent of short-lived •OH, sodium salicylate (1 m*M*), is used to monitor the generation of •OH in the culture media and also to protect cells from oxidative damage. A previously published intracerebral microdialysis of salicylate and a sensitive HPLC–EC procedure (Chiueh, 1994; Chiueh *et al.*, 1992, 1993) are employed for assaying hydroxyl adducts of salicylate, 2,3- and 2,5-dihydroxybenzoic acid. Withdrawal of 10% serum from the human SHSY-5Y cell culture media immediately increases the trapping of •OH which is accumulated time-dependently in the culture media (Andoh *et al.*, 2000) or brain perfusates (Rauhala *et al.*, 1998). The addition of fetal bovine serum to cell culture media provides not only trophic and nutritional supports but also scavenging reactive and mostly harmful ROS such as •OH. Consistently, pretreatment of cells with salicylate scavenges •OH due to serum withdrawal and prevents 24 h lethal stress-induced cell death. It is noteworthy that scavenging of •OH during the preconditioning period would stop the consequent hormetic cell signaling and preconditioning protection. Infusion of MPP^+ increases the sustained dopamine outflow and the generation of •OH radicals trapped by the intracerebral microdialysis of salicylate *in vivo* (Chiueh *et al.*, 1992).

5.2.2. NOS1 activity: Formation of •NO and activation of cGMP

An aliquot of supernatant of the culture media of experimental cells was mixed with Griess reagent for detecting nitrite, one of the major oxidized products of •NO. The amount of •NO reflected by the optical density of the nitrite-azo dye formed in the sample was measured with a spectrophotometer at 540 nm. The intracellular cGMP levels were measured using an enzyme immunoassay kit (Amersham). Preconditioning induction of nNOS or NOS1 leads to the generation of •NO and the release of •NO results in the formation of cGMP and the activation of PKG for cell signaling for upregulation of the survival protein Trx. Both sublethal and lethal oxidative stress caused by serum withdrawal increase the expression of nNOS/NOS1 and the production/release of •NO levels to a similar extent. Therefore, the underlying cell death mechanism is not caused by •NO; these findings do not support early reports that •NO-mediated neurotoxicity due to the formation of peroxylnitrite all of which can be detoxified by CO_2. Therefore, the induction of nNOS does not lead to expected cytotoxicity since peroxylnitrite do not cause anticipated neurotoxicity in the rat brain in human SHSY-5Y cell cultures. In addition to free radical scavengers, inhibition of nNOS, guanylyl cyclase, and PKG blocks hormesis induced by preconditioning. GSNO treatment activates cGMP and thus the hormetic cascade through the induction of Trx while 6-bromo-cGMP produces a cytoprotection mimicking the action of preconditioning tolerance. Theoretically, inhibitors of phosphodiesterase 5 (i.e., sildenafil) decrease the degradation of cGMP and thus increase this hormetic signaling and endogenous Trx levels for survival against subsequent severe oxidative stress and injury. It is postulated that GSNO, sildenafil, and Trx may enhance pulmonary function in acute respiratory distress syndrome and/or sepsis.

5.2.3. Lipid peroxidation

At end of the *in vitro* preconditioning experiments, cells are collected 24 h after culturing in the presence or absence of Trx (<1 μM). The collected cells are washed twice with ice-cold phosphate-buffered saline (PBS) and then sonicated in 200 μl of PBS. After a brief centrifuge, the protein concentration is determined using a protein assay kit (Bio-Rad). Fluorescent products of lipid peroxidation (fluorescent adduct products of malondialdehyde and amino acid with the excitation/emission wave lengths of 356/426 nm, respectively) are measured using a luminescent spectrometer (PerkinElmer Life Sciences) after solvent extraction (Mohanakumar *et al.*, 1994, 1998; Rauhala *et al.*, 1998). The results are presented as relative fluorescent intensity and defined as relative fluorescent units per milligram. This method can also be used to measure lipid peroxidation in rat brain homogenates *ex vivo*. Rat brain tissues are homogenized in ice-cold Ringer solution using an ultrasonic cell-disrupter and assayed for brain lipid peroxidation using a fluorescence assay procedure. Brain homogenates (25 mg) are incubated for 3, 4, or 24 h at 37 °C in 0.5 ml reaction mixtures containing ferrous citrate (0–5 μM) or MPP^+ (0–100 μM) with or without GSNO (0–20 μM). Malondialdehyde cross-linked with amino group of amino acids is extracted and its fluorescence levels are measured in relative fluorescence intensity units (Activation/emission wavelength: 356/426 nm, PerkinElmer LS50B) using aliquots equivalent to 3.75 mg brain tissue; this is used as an index of brain lipid peroxidation. Peroxynitrite is ineffective at 37.5 μM concentration whereas 25 μM iron produces a maximal lipid peroxidation *in vitro*. Greater concentrations of peroxynitrite (75–600 μM) are needed to significantly increase peroxidation of brain lipids which can be blocked concentration dependently by GSNO both *in vitro* (Rauhala *et al.*, 1996) and *in vivo* (Rauhala *et al.*, 1996, 1998). Nanomolar ferrous citrate iron complex but not peroxynitrite caused neurodegeneration of brain dopaminergic nigral neurons *in vivo*.

5.3. Methods for studying the redox function of endogenous thioredoxin

Trx antisense and sense oligonucleotides are designed by Saitoh et al. (1998) for investigating the role of endogenous Trx in mediating cytoprotection. Sequences of antisense oligonucleotide designed as 5'-TCTGCTTCAC-CATCTTGGCTGCT-3' for hybridizing with human Trx mRNA are designed to reduce the protein synthesis; sense and mutant antisense oligonucleotide sequences as 5'-AGCAGCCAAGATGGTGAAGCAGA-3' and 5'-TCGTTCTCACCATCTTGGTCCGT-3', respectively, for control groups and they are synthesized as phosphorthionate oligonucleotides (S-oligo). For transfection of SHSY-5Y cells, 2 μM of each S-oligo are mixed with 2 μl of transfection reagent TM-50 (Promega, Madison, WI) in medium (800 μl) with or without 10% fetal bovine serum for 15 min at room temperature, and then the mixture is added to cell culture before the beginning of preconditioning experiments (Andoh et al., 2003). Preconditioning procedure using serum withdrawal for 2 h significantly increases the induction of Trx 12–24 h later that can be blocked by transfection of cells with antisense S-oligo but not sense and antisense mutant S-oligo.

5.4. Molecular biology method for redox-sensitive genes, mRNA, and proteins

Other Western and slot blotting and molecular biological procedures are also employed to detect mRNA and the expression of stress proteins. The redox-sensitive genes include redox effector factor-1 (Ref-1), antioxidative responsive elements (Are), and NF-E2-related factor-2 (Nrf-2) all of which may participate in the initial response to preconditioning stress.

Cells are homogenized in cell lysis buffer consisting of 20 mM HEPES–KOH, pH 7.5, 10 mM KCl, 1.5 mM MgCl$_2$, 1 mM EDTA, 1 mM EGTA, 1 mM dithiothreitol, and phenylmethylsulfonyl fluoride. The protein concentration is quantified using a Bio-Rad protein assay kit. Cell protein (20 μg) in lysis buffer is then separated by electrophoresis using a 4–20% gradient SDS–polyacrylamide gel and then transferred to a polyvinylidene difluoride membrane (Millipore, Bedford, MA). For the slot blotting experiments, cell lysates (1 μg) are blotted on the polyvinylidene difluoride membrane. The blots are blocked with a 5% skim milk solution for 1 h at room temperature and then probed with a 1/2000 dilution of antibody such as anti-Trx, anti-cytochrome c, anti-Bcl-2, anti-MnSOD, anti-Cu/ZnSOD, or anticatalase, at 4 °C overnight. Subsequently, it is incubated with a horseradish peroxidase-linked antibody against mouse or rabbit IgG (1:2000) for 1 h. Membrane-bound horseradish peroxidase-labeled protein bands are monitored with chemiluminescent reagents (Amersham Biosciences). Chemiluminescent signals are detected using X-ray film, and the area and intensity of the signal are integrated using the NIH Image program (Wayne Rasband, NIMH, NIH).

5.5. Measurement of apoptosis

5.5.1. Fluorescent staining of condensed or fragmented nuclei by Hoechst 33258

Cell death and apoptosis are assessed microphotographically using trypan blue dye and Hoechst 33258 fluorescent dye, respectively. At the end of the serum deprivation experiments, control and treated cells are harvested and fixed with 4% paraformaldehyde in PBS at 4 °C for 30 min. After rinsing with saline, the nuclear DNAs are stained with 1 μM Hoechst 33258 fluorescent dye for 5 min at room temperature and observed with a fluorescent microscope (excitation/emission wavelength: 365/420 nm) for apoptotic fragmented nuclei.

5.5.2. Cytochrome c release assay

The cytochrome c assay procedure described by Wang (2001) is adopted (Andoh et al., 2003). Cells are harvested and separated into cytosolic fraction and mitochondrial fraction after differential centrifugation. The cytochrome c levels in the mitochondrial and cytoplasmic fractions are measured using Western blotting procedures with anti-cytochrome c antibody.

5.5.3. Measurement of caspase activity

Cells are homogenized in cell lysis buffer and centrifuged at $10,000 \times g$ for 1 min. After determining the protein concentration, one-half of the cell lysates are used for measuring the catalytic activities of caspase-9 and -3 using a colorimetric assay (R & D systems). LEHD-pNA and DEVD-pNA are the substrate of caspase-9 and caspase-3, respectively, which are incubated with cell lysates at 37 °C for 2 h. Caspase-9 and -3 activities are monitored by the pNA chromophore cleaved from the substrates and quantified at 405 nm with a spectrophotometer. To the best of our knowledge Trx is the most potent inhibitor against caspases *in vitro* (Andoh et al., 2002, 2003).

5.6. *Ex vivo* autoradiographic image of nigrostriatal dopamine terminals in the rat brain

Sprague-Dawley rats (male 250–350 g, Taconic Farms) are anesthetized with chloral hydrate (400 mg/kg, i.p.) and prepared for stereotaxic infusion of drug (0–10 nmol ferrous citrate and/or 0–16.8 nmol GSNO in 1 μl sterilized Ringer's solution or 0–33.6 nmol peroxynitrite in 1.05 μl 0.3 M NaOH solution) into right or left substantia nigra compacta area (i.n.; Paxinos and Watson stereotaxic coordinates: AP: 3.2 mm, RL: 2.1 mm,

H: 2.0 mm, mouth bar: 03.5 mm), as reported previously. Sham control groups are treated intranigrally with either 1 μl Ringer's solution or 1.05 μl 0.3 M NaOH. Eight days after intracerebral infusion of ferrous citrate (4.2 nmol, i.n.) without or with GSNO (16.8 nmol, i.n.) cotreatment in the right-side midbrain substantia nigra, dopamine uptake sites in nerve terminals of the caudate nucleus and nucleus accumbens are visualized autoradiographically using ^{125}I-labeled RTI-55 (100 μCi, i.v.; Boja et al., 1992; Rauhala et al., 1998), 10 min after blocking serotonin uptake sites by fluoxetine (10 mg/kg, i.v.). Striatal RTI-55 radioactivities in the caudate nucleus are estimated using Amersham autoradiographic ^{125}I-labeled microscales as the exposure standards. Enlarged autoradiographic images are analyzed using the NIH Image software developed by Wayne Rasband (NIMH, NIH) to measure optical densities; gray scales are then converted to rainbow color scales (orange to yellow show high bindings; green to blue depict low bindings). Normal dopamine uptake sites in dopaminergic nerve terminals of the control sides are visualized as orange areas.

5.7. Materials

The human neuroblastoma SHSY-5Y cells were kindly provided by Dr. Carol Thiele (National Cancer Institute, National Institutes of Health, Bethesda, MD). The NIH Image program was kindly provided by Dr. Wayne Rasband, NIMH, NIH, (http://rsb.info.nih.gov/nih-image/). DMEM, penicillin/streptomycin, and heat-inactivated bovine serum were purchased from Invitrogen (Carlsbad, CA). Hoechst 33258 (bisbenzimide), 1-chloro-2,4-dinitrobenzene (DNCB), and bisindolylmaleimide were ordered from Sigma Chemical Co. (St. Louis, MO). Oxidized *Escherichia coli* Trx and human catalase antibody were purchased from Calbiochem (San Diego, CA). Human Trx and mouse Trx antibody were obtained from MBL International (Watertown, MA) and Redox Bioscience (Kyoto, Japan), respectively. Sense, antisense, and antisense mutant for human Trx mRNA (nucleotide sequences: antisense, 5′-TCTGCTTCACCATCTTGGCTGCT-3′; sense, 5′-AGCAGCCAAGATGGTGAAGCAGA-3′; mutant antisense, 5′-TCGTTCTCACCATCTTGGTCCGT-3′) and mouse Trx mRNA (nucleotide sequences: antisense, 5′-TCAGCTTCACCATTTTGGCTGTT-3′; sense, 5′-AACAGCCAAAATGGTGAAGCTGA-3′; mutant antisense, 5′-TCCATGTCACCATTTTGGTGTCT-3′) were synthesized as S-oligonucleotides by Invitrogen and by Hokkaido Bioscience (Hokkaido, Japan), respectively. Antibodies against phosphorylated and nonphosphorylated c-Jun, MEK1/2, MAPK/Erk1/2, and c-Myc were obtained from Cell Signaling Technology, Inc. (Beverly, MA) and BD Pharmingen (San Diego, CA). A horseradish peroxidase-linked antibody against IgG was obtained from GE Healthcare (Little Chalfont, Buckinghamshire, UK). [^{125}I]-RTI-55 (NEX272250UC) was ordered from PerkinElmer (Waltham, MA).

5.8. Statistical analysis

Data are presented as mean ± S.E. of the results obtained from the average of four to six repeated experiments. Results are analyzed by one-way analysis of variance, and p values are assigned using the Newman–Keuls test and Student's t-test. Differences among means are considered statistically significant when the p value is less than 0.05 (n of at least 3 measurements).

REFERENCES

Andoh, T., Lee, S. Y., and Chiueh, C. C. (2000). Preconditioning regulation of bcl-2 and p66[shc] by human NOS1 enhances tolerance to oxidative stress. *FASEB J.* **14,** 2144–2146.

Andoh, T., Chock, P. B., and Chiueh, C. C. (2002a). Preconditioning-mediated neuroprotection: Role of nitric oxide, cGMP and new protein expression. *Ann. N.Y. Acad. Sci.* **962,** 1–7.

Andoh, T., Chock, P. B., and Chiueh, C. C. (2002b). The roles of thioredoxin in protection against oxidative stress-induced apoptosis in SH-SY5Y cells. *J. Biol. Chem.* **277,** 9655–9660.

Andoh, T., Chiueh, C. C., and Chock, P. B. (2003). Cyclic GMP-dependent protein kinase regulates the expression of thioredoxin and thioredoxin peroxidase-1 during hormesis in response to oxidative stress-induced apoptosis. *J. Biol. Chem.* **278,** 885–890.

Andoh, T., Chock, P. B., Murphy, D. L., and Chiueh, C. C. (2005). Role of the redox protein thioredoxin in cytoprotective mechanism evoked by (-)-deprenyl. *Mol. Pharmacol.* **68,** 1408–1414.

Bai, J., Nakamura, H., Kwon, Y. W., Tanito, M., Ueda, S., Tanaka, T., Hattori, I., Ban, S., Momoi, T., Kitao, Y., Ogawa, S., and Yodoi, J. (2007). Does thioredoxin-1 prevent mitochondria- and endoplasmic reticulum-mediated neurotoxicity of 1-methyl-4-phenyl-1,2,3,6-tetrahydropyridine? *Antioxid. Redox Signal.* **9,** 603–608.

Berndt, C., Lillig, C. H., and Holmgren, A. (2007). Thiol-based mechanisms of the thioredoxin and glutaredoxin systems: Implications for diseases in the cardiovascular system. *Am. J. Physiol. (Heart Circ. Physiol.)* **292,** H1227–H1236.

Boja, J. W., Mitchell, W. M., Patel, A., Kopajtic, T. A., Carroll, F. I., Lewin, A. H., Abraham, P., and Kuhar, M. J. (1992). High-affinity binding of [125I]RTI-55 to dopamine and serotonin transporters in rat brain. *Synapse* **12,** 27–36.

Chen, W. L., Luan, Y. C., Shieh, M. C., Chen, S. T., Kung, H. T., Soong, K. L., Yeh, Y. C., Chou, T. S., Mong, S. H., Wu, J. T., Sun, C. P., Deng, W. P., Wu, M. F., and Shen, M. L. (2004). Is chronic radiation an effective prophylaxis against cancer? *J. Am. Phys. Surg.* **9,** 6–10.

Chen, T. Y., Tsai, K. L., Lee, T. Y., Chiueh, C. C., Lee, W. S., and Hsu, C. (2009). Sex-specific role of thioredoxin in neuroprotection against iron-induced brain injury conferred by estradiol. *Stroke* [Epub ahead of print].

Chiueh, C. C. (1994). Neurobiology of NO• and •OH: Basic research and clinical relevance. *Ann. N.Y. Acad. Sci.* **738,** 279–281.

Chiueh, C. C. (2001). Iron overload, oxidative stress, and axonal dystrophy in brain disorders. *Pediatr. Neurol.* **25,** 138–147.

Chiueh, C. C., and Rauhala, P. (1999). The redox pathway of S-nitrosoglutathione, glutathione and nitric oxide in cell to neuron communications. *Free Radic. Res.* **31,** 641–650.

Chiueh, C. C., Krishna, G., Tulsi, P., Obata, T., Lang, K., Huang, S. J., and Murphy, D. L. (1992). Intracranial microdialysis of salicylic acid to detect hydroxyl radical generation though dopamine autooxidation in the caudate nucleus: Effects of MPP$^+$. *Free Radic. Biol. Med.* **13**, 581–583.

Chiueh, C. C., Miyake, H., and Peng, M. T. (1993). Role of dopamine autooxidation, hydroxyl radical generation, and calcium overload in underlying mechanisms involved in MPTP-induced Parkinsonism. *Adv. Neurol.* **60**, 251–258.

Chiueh, C. C., Andoh, T., Lai, A. R., Lai, E., and Krishna, G. (2000). Neuroprotective strategies in Parkinson's disease: Protection against progressive nigral damage induced by free radicals. *Neurotox. Res.* **2**, 293–310.

Chiueh, C. C., Andoh, T., and Chock, P. B. (2005a). Roles of thioredoxin in nitric oxide-dependent preconditioning-induced tolerance against MPTP neurotoxin. *Toxicol. Appl. Pharmacol.* **207**, 96–102.

Chiueh, C. C., Andoh, T., and Chock, P. B. (2005b). Induction of thioredoxin and mitochondrial survival proteins mediates preconditioning-induced cardioprotection and neuroprotection. *Ann. N.Y. Acad. Sci.* **1042**, 403–418.

Das, D. (2004). Thioredoxin regulation of ischemic preconditioning. *Antioxid. Redox. Signal.* **6**, 405–412.

Das, K. C., Guo, X.-L., and White, C. W. (1999). Induction of thioredoxin and thioredoxin reductase gene expression in lungs of newborn primates by oxygen. *Am. J. Physiol. (Lung Cell. Mol. Physiol.)* **276**, L530–L539.

Gidday, J. M. (2006). Cerebral preconditioning and ischaemic tolerance. *Nat. Rev. (Neurosci.)* **7**, 437–448.

Grippo, J. F., Holmgren, A., and Pratt, W. B. (1985). Proof that the endogenous heatstable glucocorticoid receptor-activating factor is thioredoxin. *J. Biol. Chem.* **260**, 93–97.

Hashemy, S. I., and Holmgren, A. (2008). Regulation of the catalytic activity and structure of human thioredoxin 1 via oxidation and S-nitrosylation of cysteine residues. *J. Biol. Chem.* **283**, 21890–21898.

Hayashi, S., Hajiro-Nakanishi, K., Makino, Y., Eguchi, H., Yodoi, J., and Tanaka, H. (1997). Functional modulation of estrogen receptor by redox state with reference to thioredoxin as a mediator. *Nucleic Acids Res.* **25**, 4035–4040.

Hofer, S., Rosenhagen, C., Nakamura, H., Yodoi, J., Bopp, C., Zimmermann, J. B., Goebel, M., Schemmer, P., Hoffmann, K., Schulze-Osthoff, K., Breitkreutz, R., and Weigand, M. A. (2009). Thioredoxin in human and experimental sepsis. *Crit. Care Med.* **37**, 2155–2159.

Holmgren, A., and Bjornstedt, M. (1995). Thioredoxin and thioredoxin reductase. *Methods Enzymol.* **252**, 199–208.

Holmgren, A., Johansson, C., Berndt, C., Lönn, M. E., Hudemann, C., and Lillig, C. H. (2005). Thiol redox control via thioredoxin and glutaredoxin systems. *Biochem. Soc. Trans.* **33**, 1375–1377.

Huang, P. L. (2004). Nitric oxide and cerebral ischemic preconditioning. *Cell Calcium* **36**, 232–329.

Ingram, D. K., Zhu, M., Mamczarz, J., Zou, S., Lane, M. A., Roth, G. S., and deCabo, R. (2006). Calorie restriction mimetics: An emerging research field. *Aging Cell.* **5**, 97–108.

Jones, D. P. (2008). Radical-free biology of oxidative stress. *Am. J. Physiol. (Cell Physiol.)* **295**, C849–C868.

Kapinya, K. J., Lowl, D., Futterer, C., Maurer, M., Waschke, K. F., Isaev, N. K., and Dirnagl, U. (2002). Tolerance against ischemic neuronal injury can be induced by volatile anesthetics and is inducible NO synthase dependent. *Stroke* **33**, 1889–1898.

Karck, M., Rahmanian, P., and Haverich, A. (1996). Ischemic preconditioning enhances donor heart preservation. *Transplantation* **62**, 17–22.

Kitano, H., Kirsch, J. R., Hurn, P. D., and Murphy, S. J. (2007). Inhalational anesthetics as neuroprotectants or chemical preconditioning agents in ischemic brain. *J. Cereb. Blood Flow Metab.* **27**, 1108–1128.

Kondo, N., Nakamura, H., Masutani, H., and Yodoi, J. (2006). Redox regulation of human thioredoxin network. *Antioxid. Redox Signal.* **8**, 1881–1890.

Lee, S. Y., Andoh, T., Murphy, D. L., and Chiueh, C. C. (2003). 17beta-estradiol activates ICI 182, 780-sensitive estrogen receptors and cyclic GMP-dependent thioredoxin expression for neuroprotection. *FASEB J.* **17**, 947–948.

Lu, J., Berndt, C., and Holmgren, A. (2009). Metabolism of selenium compounds catalyzed by the mammalian selenoprotein thioredoxin reductase. *Biochim. Biophys. Acta* **1790**, 1513–1519.

Maalouf, M., Rho, J. M., and Mattson, M. P. (2009). The neuroprotective properties of calorie restriction, the ketogenic diet, and ketone bodies. *Brain Res. Rev.* **59**, 293–315.

Matsuo, Y., Akiyama, N., Nakamura, H., Yodoi, J., Noda, M., and Kizaka-Kondoh, S. (2001). Identification of a novel thioredoxin-related transmembrane protein. *J. Biol. Chem.* **276**, 10032–10038.

Mohanakumar, K. P., de Bartolomeis, A., Wu, R. M., Yeh, K. J., Sternberger, L. M., Peng, S. Y., Murphy, D. L., and Chiueh, C. C. (1994). Ferrous-citrate complex and nigral degeneration: Evidence for free-radical formation and lipid peroxidation. *Ann. N. Y. Acad. Sci.* **738**, 392–399.

Mohanakumar, K. P., Hanbauer, I., and Chiueh, C. C. (1998). Neuroprotection by nitric oxide against hydroxyl radical-induced nigral neurotoxicity. *J. Chem. Neuroanat.* **14**, 195–205.

Mortier, E., Struys, M., and Herregods, L. (2000). Therapeutic coma or neuroprotection by anaesthetics. *Acta Neurol. Belg.* **100**, 225–228.

Rauhala, P., Sziraki, I., and Chiueh, C. C. (1996). Peroxidation of brain lipids in vitro: Nitric oxide versus hydroxyl radicals. *Free Radic. Biol. Med.* **21**, 391–394.

Rauhala, P., Lin, A. M., and Chiueh, C. C. (1998). Neuroprotection by S-nitrosoglutathione of brain dopamine neurons from oxidative stress. *FASEB J.* **12**, 165–173.

Rhee, S. G., Yang, K. S., Kang, S. W., Woo, H. A., and Chang, T. S. (2005). Controlled elimination of intracellular H_2O_2: Regulation of peroxiredoxin, catalase, and glutathione peroxidase via post-translational modification. *Antioxid. Redox Signal.* **7**, 619–626.

Rouhier, N., Lemaire, S. D., and Jacquot, J. P. (2008). The role of glutathione in photosynthetic organisms: Emerging functions for glutaredoxins and glutathionylation. *Annu. Rev. Plant Biol.* **59**, 143–166.

Saitoh, M., Nishitoh, H., Fujii, M., Takeda, K., Tobiume, K., Sawada, Y., Kawabata, M., Miyazono, K., and Ichijo, H. (1998). Mammalian thioredoxin is a direct inhibitor of apoptosis signal-regulating kinase (ASK) 1. *EMBO J.* **17**, 2596–2606.

Sano, T., Drummond, J. C., Patel, P. M., Grafe, M. R., Watson, J. C., and Cole, D. J. (1992). A comparison of the cerebral protective effects of isoflurane and mild hypothermia in a model of incomplete forebrain ischemia in the rat. *Anesthesiology* **76**, 221–228.

Sedlic, F., Pravdic, D., Ljubkovic, M., Marinovic, J., Stadnicka, A., and Bosnjak, Z. J. (2009). Differences in production of reactive oxygen species and mitochondrial uncoupling as events in the preconditioning signaling cascade between desflurane and sevoflurane. *Anesth. Analg.* **109**, 405–411.

Shahid, M., Tauseef, M., Sharma, K. K., and Fahim, M. (2008). Brief femoral artery ischaemia provides protection against myocardial ischaemia–reperfusion injury in rats: The possible mechanisms. *Exp. Physiol.* **93**, 954–968.

Takagi, Y., Mitsui, A., Nishiyama, A., Nozaki, K., Sono, H., Gon, Y., Hashimoto, N., and Yodoi, J. (1999). Overexpression of thioredoxin in transgenic mice attenuates focal ischemic brain damage. *Proc. Natl. Acad. Sci. USA* **96**, 4131–4136.

Tapuria, N., Kumar, Y., Habib, M. M., Abu Amara, M., Seifalian, A. M., and Davidson, B. R. (2008). Remote ischemic preconditioning: A novel protective method from ischemia reperfusion injury—A review. *J. Surg. Res.* **150,** 304–330.

Turoczi, T., Chang, V. W., Engelman, R. M., Maulik, N., Ho, Y. S., and Das, D. K. (2003). Thioredoxin redox signaling in the ischemic heart: An insight with transgenic mice overexpressing Trx1. *J. Mol. Cell. Cardiol.* **35,** 695–704.

Wang, X. (2001). The expanding role of mitochondria in apoptosis. *Genes Dev.* **15,** 2922–2933.

Yamawaki, H., Haendeler, J., and Berk, B. C. (2003). Thioredoxin: A key regulator of cardiovascular homeostasis. *Circ. Res.* **93,** 1029–1033.

Zhao, P., and Zuo, Z. (2004). Isoflurane preconditioning induces neuroprotection that is inducible nitric oxide synthase dependentin neonatal rats. *Anesthesiology* **101,** 695–702.

Zhou, F., Gomi, M., Fujimoto, M., Hayase, M., Marumo, T., Masutani, H., Yodoi, J., Hashimoto, N., Nozaki, K., and Takagi, Y. (2009). Attenuation of neuronal degeneration in thioredoxin-1 overexpressing mice after mild focal ischemia. *Brain Res.* **1272,** 62–70.

CHAPTER THIRTEEN

OXIDATIVE STRESS, THIOL REDOX SIGNALING METHODS IN EPIGENETICS

Isaac K. Sundar, Samuel Caito, Hongwei Yao, *and* Irfan Rahman

Contents

1. Introduction	216
2. Histone Acetylation Assays Using [^{3}H]-Acetate Incorporation	219
2.1. [^{3}H]-Acetate incorporation in cell culture	219
2.2. Histone protein extraction from cells and tissues	219
2.3. Histone acetylation assay	220
3. Histone Acetylation by Immunoblotting	221
4. HAT Activity Assay	222
4.1. HAT assay using commercial kits	222
4.2. HAT activity assay using kits	222
5. HDAC Activity Assay Using [^{3}H]-Labeled Histones	223
5.1. Preparation of [^{3}H]-labeled histones	223
5.2. HDAC assay sample preparation	223
5.3. HDAC activity assay	224
6. HDAC Activity Assay	225
6.1. HDAC assay using commercial kits	225
6.2. HDAC immunoprecipitation and specific activity assay for various deacetylases using commercial kits	226
7. HDACs Levels by Immunoblotting	227
8. Posttranslational Modifications of HDACs and SIRTs (Sirtuins 1–7) by Immunoprecipitation	228
9. Preparation of Whole Cell Lysate	229
9.1. From cells	229
9.2. From tissues	229
10. Preparation of Cytoplasmic and Nuclear Proteins	229
10.1. From cells	229
10.2. From tissues	230
11. Redox-Mediated Posttranslational Modification Assays	230
11.1. Protein carbonylation	230
11.2. Biotin-switch assay	232

Lung Biology and Disease Program, Department of Environmental Medicine, University of Rochester Medical Center, Rochester, New York, USA

12. Chromatin Immunoprecipitation (ChIP) Assay   233
    12.1. Methods for ChIP assay using cells   234
    12.2. Methods for ChIP assay using tissues   236
    12.3. General considerations for ChIP assay   237
13. Conclusions   238
Acknowledgments   239
References   239

Abstract

Epigenetics is referred to as heritable changes in gene expression but not encoded in the DNA sequence itself which occurs during posttranslational modifications in DNA and histones. These epigenetic modifications include histone acetylation, deacetylation, and methylation. Acetylation by histone acetyltransferases (HATs) of specific lysine residues on the N-terminal tail of core histones results in uncoiling of the DNA and increased accessibility to transcription factor binding. In contrast, histone deacetylation by histone deacetylases (HDACs) represses gene transcription by promoting DNA winding thereby limiting access to transcription factors. Reactive oxygen species (ROS) are involved in cellular redox alterations, such as amplification of proinflammatory and immunological responses, signaling pathways, activation of transcription factors (NF-κB and AP-1), chromatin remodeling (histone acetylation and deacetylation), histone/protein deacetylation by sirtuin 1 (SIRT1) and gene expression. The glutathione redox status plays an important role in protein modifications and signaling pathways, including effects on redox-sensitive transcription factors. Protein S-glutathiolation and mixed disulfide formation as candidate mechanisms for protein regulation during intracellular oxidative stress have gained a renewed impetus in view of their involvements in redox regulation of signaling proteins. A variety of methods are applied to study the epigenetic processes to elucidate the molecular mysteries underlying epigenetic inheritance. These include chromatin immunoprecipitation (ChIP), which is a powerful tool to study protein–DNA interaction and is widely used in many fields to study protein associated with chromatin, such as histone and its isoforms and transcription factors, across a defined DNA domain. Here, we describe some of the contemporary methods used to study oxidative stress and thiol redox signaling involved in epigenetic (histone acetylation, deacetylation, and methylation) and chromatin remodeling (HAT, HDAC, SIRT1) research.

ABBREVIATIONS

AP-1   activator protein-1
APS   ammonium persulfate
BCA   bicinchoninic acid
CBP   CREB binding protein

ChIP	chromatin immunoprecipitation
CoA	coenzyme A
COPD	chronic obstructive pulmonary disease
CS	cigarette smoke
2DGE	2-D gel electrophoresis
DEAE	diethylaminoethyl
DMEM	Dulbecco's modified Eagle's medium
DNP	dinitrophenyl
DNPH	dinitrophenylhydrazine
dpm	disintegrations per minute
DTT	dithiothreitol
ECL	enhanced chemiluminescence
EDTA	ethylenediaminetetraacetic acid
ERK	extracellular signal-regulated kinase
ESI-MS	electrospray ionization – mass spectrometry
FACS	fluorescence activated cell sorter
FBS	fetal bovine serum
FOXO3	Forkhead box O (Fox O)$_3$
GNATs	(general control nonderepressible 5)-related acetyltransferases
H_2O_2	hydrogen peroxide
HAT	histone acetyltransferase
HBSS	Hank's balanced saline solution
HDAC	histone deacetylase
HEPES	4-(2-hydroxyethyl)-1-piperazineethanesulfonic acid
4-HNE	4-hydroxy-2-nonenal
ICAT	isotope – coded affinity tags
IP	immunoprecipitation
iTRAQ	isobaric tag for relative and absolute quantitation
JNK	Jun N-terminal kinase
MALDI-TOF MS	matrix – assisted laser desorption ionization – time of flight tandem mass spectrometry
MDA	malondialdehyde
ME	mercaptoethanol
NAD	nicotinamide adenine dinucleotide
NEM	N-ethylmaleimide
NF-κB	nuclear factor κB
PAGE	polyacrylamide gel electrophoresis
PBS	phosphate-buffered saline
PGC-1α	peroxisome proliferators – activated receptor-γ coactivator-1 α
PM	particulate matter
PMSF	phenylmethylsulfonyl fluoride

RIPA	Radioimmunoprecipitation assay
ROS	reactive oxygen species
SDS	sodium dodecyl sulfate
SIRT1	Sirtuin1
TBS	Tris-buffered saline
TCA	trichloroacetic acid
TEMED	$N,N,N'N'$-tetramethylethylenediamine
TSA	trichostatin A

1. INTRODUCTION

In Eukaryotes, the DNA is packed into chromatin by histones, which assembles the DNA into a tightly coiled structure. DNA carries out a series of processes such as replication, transcription, DNA repair as well as recombination which includes organization of chromatin through various posttranslational modifications (i.e., acetylation, deacetylation, methylation, phosphorylation, carbonylation, ubiquitination, SUMOylation, and poly-ADP-ribosylation) (Fukuda et al., 2006). Recent research has focused in understanding these processes with a view that these biochemical events can potentially be used as therapeutic targets involving chromatin modifications in various chronic diseases. For example, enzymes such as the histone acetyltransferases (HATs) and histone deacetylases (HDACs) which mediate many of the epigenetic processes have been the recent focus of novel drug-development strategies and potential targets of anticancer agents as well as other chronic diseases of heart, lung, liver, and brain, and aging processes (Roth et al., 2001). The analysis on the relationship of histone acetylation and gene expression showed that not only transcription but other DNA-mediated reactions are also regulated by histone acetylation (Gregory and Horz, 1998; Grunstein, 1997; Kuo and Allis, 1998). It is becoming clearer now that site-specific modifications and recognition of acetylated histone/DNA complexes are important in flow of genetic information via histone acetylation (Kouzarides, 1999; Wolffe, 1997; Workman and Kingston, 1998). Therefore, elucidation of the downstream effects of histone modification as well as the identification, isolation, and characterization of the relevant factors involved will aid in understanding the mechanisms of gene regulation by histone acetylases and deacetylases.

Acetylation of the amino-terminal (lysine) of core histone proteins has been linked to many cellular processes including cell-cycle progression, DNA replication, chromatin assembly, and the regulation of gene expression (Roth et al., 2001). Specific lysine residues in the N-terminal tails of the core histone can be posttranslationally modified by acetylation of the ε-amino group. The amino-terminal region of histones is a hotspot for posttranslational modifications that affect the interactions of histones with DNA, each other, and chromatin remodeling proteins (Fischle et al., 2003). The balance in histone acetylation/deacetylation is maintained by HATs and HDACs/SIRTUINS which affect the binding of DNA sequence-specific transcription factors and subsequently leading to recruitment of coactivators on specific regions in the genes which forms either the coactivator or corepressor complex (Wang et al., 2001). Several transcriptional regulators possess intrinsic HAT (mainly by CBP/p300) and HDAC activities, suggesting that histone acetylation and deacetylation play a causal role in regulating transcription (Gregory and Horz, 1998; Grunstein, 1997; Kouzarides, 1999; Kuo and Allis, 1998; Wolffe, 1997; Workman and Kingston, 1998). There are several evidences showing increased gene transcription that leads to increase in histone acetylation, and histone hypoacetylation which is further linked to decreased gene transcription (Kuo and Allis, 1998; Wolffe, 1997; Workman and Kingston, 1998). A variety of stimuli in air such as airborne particulate matter with a diameter of <10 μm diameter (PM_{10}) and cigarette smoke (CS), which are associated with chronic lung and cardiovascular diseases, increased histone acetylation by reactive oxygen species (ROS), and redox modifications of proteins involved in chromatin remodeling in lung epithelial cells, human monocytes/macrophages, and mouse lungs (Gilmour et al., 2003; Tomita et al., 2003; Yang et al., 2008, 2009). Furthermore, histone acetylation and deacetylation, and SIRT1 are linked to cell-cycle progression, DNA repair, and recombination events as well as inflammatory gene transcription which can be affected by redox signaling (Kouzarides, 1999; Taplick et al., 1998). Hence, analysis of histone acetylation and deacetylation status of a cell gives an indication on the activation status of the cell or a possible marker of cancer and oxidants-mediated chronic inflammatory diseases, such as cancer, chronic obstructive pulmonary disease (COPD), asthma, idiopathic fibrosis, and cardiovascular diseases.

Eighteen mammalian HDACs have been identified to date. These can be classified into one of four classes based upon their homology to a prototypical HDAC found in yeast. Class I HDACs (HDACs 1, 2, 3, and 8) are ubiquitously expressed ($\sim$50 kDa in size), with the possible exception of HDAC3 and HDAC10 are also predominantly nuclear in localization (Taplick et al., 2001). Class II HDACs (HDACs 4, 5, 6, 7, 9, and 10) are larger in size (120–150 kDa) and are expressed in a tissue-specific manner (Bertos et al., 2001; Grozinger et al., 1999). Class II HDACs exist in both the nucleus and cytoplasm, and the shuttling of class II HDACs out of the nucleus, which is regulated by nuclear export signaling 14-3-3 proteins, is a major mechanism by which their activity is regulated (Bertos et al., 2001). Class II HDACs can be further classified into class IIa (HDACs 4, 5, 7, and 9) and IIb (HDACs 6 and 10) based upon the existence of tandem deacetylase domains in HDACs 6 and 10 (Johnstone, 2002). Class I and II HDACs share significant homology at the deacetylase domain but differ in their N-terminal sequence.

HDAC11, which shares some but not sufficient homology to both class I and II HDACs, is assigned to its own class, class IV (Gao et al., 2002). Previous studies showed that the levels and activities of HDACs were decreased in response to oxidative stress (e.g., H_2O_2), PM_{10}, redox status, and CS exposure which is associated with severity of COPD, asthma, and steroids resistance in asthmatics who smoke tobacco (Adenuga et al., 2009; Ito et al., 2005; Moodie et al., 2004; Yang et al., 2006). However, overexpression of HDAC1 correlated with advanced stage of disease and adverse outcome in lung cancer patients (Osada et al., 2004; Sasaki et al., 2004). Further studies have been shown that a decrease in HDACs levels is due to its posttranslational modifications such as oxidation/carbonylation, nitrosylation, acetylation, and phosphorylation in response to oxidants (Adenuga et al., 2009; Ito et al., 2004; Moodie et al., 2004; Yang et al., 2006). Therefore, determination of HDAC levels/activities and its posttranslational modifications give an implication of chromatin status which drives the expression or repression of genes involved in chronic inflammatory diseases and cancer.

The class III HDAC, or Sir2 family named for their homology to the yeast Sir2 gene, is a highly conserved gene family which in humans presently comprises seven members, sirtuin (SIRT) 1–7 (Michan and Sinclair, 2007). Among these, SIRT1, 2, 3, and 5 have an NAD^+-dependent deacetylase domain (distinct from the zinc-dependent deacetylase domains of class I and II HDACs), and catalyze the deacetylation of a number of nonhistone proteins, such as p53, RelA/p65, Ku70, PGC-1α and FOXO3. In contrast, SIRT4 and SIRT6 have an NAD^+-dependent ADP ribosylation domain and catalyze protein ribosylation. Sirtuin family members have well-defined subcellular localizations, with SIRT1, SIRT6, and SIRT7 localized to the nucleus; SIRT2 to the cytoplasm; and SIRT3, SIRT4, and SIRT5 localized to the mitochondria. Class III HDACs share little homology to the first two classes, and are not inhibited by the widely used HDAC inhibitors butyrate, valproic acid, trichostatin A, or suberoylanilide hydroxamic acid (Xu et al., 2007), but are inhibited by nicotinamide, splitomicin, and sirtinol. Our recent studies showed that the levels/activities of SIRT1 were decreased in monocytes (MonoMac6 cells), lung epithelial cells (Beas-2B), and mouse lungs exposed to CS as well as in lungs of patients with COPD, which was associated with increased inflammatory response (Rajendrasozhan et al., 2008; Yang et al., 2007). A reduction in SIRT1 level attributed to the posttranslational modifications such as phosphorylation and nitrosylation, and oxidation/carbonylation by ROS (Rajendrasozhan et al., 2008). Using SIRT1 activators or caloric restriction may attenuate inflammatory response induced by oxidative stress in type 2 diabetes, and neurodegenerative disorders (Banks et al., 2008; Lavu et al., 2008; Yang et al., 2007) cardiovascular and chronic lung diseases. Hence, study in the level/activity and posttranslational modifications of SIRT1 is critical in understanding the mechanisms underlying chronic inflammatory and oxidative stress-mediated diseases.

2. Histone Acetylation Assays Using [³H]-Acetate Incorporation

The source of the acetyl group in histones acetylation is acetyl-CoA whereas in histone deacetylation the acetyl group is transferred to coenzyme A (CoA). Acetyl-CoA cannot enter into the living cells, and radioactive acetate should be supplied to cells, and allow cells convert it into radioactive acetyl-CoA. Therefore, the radioactivity of reactive products of [³H]-labeled histones reflects the degree of histones being acetylated which can be determined by liquid scintillation as described previously (Ito *et al.*, 2000a,b; Kagoshima *et al.*, 2001) with some modifications. Direct hydrolysis of the acetyl group present in the acetylated protein substrate is catalyzed by class I and II HDACs via activating a water molecule to form deacetylated protein and free acetate products (Eq. (13.1)):

$$\text{Acetyl-protein substrate} \xrightarrow{\text{HDAC (Class I and II)}} \text{Deacetylated protein} + \text{Acetate} \tag{13.1}$$

2.1. [³H]-Acetate incorporation in cell culture

1. Fetal bovine serum (FBS)-free Dulbecco's modified Eagle's medium (DMEM) will be added to 50% subconfluent (exponentially growing) cells (e.g., A549, HUVECs, H292, Beas-2B, MonoMac6, THP-1, and HepG2 cells) in a 6-well plate ($\sim 0.25 \times 10^6$ cells/well) at 72 h before the experiment.
2. Synchronization of the cells is an important step which requires 60–72 h. Before the start of experiments, simple DNA content should be checked by FACS using propidium iodide staining to synchronize cells.
3. The cells are incubated with 5 μCi/ml of [³H]-acetic acid (Amersham Pharmacia Biotech, Piscataway, NJ) in FBS-free medium for 10 min at 37 °C before treatments (e.g., CS extract, H_2O_2 redox recycling agents, and aldehydes).

2.2. Histone protein extraction from cells and tissues

1. The cells are washed twice with ice-cold PBS, and 500 μl of ice-cold lysis buffer (10 mM Tris–HCl [pH 6.5], 50 mM sodium bisulfite, 10 mM MgCl$_2$, 8.6% sucrose, 2% Triton X-100) is added into each well to scrape the cells off quickly with a rubber policeman at 4 °C after abovementioned treatments.
2. Cells are collected into 1.5-ml Eppendorf tube, and kept on ice for 10 min. The cells are washed three times with 200 μl of ice-cold lysis buffer and the

pellets are washed once with nuclei wash buffer (10 mM Tris–HCl [pH 7.4], 13 mM EDTA, 10,000 × g, 4 °C, 5 min), if the nuclear pellet is difficult to resuspend then the pellets are sonicated for 5 s on ice and then resuspended in 150 μl of ice-cold distilled water and vortex mixed.
3. Six microliters of 5 N HCl and 3 μl of 18 N H$_2$SO$_4$ solutions are added into nuclear suspension to reach final concentration at 0.2 and 0.36 N, respectively, for histone protein extraction (Ito et al., 2000a,b; Yang et al., 2009).
4. The mixture is rotated on a rocker at 4 °C for 6–18 h. If necessary, the pellet is sonicated for 2 s on ice and centrifuged at 14,000×g for 10 min at 4 °C to remove acid-insoluble material. A 140 μl of supernatant is transferred into a new tube and 1.1 ml of ice-cold acetone is added to precipitate the histones.
5. The suspension is kept at -20 °C for overnight, centrifuged at 14,000×g for 10 min at 4 °C, and washed with cold acetone by centrifugation to remove acid.
6. The pellets are dried under vacuum at room temperature, redissolve in 50 μl of distilled water or HDAC buffer C containing 15 mM Tris–HCl (pH 7.9), 10 mM NaCl, 0.25 mM EDTA, 10 mM β-mercaptoethanol, 10% glycerol, and complete protease inhibitors (Cat. No. P8340; Sigma, St. Louis, MO), and finally assess the protein concentration by bicinchoninic acid (BCA), Bradford, or Lowry protein assays.

2.3. Histone acetylation assay

1. Fifty micrograms of acid-extracted protein is mixed with 4× SDS–PAGE sample buffer (2.5 ml of 1 M Tris–HCl [pH 6.8], 2 ml of glycerol, 2 ml of 10% [w/v] SDS solution, 20 mg of bromophenol blue, and 3.5 ml of distilled water). Add 20 μl of β-mercaptoethanol to 1 ml of 4× sample buffer just before use for SDS–PAGE (Ito et al., 2000b). If the blue SDS sample buffer turns brown or yellow, then the sample is still acidic.
2. Add about 2–3 μl of 1 M Tris base until the sample turns blue and boil for 4–5 min, and then centrifuge at 14,000×g for 20 s.
3. SDS–PAGE is used to separate [^{3}H]-labeled and -acetylated histones. Separating gel mix for one mini gel consists of 2.0 ml of 40% acrylamide/bisacrylamide solution, 1.88 ml of 1 M Tris–HCl (pH 8.8), 50 μl of 10% (w/v) SDS, and 2 ml of distilled water. Add 5 μl of N,N,N',N'-tetramethylethylenediamine (TEMED) and 50 μl of 10% (w/v) ammonium persulfate (APS) just prior to pouring the gel. A 50% Stacking gel mix for one mini gel is 0.32 ml of 40% acrylamide/bisacrylamide solution, 0.32 ml of 1 M Tris–HCl (pH 6.8), 50 μl of 10% (w/v) SDS solution, and 1.82 ml of distilled water. Add 2.5 μl of TEMED and 25 μl of 10% (w/v) APS solution just prior to pouring the gel. Prepare a stock of 10× SDS–PAGE running buffer (30.3 g Tris-base, 144 g glycine, 10 g SDS in 1 l of distilled water).

4. Samples are loaded into appropriate wells on the 16% SDS-gel and run at 40 mA/gel for 30 min then stained for 20 min with Coomassie brilliant blue staining solution (0.2% (w/v) Coomassie brilliant blue in 50% (v/v) methanol and 10% (v/v) acetic acid, store at room temperature).
5. Destain the gel in destaining solution [33% (v/v) methanol and 10% (v/v) acetic acid can be prepared and stored at room temperature] for over 6 h and cut out the band of interest, that is, core histones (e.g., histones H3 or H4) using a scalpel blade and put into 3 ml scintillation, mix by vortex and leave at room temperature for more than 30 min, and determine the radioactivity in excised core histone bands by liquid scintillation counting (Ito et al., 2000a,b).
6. The radioactivity of acetylated histones is counted by liquid scintillation counter, and expressed as disintegrations per minute (dpm) per μg protein.
7. For visualization of histone acetylation, immerse and soak the gel in [^{3}H]-amplifier solution for 15–30 min, stain with Coomassie brilliant blue solution and destain, then dry the gel onto Whatmann 3MM filter paper using a gel drier.
8. Transfer the dry gel to an X-ray film cassette and allow to autoradiograph at $-80\,^\circ$C for an appropriate time (6–24 h).

3. Histone Acetylation by Immunoblotting

With the development of antibodies for specific lysine residues in histones such as H3K9, H3K14, H3K18, H3K27, H4K5, H4K8, H4K12, and H4K16, specific histone acetylation assay is now possible by Western blotting (Yang et al., 2008, 2009). Histone protein in cells or tissues is extracted as mentioned in histone extraction procedure. In tissues, cell pellets are likewise prepared to extract nuclear proteins. Pellets are washed once with nuclei wash buffer (10 mM Tris–HCl [pH 7.4], 13 mM EDTA, 10,000×g, 4 $^\circ$C, 5 min). The pellets are resuspended with 150 μl of ice-cold distilled water for further histone extraction as per the histone extraction procedure.

1. Acid-extracted histone proteins are subjected to 15% SDS–PAGE, and separated proteins are electroblotted onto nitrocellulose membranes (Amersham, Arlington Heights, IL), and blocked for 1 h at room temperature with 5% nonfat dry milk.
2. The membranes are then probed with specific anti-acetylated or antimethylated, anti-histone H3 or anti-histone H4 antibodies (Upstate, Charlottesville, VA; Cell Signaling Technology, Inc., Danvers, MA, or Abcam, Cambridge, MA) at room temperature for 1 h.
3. After three washing steps, the levels of proteins are detected using anti-rabbit, anti-mouse, or anti-goat secondary antibody linked to horseradish peroxidase (DAKO), and bound complexes are detected using the enhanced chemiluminescence (ECL) method (Perkin Elmer, Waltham, MA).

4. HAT Activity Assay

Over 30 HATs including transcription factors, coactivators, and other signaling molecules are discovered to date, which display distinct substrate specificities for histone and nonhistone proteins (Spange *et al.*, 2009). CBP/p300 is the most extensively studied among the HATs, and it is vital for the coactivation of several transcription factors, including NF-κB and AP-1. CBP and p300 (referred to CBP/p300 because of their mutual interaction) are transcriptional coactivators with intrinsic HAT activity, and are regulated by MAP kinase (Rahman and Adcock, 2006). Specific core histone lysine residues can be acetylated by CBP/p300 coactivator. Both p300 and CBP are also known to involve in the regulation of various DNA-binding transcriptional factors. For example, lysine acetylation of histones by CBP/p300-HAT causes DNA uncoiling, and allows accessibility of NF-κB (RelA/p65) to bind the promoters of genes (Chen *et al.*, 2001). Thus, histone acetylation via CBP/p300 has a significant role in the activation of NF-κB-mediated proinflammatory gene expression. Recently, Yang *et al.* (2008) showed that IKKα mediates chromatin remodeling (by increasing intrinsic HAT activity) via the activation of NF-κB-inducing kinase (NIK) in response to CS-mediated oxidative stress in human lung epithelial cells, macrophages, and mouse lungs.

4.1. HAT assay using commercial kits

HAT activity assay can be performed using commercially available kits of different companies (Cat. No. K322-100, BioVision Research Products, Mountain View, CA; Cat. No. 56100, Active Motif, Carlsbad, CA; Cat. No. KT-146, Kamiya Biomedical Company, Seattle, WA; Cat. No. PK-CA577-K332-100; PromoCell GmbH, Heidelberg, Germany).

4.2. HAT activity assay using kits

An easy and nonradioactive system for rapid and sensitive detection of HAT activity using cell and tissue extracts have been developed as commercial available kits (see under HAT assay kits—BioVision or Active Motif). The kit utilizes active nuclear extracts as a positive control and acetyl-CoA as a cofactor. Acetylation of peptide substrate by active HAT releases the free form of CoA which then serves as an essential coenzyme for producing NADH. This NADH can be easily detected spectrophotometrically upon reacting with a soluble tetrazolium dye.

1. Briefly, the procedure includes preparation of nuclear extracts of the test sample (consisting of 50 μg protein) in 40 μl water (final volume) for each assay from a cell culture 96-well plate.
2. For positive control, 10 μl of nuclear extracts are added with 30 μl of water. Assay mix should be prepared using the HAT assay buffer, substrates, and enzyme provided along with the kit to a total volume of 65 μl/well.
3. Assay cocktail (65 μl) is mixed with each test sample in wells and allowed to incubation at 37 °C for 1–4 h depending on the color development.
4. Finally, the samples should be read using a microplate reader at 440 nm according to the manufacturer's instructions provided in the kit.

5. HDAC Activity Assay Using [^{3}H]-Labeled Histones

HDAC can deacetylate acetyl-lysine present in histones thereby generates free acetic acid. [^{3}H]-labeled acetylated histones are used as substrate to assess the ability of samples (HDAC activity) to remove acetyl moieties from acetylated histones and form [^{3}H]-acetic acid in this approach. Therefore, radioactivity of [^{3}H]-acetic acid in the process of histone deacetylation reflects the activity of HDAC in samples which is described previously (Ito et al., 2000a,b, 2001; Kolle et al., 1998) with some modifications.

5.1. Preparation of [^{3}H]-labeled histones

Eighty percentage of subconfluent epithelial or endothelial cells in 150 cm^2 flask are cultured in DMEM without FBS for 24 h before incubation with 100 μCi/ml of [^{3}H]-acetic acid (Amersham Pharmacia Biotech) in FBS-free medium for 10 min (Ito et al., 2001). TSA (100 ng/ml) can be added, and incubated at 37 °C for 6–8 h as positive control. Extraction of histones is performed as mentioned earlier and diluted with distilled water or HDAC buffer C to give a concentration of 1.5 μg/μl protein.

5.2. HDAC assay sample preparation

1. After treatment of cells (e.g., MonoMac6, A549, H292, and Beas-2B cells) with oxidative stress imposed by H_2O_2, aldehydes, or CS extracts, cells are washed twice with ice-cold PBS, and collected by scraping into 1 ml ice-cold HBSS into Eppendorf tube.
2. The cells are spun at $10,000 \times g$ for 5 min at 4 °C and extract nuclei by cell lysis buffer (10 mM Tris–HCl [pH 6.5], 50 mM sodium bisulfite, 10 mM $MgCl_2$, 8.6% sucrose, 2% Triton X-100).

3. Twenty microliters of HDAC buffer A (15 mM Tris–HCl [pH 7.9], 450 mM NaCl, 0.25 mM EDTA, 10 mM β-mercaptoethanol, 10% glycerol, and complete protease inhibitor cocktail tablet without EDTA is added just before use) is added, vortexed for 15 s vigorously and kept on ice for 20 min, and spun at 10,000 × g for 5 min at 4 °C.
4. Finally, the supernatant is collected and 180 μl of HDAC buffer B (15 mM Tris–HCl [pH 7.9], 0.25 mM EDTA, 10 mM β-mercaptoethanol, 10% glycerol, and complete protease inhibitor cocktail tablet without EDTA is added just before use) is added.
5. Salt concentrations above 200 mM increasingly inhibit HDAC activity. Therefore, dialysis may be necessary. Final salt concentration in assay is recommended to 10–45 mM. In this protocol, final 30 mM of salt is used. This is a crude HDAC extraction. For partial purification of deacetylases, the deacetylase is precipitated by raising the concentration of $(NH_4)_2SO_4$ to 3.5 M.
6. The dialysate is then loaded onto a DEAE-cellulose (Whatman DE52) column equilibrated with the HDAC buffer C (15 mM Tris–HCl [pH 7.9], 10 mM NaCl, 0.25 mM EDTA, 10 mM β-mercaptoethanol, 10% glycerol, and complete protease inhibitor cocktail mixture without EDTA is added just before use) and eluted with a linear gradient (0–0.6 M) of NaCl. A single peak of HDAC activity is eluted with 0.15 and 0.2 M NaCl.

Extraction of HDAC samples from tissues, cell pellets from nuclear fraction, which is described earlier are dissolved in 20 μl of HDAC buffer A, vortexed for 15 s vigorously, and kept on ice for 20 min, finally supernatant is collected and 180 μl of HDAC buffer B is added. This is a crude HDAC extraction. Partial purification of deacetylase in tissues can be followed by the same procedure as described earlier.

5.3. HDAC activity assay

1. Fifteen microliters of [^{3}H]-labeled histone (1.5 μg/μl) is mixed with 25 μl of crude HDAC extraction, and incubated at 30 °C for 30 min (Ito *et al.*, 2000b).
2. Ten microliters of acid mixture (2.5 N HCl and 1 N acetic acid) is added to stop the reaction. Then add 900 μl of ethylacetate into reaction mixture, vortex vigorously, keep for 5 min, and mix again.
3. Centrifuge at 13,000×g for 5 min at 4 °C and aliquot 600 μl of the upper organic phase and transfer into 3 ml of liquid scintillation cocktail.
4. Determine the radioactivity of sample by liquid scintillation counter. Count the release of [^{3}H]-acetic acid from acetylated histones by HDAC samples with liquid scintillation counter (Ito *et al.*, 2000a,b, 2001; Kolle *et al.*, 1998), and express the unit as dpm/μg protein.

Notes

1. For macrophage/monocyte cell lines, such as MonoMac6, THP-1, U937, mononuclear cells, endothelial cells HUVECs and bronchial epithelial cells (e.g., NIH-H292 and Beas-2B cells or other epithelial cells), 0.5–1% Triton X-100 is needed. These concentrations are relatively high so it should not be left for more than 10 min. Nonidet P-40 (0.2–0.5%) can be used instead of Triton X-100.
2. Complete protease inhibitor cocktail (Boehringer Mannheim) can be used for the preparation of lysis buffer, the tablet with EDTA is used. For other buffers used in this protocol containing EDTA, the tablet without EDTA is used.
3. Trichostatin A (Sigma) stock is prepared at a concentration of 100 μg/ml in 100% EtOH and stored at $-20\,°C$ and added to lysis buffer just before use.
4. These reagents (HDAC buffer A, B, and C) can be stored at 4 °C for up to 1 month.

6. HDAC Activity Assay

Among various HDACs, HDAC2 is an important corepressor protein which is redox-sensitive and prone to alterations by oxidants and/or free radicals, which is associated with the reduced levels/activity of these deacetylases in lungs of smokers and patients with COPD compared to healthy nonsmokers (Barnes, 2009; Marwick et al., 2004; Yang et al., 2006). The decreased levels/activity of HDAC2 is also observed in response to CS extract *in vitro* exposure to macrophage–monocytes (MonoMac6 cells), human airway and bronchial epithelial cells, and in lungs of mice exposed to CS showing the modifications by aldehydes, and by protein nitration (Adenuga et al., 2009; Yang et al., 2006). Oxidative stress plays an important role in reducing the activity of HDAC2 by oxidative/carbonyl posttranslational modifications and kinase-dependent signaling mechanisms. However, the mechanism underlying reduced level/activity of HDAC2 by oxidative modification is not clearly understood.

6.1. HDAC assay using commercial kits

HDAC activity assay can be performed using commercially available kits which are available in different companies (Cat. No. K332-100, BioVision Research Products; AK-501, Biomol International, Plymouth Meeting, PA; Cat No. 56200, Cat. No. 56210, Active Motif) and SIRT1 and SIRT2 fluorimetric drug discovery kit (AK-555, AK-556; Biomol International).

6.2. HDAC immunoprecipitation and specific activity assay for various deacetylases using commercial kits

An efficient and simple method to perform HDAC activity assay is by using a colorimetric assay kit (Yang et al., 2006). The procedure involves the use of the HDAC colorimetric substrate (Color *de* Lys substrate, 500 μM), which comprises an acetylated lysine side chain and is incubated with a sample containing nuclear extract. Deacetylation sensitizes the substrate, and treatment with the lysine developer produces a chromophore, which can be analyzed with a colorimetric plate reader at 405 nm. HeLa or A549 cell nuclear extract can be used as a positive control. A standard curve is prepared, using the known amount of the deacetylated standard (Boc-Lys-pNA) included in the Biomol kit.

6.2.1. HDAC activity assay

1. The different fractions isolated from cells or tissues are immunoprecipitated for HDAC1, HDAC2, and HDAC3. HDAC1, HDAC2, or HDAC3 antibody (Santa Cruz Biotechnology, Santa Cruz, CA; 1:1000 dilution) is added to 200 μg of protein sample in a final volume of 500 μl with RIPA buffer [50 mM Tris–HCl, 150 mM NaCl, 1 mM EDTA, 0.25% deoxycholate, 1 mM Na$_3$VO$_4$ or sodium orthovandate, 1 mM NaF, 1 μg of leupeptin/ml, 1 μg of aprotinin/ml, 1 mM phenylmethylsulfonyl fluoride (PMSF) or containing complete protease inhibitor cocktail (Sigma)] and incubated for overnight.
2. Protein A/G agarose beads (40 μl; Santa Cruz Biotechnology) are added to each sample and kept for 1–2 h at 4 °C on a rotor. The samples are then centrifuged at 2500 rpm at 4 °C for 5 min.
3. The supernatant is discarded, and the beads are washed twice using 500 μl of RIPA buffer and then resuspended in 150 μl of 1 mM Color *de* Lys substrate (Adenuga et al., 2009). The assay is allowed to continue at 37 °C for 90 min with continuous mixing.
4. Then the agarose beads are pelleted by centrifugation, 30 μl of supernatant is transferred to a 96-well plate and made up to 50 μl with HDAC-specific buffer. Finally, 50 μl Color *de* Lys developer containing 2 μM TSA is added and the plate is incubated for 20 min at 37 °C. Color development is monitored spectrophotometrically at 405 nm.

6.2.2. SIRT1 activity assay

1. SIRT1 is immunoprecipitated as described earlier from whole cell or tissue extracts and 100 μg of protein is used for SIRT1 activity assay.

2. The assay involved the use of the SIRT1 fluorometric substrate (Fluor de Lys-SIRT1 substrate-developer combination, 0.2 U/μl), which comprised amino acids 379–382 of human p53 (Arg-His-Lys-Lys AC), and incubation with human recombinant SIRT1 and the cosubstrate NAD^+ (Biomol). SIRT1 activity assay kit employs an indirect method using fluorescent substrate (acetylated p53) (Pacholec et al., 2010).
3. Deacetylation of SIRT1 sensitizes the substrate, and treatment with the developer produces a fluorophore, which can be analyzed with a fluorescence plate reader at 460 nm.
4. Resveratrol (200 μM) may be used as a positive control, and 1 mM nicotinamide and 200 μM suramin, 10 μM sirtinol or 150–600 μM splitomicin served as negative controls. Resveratrol is an indirect activator of SIRT1 (Pacholec et al., 2010).
5. A standard calibration curve is prepared with a known amount of the deacetylated standard. SIRT1 activity is assayed using SIRT1 activity assay kit according to the manufacturer's instructions.

7. HDACs Levels by Immunoblotting

In addition to HDAC activity, the levels of HDACs are also regulated by redox pathway via modulating their nucleocytoplasmic shuttling and posttranslational modifications, such as proteasome-dependent degradation (Adenuga et al., 2009; Rahman et al., 2002). Therefore, protein level of HDACs is an important marker of oxidant-induced cell dysfunction or diseases such as COPD (Ito et al., 2005).

1. In general, different cells fractions, such as whole lysates, and both cytoplasmic and nuclear protein are used to investigate the degradation and shuttling of HDACs (Adenuga et al., 2009; Yang et al., 2006).
2. Protein estimation is done by the BCA, Bradford, or Lowry protein assay, following the manufacturer's instructions.
3. For HDAC1, HDAC2, HDAC3, SIRT1, and SIRT2 assays, 20 μg of isolated soluble proteins are electrophoresed on 7.5% SDS–PAGE gels, transferred onto nitrocellulose membranes (Amersham), and blocked with 10% nonfat dry milk in Tris-buffered saline (TBS) with 0.1% Tween-20 at 4 °C overnight.
4. Membranes are incubated with goat polyclonal anti-human anti-HDAC or anti-SIRT1 (1:1000 dilution in 5% nonfat dry milk in TBS) antibodies (HDAC1, SC-6298; HDAC2, SC-6296; HDAC3, SC-8138 from Santa Cruz Biotechnology; SIRT1, Ab7343; SIRT2, Ab10659 from Abcam).
5. After washing, the levels of HDAC proteins are detected with rabbit anti-goat or anti-mouse antibody (1:20,000 dilution in 2.5% nonfat dry

milk in TBS for 1 h) linked to horseradish peroxidase (Dako, Santa Barbara, CA), and bound complexes are detected with ECL (Jackson Immunology Research, West Grove, PA).

8. Posttranslational Modifications of HDACs and SIRTs (Sirtuins 1–7) by Immunoprecipitation

Immunoprecipitation is used to assess the posttranslational modifications, such as phosphorylation, nitrosylation, and SUMOylation since the specific phosphorylated, carbonylated, or nitrated HDAC antibody is not available (Adenuga et al., 2009; Yang et al., 2006).

1. After extraction of cytoplasmic, nuclear proteins, or whole lysates from cells or tissues, HDAC1, HDAC2, HDAC3, or SIRT1 antibodies (1:80 dilution) are added to 100–200 μg of protein in a final volume of 400 μl RIPA buffer and incubated for 1 h. Protein-A/G agarose beads (20 μl) (Santa Cruz Biotechnology) are added to each sample and left overnight at 4 °C on a rotator.
2. The samples are then centrifuged at 13,000 rpm at 4 °C for 5 min. The supernatant is discarded, and the beads are washed three times and then resuspended in 40 μl of RIPA buffer.
3. For Western blots, 100 μg of the immunoprecipitated HDAC1, HDAC2, HDAC3, or SIRT1 agarose bead suspension is added to 10 μl of 4× sample buffer [2.5 ml of 1 M Tris–HCl (pH 6.8), 2 ml of glycerol, 2 ml of 10% (w/v) SDS solution, 20 mg of bromophenol blue, and 3.5 ml of distilled water].
4. Add 20 μl of β-mercaptoethanol to 1 ml of 4× sample buffer just before use, boiled at 95 °C for 5 min, and resolved by SDS–PAGE as described earlier.
5. To determine the posttranslational modifications of HDAC1, HDAC2, HDAC3, and SIRT1, blots are probed with mouse monoclonal anti-nitrotyrosine (Cat. No. 05-223; Upstate), anti-4-hydroxy-2-nonenal (4-HNE) (Cat. No. 24327; Oxis International), anti-phosphoserine (Santa Cruz Biotechnology), anti-acrolein, anti-ubiquitin (Santa Cruz Biotechnology), or anti-SUMO1 (Zymed, Carlsbad, CA or Santa Cruz Biotechnology) antibodies.
6. Recently, antibodies for phospho-SIRT1 at Ser27 and Ser47 residue (Cat. No. 2327 and 2314; Cell Signaling Technology, Inc) and phospho-HDAC2 at Ser394 (Cat. No. sc-135639; Santa Cruz Biotechnology) are commercially available which can be used to determine the phosphorylation status of SIRT1 and HDAC2 on specific residues by immunoblotting.

9. Preparation of Whole Cell Lysate

9.1. From cells

1. Harvested cells are washed in 1 ml sterile 1× PBS buffer and then resuspended in 100 μl of ice-cold RIPA buffer containing 50 mM Tris–HCl, 150 mM NaCl, 1 mM EDTA, 0.25% deoxycholate, 1 mM Na_3VO_4 or sodium orthovandate, 1 mM NaF, 1 μg of leupeptin/ml, 1 μg of aprotinin/ml, 1 mM PMSF or containing complete protease inhibitor cocktail (Sigma) and incubated on ice for 30 min, vortexed for 15 s.
2. Finally, centrifuged at 10,000×g for 5 min, the supernatant is whole cell lysate transferred to fresh tube and kept frozen at −80 °C.

9.2. From tissues

1. For tissues, 50 mg of tissue is homogenized (Pro 200 homogenizer, at maximum speed, 5th gear for 40 s) in 0.5 ml of ice-cold RIPA buffer containing complete protease inhibitor cocktail (Sigma) (Yao et al., 2008).
2. The tissue homogenate is then incubated on ice for 45 min to allow total cell lysis. The homogenate is then centrifuged at 13,000×g for 5 min at 4 °C to separate the protein fraction from the cell/tissue debris.
3. The supernatant, containing the protein is aliquoted and stored at −80 °C. The level of protein is determined by BCA, Bradford, or Lowry protein assay.

10. Preparation of Cytoplasmic and Nuclear Proteins

10.1. From cells

1. For nuclear proteins, cells are washed twice with ice-cold PBS and resuspended in 400 μl of Buffer "A" (10 mM HEPES [pH 7.9], 10 mM KCl, 0.1 mM EDTA, 0.1 mM EGTA, 1 mM DTT, and 0.5 mM PMSF). After 15 min, Nonidet P-40 is added to a final concentration of 0.6% and vortexed for 15 s.
2. Samples are centrifuged for collection of the supernatants containing cytosolic proteins. The pelleted nuclei are resuspended in 50 μl of Buffer "B" (20 mM HEPES [pH 7.9], 0.4 M NaCl, 1 mM EDTA, 1 mM EGTA, 1 mM DTT, and 1 mM PMSF). After 30 min at 4 °C, lysates are centrifuged, and supernatants containing the nuclear proteins are stored at −80 °C. The pellet after nuclear extraction from cells can be used for histone protein extraction as described earlier.

10.2. From tissues

1. One hundred milligrams of tissue is mechanically homogenized (Pro 200 homogenizer, maximum speed, 5th gear for 40 s) in 0.5 ml Buffer "A" (10 mM HEPES [pH 7.8], 10 mM KCl, 2 mM MgCl$_2$, 1 mM DTT, 0.1 M EDTA, 0.2 mM NaF, 0.2 mM Na orthovandate, 0.4 mM PMSF, and 1 µg/ml leupeptin), then 1% (v/v) Nonidet P-40 is added, and the tubes are incubated in ice for 15 min (Yao et al., 2008).
2. The homogenate is centrifuged at 2000×g in a benchtop centrifuge for 30 s at 4 °C to remove cellular debris. The supernatant is then transferred to a 1.7-ml ice-cold microtube and further centrifuged for 30 s at 13,000×g at 4 °C.
3. The supernatant is collected as a cytoplasmic extract. The pellet is resuspended in 100 µl of Buffer "B" (50 mM HEPES [pH 7.8], 50 mM KCl, 300 mM NaCl, 0.1 M EDTA, 1 mM DTT, 10% [v/v] glycerol, 0.2 mM NaF, 0.2 mM Na-orthovandate, and 0.6 mM PMSF) and placed on the rotator in the cold room for 30 min.
4. After centrifugation at 13,000×g in an Eppendorf tube for 5 min, the supernatant is collected as the nuclear extract and kept frozen at −80 °C. The pellet after nuclear extraction from tissues can be used for histone protein extraction as described earlier. BCA, Bradford, or Lowry protein assay is used for determination of protein concentration.

11. Redox-Mediated Posttranslational Modification Assays

11.1. Protein carbonylation

Protein carbonylation occurs via the introduction of carbonyl groups into proteins either directly by oxidation of amino acid side chains, particularly lysine, arginine, and histidine or by oxidation of peptide backbone leading to backbone cleavage (Ghezzi and Bonetto, 2003). Additionally, proteins can also react with glycation and lipid peroxidation products such as 4-HNE or malondialdehyde (MDA) (Stadtman and Berlett, 1997). Mostly, carbonylated proteins are analyzed by one- or two-dimensional electrophoresis and subsequent immunoblotting. All protein-carbonyls can be targeted by reaction with hydrazides coupled to fluorescent dyes, radioactive labels, or using the far more frequently used method by reacting with 2,4-dinitrophenylhydrazine (DNPH; Ahn et al., 1987; Levine et al., 1990). Dinitrophenyl-(DNP)-adducts can be detected using specific anti-DNP antibodies that may be used for immunoprecipitation of carbonylated proteins. Furthermore, a variety of specific antibodies (anti-MDA and anti-HNE) against adducts of lipid peroxidation products are available for immunoblotting (Toyokuni *et al.*, 1995; Yamada *et al.*, 2001, 2004).

11.1.1. Covalent modification of SIRT1 by carbonylation

1. SIRT1 is immunoprecipitated using whole cell lysates; SIRT1 antibody (1:80 dilution; Abcam) is added to 100 μg of protein in a final volume of 400 μl and incubated for 1 h at 4 °C.
2. Protein-A/G agarose beads (20 μl) (Santa Cruz Biotechnology) are added to each sample and left overnight at 4 °C on a rotator (Barnstead Thermolyne, Dubuque, IA).
3. The samples are then centrifuged at 13,000$\times g$ at 4 °C for 5 min. The supernatant was discarded and the beads are washed three times and then resuspended in 50 μl lysis buffer.
4. For immunoblots, 100 μg of the immunoprecipitated SIRT1 agarose bead suspension is added to 5$\times$ sample loading buffer, boiled and resolved using SDS–PAGE. Agarose beads alone were used as negative control.
5. To determine the carbonylation of SIRT1, membranes were probed first with anti-SIRT1 antibody (Abcam). After stripping, membranes were equilibrated with 20% (v/v) methanol, 80% TBS for 5 min, then incubated with 0.5 mM DNPH for 30 min at room temperature (Murtaza *et al.*, 2008).
6. The membranes were washed and then incubated overnight in anti-DNP antibody, as described earlier.

11.1.2. Protein carbonylation using proteomics approach

Protein carbonylation occurs in response to oxidative/carbonyl stress by CS and environmental pollution which may serve as a potential biomarker based on the relatively early formation and relative stability of carbonylated proteins (Dalle-Donne *et al.*, 2003; Rahman and Biswas, 2004; Reznick *et al.*, 1992). In order to assess the relative contribution of carbonylated protein to total protein, two-dimensional PAGE can be performed according to a method described previously (Lin *et al.*, 2002; Umstead *et al.*, 2009).

1. Briefly, trichloroacetic acid (TCA)–acetone precipitated samples will be loaded onto Immobiline Drystrip IPG gel (Amersham Biosciences), and proteins were separated using the Amersham Biosciences IPGpho Isoelectric Focusing System.
2. Strips will be removed and treated with 10 mM of DNPH–5% TCA, and then transferred to 12.5% horizontal SDS–PAGE gels for electrophoresis.
3. After transfer, the PVDF membrane will be stained with deep purple total protein stain (GE Healthcare).

4. Carbonylated proteins will be detected using the same membrane after imaging for total protein.
5. The membranes will be washed and incubated with 2 μg/ml Biotin-XX-rabbit anti-dinitrophenyl-KLH IgG (Invitrogen) followed by incubation with 1 μg/ml Cy5-labeled streptavidin (GE Healthcare).
6. Deep purple total protein-stained images and Cy5-stained oxidized protein images will be then imported into Progenesis Same Spots (v2.0; Nonlinear USA, Durham, NC) for quality assessment, alignment, spot matching, and analysis.
7. Protein spots of interest will be excised, digested, and processed to identify the specific/novel protein carbonylation through peptide mass fingerprinting and sequence analyses using matrix-assisted laser desorption ionization-time of flight tandem mass spectrometry (MALDI-TOF MS) (Caito et al., 2010, England and Cotter, 2004; Umstead et al., 2009).

11.2. Biotin-switch assay

Protein S-nitrosylation is a dynamic process where the covalent addition of a nitroso group to a cysteine thiol side chain occurs in a wide variety of proteins (Foster et al., 2003; Hess et al., 2005). Protein S-nitrosylation is linked to various cellular redox signaling functions in normal physiology and during pathophysiology of disease. Due to the increase in research interest on protein S-nitrosylation, most of the recent studies are relying on biotin-switch assay for the detection of endogenously S-nitrosylated proteins (protein–SNOs). Biotin-switch assay involving the study of S-nitrosylation has gained considerable interests both *in vitro* and *in vivo*, largely due to its compatibility and availability of molecular methods including SDS–PAGE, immunodetection, and mass spectrometry (Jaffrey et al., 2001).

11.2.1. Biotin-switch assay for HDACs and SIRT1

1. Modifications on cysteine residues were measured by maleimide–PEO_2–biotin labeling. After immunoprecipitation of HDACs or SIRT1, conjugated to the agarose beads were diluted in PBS and 1 μl of maleimide–PEO_2–biotin (Pierce) was added for 2 h rocking at room temperature in dark (Caito et al., 2010).
2. Samples were then boiled and electrophoresed on a polyacrylamide gel. The gel was transferred onto the nitrocellulose membrane, blocked with 5% BSA, streptavidin conjugated to horseradish peroxidase was added for 1 h, and visualized using the ECL method (Perkin Elmer).

11.2.2. Cysteine group modifications on deacetylases

To determine the aldehyde modifications on cysteines, biotin-switch labeling is performed as described above.

1. Immunoprecipitated HDAC2 or SIRT1 is labeled with maleimide–PEO_2–biotin, which reacts with free cysteine groups. After electrophoresis and transfer, the nitrocellulose membranes will be developed using streptavidin-conjugated HRP.
2. Cells and recombinant SIRT1 or HDAC2 (20 μg, Biomol) are treated with N-ethylmaleimide (NEM; 100 μM) for 30 min, and then alkylation of cysteine groups on SIRT1 is determined after labeling with maleimide–PEO_2–biotin. Alkylation of SIRT1 by NEM will lead to lower enzymatic activity by alkylation of cysteines (Caito et al., 2010).

To determine the sites of carbonylation on HDAC2 or SIRT1, cells are treated with CSE, oxidants and nitrosants, cell/tissue lysate is labeled with DNPH, separated on 2DGE, immunoblotted against anti-acrolein and anti-4-HNE antibodies and then analyzed by MALDI-TOF/TOF mass spectrometry (Autoflex III TOF/TOF) MALDI mass spectrometer (Bruker Daltonics). Data are analyzed using MASCOT version 2.1.04. Recombinant human HDAC2/SIRT1 (20 μg) modified with acrolein (30 μM) is used as a positive control (Caito et al., 2010; Grimsrud et al., 2008; Ishii et al., 2003).

To further determine the oxidation of cysteine residues to sulfenic acid (–SOH), sulfinic acid (–SO_2H), or sulfonic acid (–SO_3H), cells are treated with CSE, aldehydes, and oxidants, and whole cell extracts and lung homogenate will be made for HDAC2/SIRT1 immunoprecipitation, and HDAC2/SIRT1 activity is determined. A 100 μg of immunoprecipitated HDAC2 or SIRT1 is used for electrospray ionization-mass spectrometry (ESI-MS) as previously described (Woo et al., 2003), allowing comparison of the mass/charge ratio (m/z) of unmodified cysteines on HDAC2 or SIRT1 to their respective sulfinic and sulfonic acids (Eaton, 2006; Hurd et al., 2009). Recently developed methods, such as isotope-coded affinity tags (ICAT) and isobaric tag for relative and absolute quantitation (iTRAQ) can also be used for cysteine group modifications on deacetylases (Yang et al., 2008; Sethuraman et al., 2004)

12. Chromatin Immunoprecipitation (ChIP) Assay

ChIP assay allows the researchers to take a virtual molecular snapshot of protein–DNA interactions in the context of the living cells (Ooi and Wood, 2009). By this method, living cells are fixed with formaldehyde to cross-link proteins which are in close association with the DNA. This is an important property allowing for the analysis of proteins that are recruited to the DNA through association with another DNA-binding factor. Immunoprecipitation of the sample with a transcription factor-specific antibody will allow the isolation of DNA fragments that are specifically associated with the protein of interest. In ChIP assay, a protein interacts with the genome, and the analysis finally depends on PCR amplification of DNA using gene-specific primers. The enrichment of the region that is specifically bound to the protein of interest is confirmed based on the fragment that is specifically immunoprecipitated in the sample but not in the nonspecific antibody control (Ooi and Wood, 2009; Orlando, 2000; Orlando and Paro, 1993). Furthermore, ChIP assay is also introduced to study protein–DNA interactions through formaldehyde cross-linking of proteins to DNA, followed by immunoprecipitation with specific antibodies. Using this approach, the *in vivo* DNA binding site of any protein can be detected. Here, we describe the stepwise protocol (Fig. 13.1) currently used to immunoprecipitate the formaldehyde cross-linked chromatin and further analyze the immunoprecipitated DNA by semiquantitative PCR (Yan *et al.*, 2004; Yang *et al.*, 2008, 2009). This technique is elegantly described by several investigators (Ooi and Wood, 2009; Weinmann, 2004; Yan *et al.*, 2004).

12.1. Methods for ChIP assay using cells

1. Day 1: The cells are washed once with 10 ml 1× PBS, and then resuspended in 10 ml of 1× PBS, 1 ml of cross-linking solution containing 100 mM NaCl, 1 mM EDTA, 0.5 mM EGTA, 50 mM HEPES (pH 8.0), and 11% formaldehyde (final concentration 1% formaldehyde) is added and incubated at room temperature for 10 min.

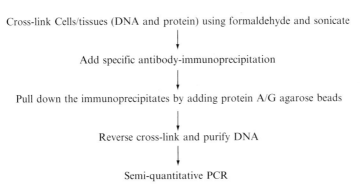

Figure 13.1 Flow chart describing the steps involved in chromatin immunoprecipitation assay.

2. Glycine 0.5 ml (2.5 M) is added to neutralize it and centrifuged the cells at 1500 rpm for 5 min. Further, the cells are washed once with 10 ml of 1× PBS, resuspended in 3 ml lysis buffer (50 mM Tris–HCl [pH 8.0], 1% SDS, 5 mM EDTA, 5 mM Na-butyrate, and 1× protease inhibitor) and incubated on ice for 5 min (see Note 5).
3. The cells are sonicated on Misonix Sonicator 3000 with microtip setting "8" (i.e., 3× 30 s for MonoMac6, U937, THP-1, Beas-2B, HUVECs, A549, H292, HepG2, Jurkat T-, and HeLa cells) (see Note 6).
4. Aliquot 150 µl of the sonicated samples as "input" fraction and store at 4 °C, the sonicated sample is diluted 10-fold (i.e., add 27 ml of dilution buffer containing 1% Triton X-100, 2 mM EDTA, 150 mM NaCl, 20 mM Tris–HCl [pH 8.0], 5 mM Na-butyrate and 1× protease inhibitor to 3 ml sonicated cells) to dilute SDS present in the lysis buffer (see Note 7).
5. Preclearing is done by adding 300 µl of normal rabbit serum (10 µl/ml of chromatin) and 600 µl of 50% slurry of protein A (and or G depending on the antibody used) sepharose beads. The tubes are incubated in a rotor at 4 °C for 3–4 h followed by centrifugation of the beads.
6. Equal volume of chromatin samples is aliquoted into new Eppendorf tubes for each antibody and incubated on a rocker at 4 °C overnight. ChIP grade antibodies are recommended for this as provided by commercial vendors (Abcam or Upstate).
7. Day 2: To each aliquot, 20 µl of 50% slurry protein A (and or G) beads per milliliter of chromatin is added and incubated on a rocker at 4 °C for 2–3 h. Beads are centrifuged, supernatants are discarded, and sequential washes using Paro buffer I (0.1% SDS, 1% Triton X-100, 2 mM EDTA, 20 mM Tris–HCl [pH 8.1], and 150 mM NaCl), Paro buffer II (0.1% SDS, 1% Triton X-100, 2 mM EDTA, 20 mM Tris–HCl [pH 8.1], and 500 mM NaCl), and Paro buffer III (0.25 M LiCl, 1% Igepal CA-630/NP-40, 1% deoxycholate, 1 mM EDTA, and 10 mM Tris–HCl [pH 8.1]) on ice are performed (see Note 8).
8. The beads are washed two times with 1× TE buffer at room temperature, 2× elution buffer (50–100 µl) containing 1% SDS, 0.1 M NaHCO$_3$, and 1 mg/ml proteinase K is added and incubated at room temperature for 10–15 min. The samples are mixed for every couple of minutes and spin down the sample.
9. Eluates are pooled, a 5 µl of proteinase K (5 mg/ml) is added to each input fractions stored at 4 °C (Step 7) and incubated overnight at 65 °C (see Note 9).
10. Day 3: DNA is purified using QIAquick PCR purification kit (Cat No. 28106; Qiagen, Valencia, CA) (see Note 10). Samples of input DNA are also prepared in the same way using Qiagen PCR purification kit according to the manufacturer's instructions.
11. PCR amplification is performed in PCR thermal cycler (PTC-200 DNA Engine, MJ Research, Waltham, MA) using optimized PCR

conditions of gene-specific primers. PCR for the input reaction is performed using 100 ng of genomic DNA and the PCR products are analyzed on a 1.5–2.0% agarose gel.

12.2. Methods for ChIP assay using tissues

CRIP assay in tissues can be performed using the commercially available kits by Epigentek [Cat. No. P-2012 and P-2013; EpiQuik tissue acetyl-H3/H4 CRIP kit] (Epigentek group Inc. Brooklyn, NY; www.epigentek.com).

1. Tissue sample (100 mg) is homogenized using PBS containing BSA (1 mg/ml) along with protease inhibitor cocktail (Sigma). Cross-linking is performed using formaldehyde containing 100 mM NaCl, 1 mM EDTA, 0.5 mM EGTA, 50 mM HEPES (pH 8.0), and 11% formaldehyde (final concentration of 1% formaldehyde) for 10 min, and washed three times with PBS, followed by addition of 0.5 ml glycine (2.5 M) to neutralize (Yao et al., 2010).
2. The samples are centrifuged at 1500×g for 5 min, and cell pellet is resuspended in SDS-lysis buffer and sonication of nuclear pellet containing chromatin is performed (four times for 30 s and one time for 15 s) at a maximum speed using Misonix-3000 Sonicator (Misonix, Inc., Farmingdale, NY).
3. Sonicated samples are centrifuged at 1500×g for 5 min, the supernatant is collected and diluted (1:10 dilution) with dilution buffer containing 1% Triton X-100, 2 mM EDTA, 150 mM NaCl, 20 mM Tris–HCl (pH 8.0), 5 mM Na-butyrate and 1× protease inhibitor (Roche) followed by preclearing the extract with 60 μl of protein A agarose/salmon sperm DNA (Cat No. 16-157, Upstate) for 3 h at 4 °C.
4. Immunoprecipitation is carried out overnight at 4 °C with 1 μg of specific antibodies (ChIP grade antibodies are recommended). After immunoprecipitation, 40 μl of protein A agarose/salmon sperm DNA is added and incubated for 2 h followed by brief centrifugation.
5. Precipitates are washed sequentially with Paro buffer I (0.1% SDS, 1% Triton X-100, 2 mM EDTA, 20 mM Tris–HCl [pH 8.1], and 150 mM NaCl), Paro buffer II (0.1% SDS, 1% Triton X-100, 2 mM EDTA, 20 mM Tris–HCl [pH 8.1], and 500 mM NaCl), and Paro buffer III (0.25 M LiCl, 1% Igepal CA-630/NP-40, 1% Deoxycholate, 1 mM EDTA, and 10 mM Tris–HCl [pH 8.1]), respectively, for 5 min at 4 °C. Finally, the precipitates are washed again with Tris-buffer twice for 5 min each.
6. The antigen–antibody complexes are extracted two times with 50 μl elution buffer (1% SDS, 0.1 M NaHCO$_3$, and 1 mg/ml proteinase K) and the eluted samples are heated at 65 °C overnight to reverse cross-linking. The recovered DNA is purified with a QIAquick PCR purification kit (Cat No. 28106, Qiagen).

7. Samples of input DNA are also prepared in the same way using Qiagen PCR purification kit according to the manufacturer's instructions. PCR amplification is performed in PCR thermal cycler (PTC-200 DNA Engine, MJ Research) using optimized PCR conditions of gene-specific primers. PCR for the input reaction is performed using 100 ng of genomic DNA and the PCR products are analyzed using a 1.5–2.0% agarose gel.

Notes

1. Before starting the ChIP assay, prepare preblocked protein A (and or G) bead slurry (Protein A-Agarose Cat. No. 16-156; Upstate).
2. When using the bead slurry, use 1000 μl tips (cut the tip ends) to avoid damage of the beads used. Wash the beads three times each with protein wash buffer (10 mM Tris–HCl, [pH 8.0], 1 mM EDTA, and 1 mg/ml BSA).
3. *Cross-linking solution*: Formaldehyde (11%) is prepared using 16% formaldehyde stock solution (Cat. No. 15710; Electron Microscopy Sciences).
4. Centrifuge at 1000 rpm for 5 min, resuspend the beads in equal volume of wash buffer.
5. Sodium butyrate is only necessary if ChIP is performed using antibodies specific for acetylated proteins.
6. Extent of sonication need to be empirically determined for each cell type used for the ChIP assay.
7. The SDS present in the lysis buffer could interfere with the antibody binding during immunoprecipitation process. Hence, diluted using dilution buffer (1:10) that lacks SDS.
8. All Paro buffers contain Na-butyrate (5 mM) and 1× protease inhibitors are completely EDTA-free. In between each Paro buffer washes tap, mix the samples, and incubate on ice for 10 min. Washes are done in 1 ml using 1.7 ml sterile Eppendorf tubes. Centrifuge at 2500 rpm for 1 min. TE buffer wash does not require incubation.
9. Elution and reverse cross-linking of the samples should be done using screw cap tubes for overnight incubation at 65 °C.
10. Once DNA is purified, check the input DNA samples (load 1, 2, and 4 μl) by performing agarose gel eletrophoresis using 1% gel in 1× TBE along with 100 bp and 1 kb DNA ladder.
11. One tablet of protease inhibitor dissolved in 1 ml of sterile double distilled water to make 50× concentration of protease inhibitors (protease inhibitor completely EDTA-free: Roche, Cat. No. 11873580001).

12.3. General considerations for ChIP assay

a. The extent of cross-linking (usually performed using 1% formaldehyde for 10 min) is crucial and it depends on the protein of interest. The conditions for cross-linking should be optimized. The concentration of formaldehyde, time duration for cross-linking, or the temperature of cross-linking should be adjusted for different cell types. Insufficient cross-linking may lead to incomplete fixation and the average size of the DNA fragment is less than 500 bp. Over-cross-linking may result in substantial loss of the material and prevent the production of small chromatin fragments even after prolonged sonication.

b. The sonicator need to be calibrated to yield the desired average length of DNA. Variables such as processing time and output control settings can be adjusted to produce optimum results. The size of the DNA fragments may be critical for high-resolution analysis. If the ChIP assay is performed to show binding of a protein to a particular site or a specific modification of histone in a particular site, fine tuning of the extent of cross-linking and sonication variables may be required.

c. Preliminary immunoprecipitation experiments should be performed to determine the optimum amount of antibody to be used. Generally, 5 μg of specific antibodies will be sufficient enough to produce immunoprecipitated (IP) DNA from 300 to 400 μg chromatin, but lower specificity of antibody lowers the relative enrichment of specific IP DNA.

d. It is important to use control to avoid nonspecific binding. Affinity-purified antibodies can reduce the amount of nonspecific binding, but IP without antibody is also needed. It is also important to include control using noncross-linked chromatin extract to avoid nonchromatin proteins and free DNA binding. This control allows monitoring whether the high salt washes are efficient. When monoclonal antibodies are used, protein A beads are not efficient to bind the antibodies, then protein A/G beads or other specific beads should be used.

e. It is very important to test the sensitivity and efficiency of the PCR before analyzing the IP DNA. The signal should be proportional to the amount of template DNA.

13. Conclusions

Epigenetic events such as histone acetylation, deacetylation, phosphorylation, and histone/DNA methylation play an important role in oxidative stress-induced chronic inflammatory processes in various diseases, such as, COPD, neurodegenerative diseases, disease of aging, aging process, and cancer (Ito *et al.*, 2005; Johnstone, 2002; Migliore and Coppede, 2009). Hence, study in epigenetics using different approaches will further unravel the mechanisms and identify the possible epigenetics-based therapies for chronic diseases. HAT and HDAC assays are useful in deciding whether a transcription factor recruits a coactivator (HAT complex) and/or a corepressor (HDAC complex). Further, as HATs and HDACs/SIRTs and transcription factors are modified by acetylation, phosphorylation, ubiquitination, sumoylation, and glycosylation, these assays will be important in deciding the function of the modification on enzymatic activity (Sun et al., 2002, 2003). ROS produced by extracellularly or by normal intracellular processes have the potential to introduce considerable chemical complexity/modifications into proteins, their principal biochemical redox targets (Sheehan, 2006). The surge of proteomics approaches in biological research has a significant impact on the study of protein oxidation and posttranslational modifications. Future advancements in developing various methods will help us to understand the extent that these modifications play in regulation of cellular redox function to understand epigenetic events in physiology and pathology (Eaton, 2006). Mass spectrometry-based proteomics methods have now begun to provide tools to understand the protein modifications which are linked to disease states, signaling systems, and age-related conditions. Further advancement in this field may provide more sensitive and reliable methods to identify site-specific modification of proteins and characterization of their stoichiometry *in vivo* (Grimsrud et al., 2008). ChIP assay monitors the ability of a protein to associate with DNA, but before further analysis, it is not known whether there is a functional outcome to the identified interaction. Hence, additional expression analysis will be required to confirm whether the identified target genes are regulated by the factor in a specific cell type. Methods described in this chapter include the techniques, assays in current use and practice involved in redox signaling, epigenetic changes and chromatin modifications associated with DNA and its associated proteins. However, several other newer methods can be employed in combination with different methods tailored to better understand the redox signaling in epigenetics and chromatin research.

ACKNOWLEDGMENTS

This study was supported by NIH R01-HL085613, 1R01HL097751, 1R01HL092842 and NIEHS Environmental Health Sciences Center grant ES01247.

REFERENCES

Adenuga, D., Yao, H., March, T. H., Seagrave, J., and Rahman, I. (2009). Histone deacetylase 2 is phosphorylated, ubiquitinated and degraded by cigarette smoke. *Am. J. Respir. Cell Mol. Biol.* **40,** 464–473.

Ahn, B., Rhee, S. G., and Stadtman, E. R. (1987). Use of fluorescein hydrazide and fluorescein thiosemicarbazide reagents for the fluorometric determination of protein carbonyl groups and for the detection of oxidized protein on polyacrylamide gels. *Anal. Biochem.* **161,** 245–257.

Banks, A. S., Kon, N., Knight, C., Matsumoto, M., Gutierrez-Juarez, R., Rossetti, L., Gu, W., and Accili, D. (2008). SirT1 gain of function increases energy efficiency and prevents diabetes in mice. *Cell Metab.* **8,** 333–341.

Barnes, P. J. (2009). Role of HDAC2 in the pathophysiology of COPD. *Annu. Rev. Physiol.* **71,** 451–464.

Bertos, N. R., Wang, A. H., and Yang, X. J. (2001). Class II histone deacetylases: Structure, function, and regulation. *Biochem. Cell Biol.* **79,** 243–252.

Caito, S., Rajendrasozhan, S., Cook, S., Chung, S., Yao, H., Friedman, A. E., Brookes, P. S., and Rahman, I. (2010). SIRT1 is a redox-sensitive deacetylase that is post-translationally modified by oxidants and carbonyl stress. *FASEB J.* (in press).

Chen, L., Fischle, W., Verdin, E., and Greene, W. C. (2001). Duration of nuclear NF-kappaB action regulated by reversible acetylation. *Science* **293,** 1653–1657.

Dalle-Donne, I., Rossi, R., Giustarini, D., Milzani, A., and Colombo, R. (2003). Protein carbonyl groups as biomarkers of oxidative stress. *Clin. Chim. Acta* **329,** 23–38.

Eaton, P. (2006). Protein thiol oxidation in health and disease: Techniques for measuring disulfides and related modifications in complex protein mixtures. *Free Radic. Biol. Med.* **40,** 1889–1899.

England, K., and Cotter, T. (2004). Identification of carbonylated proteins by MALDI-TOF mass spectroscopy reveals susceptibility of ER. *Biochem. Biophys. Res. Commun.* **320,** 123–130.

Fischle, W., Wang, Y., and Allis, C. D. (2003). Histone and chromatin cross-talk. *Curr. Opin. Cell Biol.* **15,** 172–183.

Foster, M. W., McMahon, T. J., and Stamler, J. S. (2003). S-Nitrosylation in health and disease. *Trends Mol. Med.* **9,** 160–168.

Fukuda, H., Sano, N., Muto, S., and Horikoshi, M. (2006). Simple histone acetylation plays a complex role in the regulation of gene expression. *Brief Funct. Genomic Proteomic* **5,** 190–208.

Gao, L., Cueto, M. A., Asselbergs, F., and Atadja, P. (2002). Cloning and functional characterization of HDAC11, a novel member of the human histone deacetylase family. *J. Biol. Chem.* **277,** 25748–25755.

Ghezzi, P., and Bonetto, V. (2003). Redox proteomics: Identification of oxidatively modified proteins. *Proteomics* **3,** 1145–1153.

Gilmour, P. S., Rahman, I., Donaldson, K., and MacNee, W. (2003). Histone acetylation regulates epithelial IL-8 release mediated by oxidative stress from environmental particles. *Am. J. Physiol. Lung Cell Mol. Physiol.* **284,** L533–L540.

Gregory, P. D., and Horz, W. (1998). Chromatin and transcription—How transcription factors battle with a repressive chromatin environment. *Eur. J. Biochem.* **251,** 9–18.

Grimsrud, P. A., Xie, H., Griffin, T. J., and Bernlohr, D. A. (2008). Oxidative stress and covalent modification of protein with bioactive aldehydes. *J. Biol. Chem.* **283,** 21837–21941.

Grozinger, C. M., Hassig, C. A., and Schreiber, S. L. (1999). Three proteins define a class of human histone deacetylases related to yeast Hda1p. *Proc. Natl. Acad. Sci. USA* **96,** 4868–4873.

Grunstein, M. (1997). Histone acetylation in chromatin structure and transcription. *Nature* **389,** 349–352.

Hess, D. T., Matsumoto, A., Kim, S. O., Marshall, H. E., and Stamler, J. S. (2005). Protein S nitrosylation: Purview and parameters. *Nat. Rev. Mol. Cell Biol.* **6,** 150–166.

Hurd, T. R., James, A. M., Lilley, K. S., and Murphy, M. P. (2009). Measuring redox changes in mitochondrial protein thiols with redox difference gel electrophoresis (REDOX-DIGE). *Methods Enzymol.* **456,** 344–361.

Ishii, T., Tatsuda, E., Kumazawa, S., Nakayama, T., and Uchida, K. (2003). Molecular basis of enzyme inactivation by an endogenous electorphile 4-hydroxy-2-nonenal:

Identification of modification sites in glyceraldehyde-3-phosphate dehydrogenase. *Biochemistry* **42,** 3474–3480.

Ito, K., Barnes, P. J., and Adcock, I. M. (2000a). Glucocorticoid receptor recruitment of histone deacetylase 2 inhibits interleukin-1beta-induced histone H4 acetylation on lysines 8 and 12. *Mol. Cell. Biol.* **20,** 6891–6903.

Ito, K., Barnes, P. J., and Adcock, I. M. (2000b). Histone acetylation and histone deacetylation. *In* "Methods in Molecular Medicine, Vol. 44, Asthma Mechanisms and Protocols," (K. F. Chung and I. M. Adcock, eds.), pp. 309–319. Human Press, Totowa, NJ.

Ito, K., Lim, S., Caramori, G., Chung, K. F., Barnes, P. J., and Adcock, I. M. (2001). Cigarette smoking reduces histone deacetylase 2 expression, enhances cytokine expression, and inhibits glucocorticoid actions in alveolar macrophages. *FASEB J.* **15,** 1110–1112.

Ito, K., Hanazawa, T., Tomita, K., Barnes, P. J., and Adcock, I. M. (2004). Oxidative stress reduces histone deacetylase 2 activity and enhances IL-8 gene expression: Role of tyrosine nitration. *Biochem. Biophys. Res. Commun.* **315,** 240–245.

Ito, K., Ito, M., Elliott, W. M., Cosio, B., Caramori, G., Kon, O. M., Barczyk, A., Hayashi, S., Adcock, I. M., Hogg, J. C., and Barnes, P. J. (2005). Decreased histone deacetylase activity in chronic obstructive pulmonary disease. *N. Engl. J. Med.* **352,** 1967–1976.

Jaffrey, S. R., Erdjument-Bromage, H., Ferris, C. D., Tempst, P., and Snyder, S. H. (2001). Protein S-nitrosylation: A physiological signal for neuronal nitric oxide. *Nat. Cell Biol.* **3,** 193–197.

Johnstone, R. W. (2002). Histone-deacetylase inhibitors: Novel drugs for the treatment of cancer. *Nat. Rev. Drug Discov.* **1,** 287–299.

Kagoshima, M., Wilcke, T., Ito, K., Tsaprouni, L., Barnes, P. J., Punchard, N., and Adcock, I. M. (2001). Glucocorticoid-mediated transrepression is regulated by histone acetylation and DNA methylation. *Eur. J. Pharmacol.* **429,** 327–334.

Kolle, D., Brosch, G., Lechner, T., Lusser, A., and Loidl, P. (1998). Biochemical methods for analysis of histone deacetylases. *Methods* **15,** 323–331.

Kouzarides, T. (1999). Histone acetylases and deacetylases in cell proliferation. *Curr. Opin. Genet. Dev.* **9,** 40–48.

Kuo, M. H., and Allis, C. D. (1998). Roles of histone acetyltransferases and deacetylases in gene regulation. *Bioessays* **20,** 615–626.

Lavu, S., Boss, O., Elliott, P. J., and Lambert, P. D. (2008). Sirtuins-novel therapeutic targets to treat age-associated diseases. *Nat. Rev. Drug Discov.* **7,** 841–853.

Levine, R. L., Garland, D., Oliver, C. N., Amici, A., Climent, I., Lenz, A. G., Ahn, B. W., Shaltiel, S., and Stadtman, E. R. (1990). Determination of carbonyl content in oxidatively modified proteins. *Methods Enzymol.* **186,** 464–478.

Lin, T. K., Hughes, G., Muratovska, A., Blaikie, F. H., Brookes, P. S., Darley-Usmar, V., Smith, R. A., and Murphy, M. P. (2002). Specific modification of mitochondrial protein thiols in response to oxidative stress: A proteomics approach. *J. Biol. Chem.* **277,** 17048–17056.

Marwick, J. A., Kirkham, P. A., Stevenson, C. S., Danahay, H., Giddings, J., Butler, K., Donaldson, K., MacNee, W., and Rahman, I. (2004). Cigarette smoke alters chromatin remodeling and induces proinflammatory genes in rat lungs. *Am. J. Respir. Cell. Mol. Biol.* **31,** 633–642.

Michan, S., and Sinclair, D. (2007). Sirtuins in mammals: Insights into their biological function. *Biochem. J.* **404,** 1–13.

Migliore, L., and Coppede, F. (2009). Environmental-induced oxidative stress in neurodegenerative disorders and aging. *Mutat. Res.* **674,** 73–84.

Moodie, F. M., Marwick, J. A., Anderson, C. S., Szulakowski, P., Biswas, S. K., Bauter, M. R., Kilty, I., and Rahman, I. (2004). Oxidative stress and cigarette smoke

alter chromatin remodeling but differentially regulate NF-kappaB activation and proinflammatory cytokine release in alveolar epithelial cells. *FASEB J.* **18,** 1897–1899.
Murtaza, I., Wang, H. X., Feng, X., Alenina, N., Bader, M., Prabhakar, B. S., and Li, P. F. (2008). Down-regulation of catalase and oxidative modification of protein kinase CK2 lead to the failure of apoptosis repressor with caspase recruitment domain to inhibit cardiomyocyte hypertrophy. *J. Biol. Chem.* **283,** 5996–6004.
Ooi, L., and Wood, I. C. (2009). Identifying transcriptional regulatory regions using reporter genes and DNA–protein interactions by chromatin immunoprecipitation. *Methods Mol. Biol.* **491,** 3–17.
Orlando, V. (2000). Mapping chromosomal proteins *in vivo* by formaldehyde-crosslinked-chromatin immunoprecipitation. *Trends Biochem. Sci.* **25,** 99–104.
Orlando, V., and Paro, R. (1993). Mapping Polycomb-repressed domains in the bithorax complex using *in vivo* formaldehyde cross-linked chromatin. *Cell* **75,** 1187–1198.
Osada, H., Tatematsu, Y., Saito, H., Yatabe, Y., Mitsudomi, T., and Takahashi, T. (2004). Reduced expression of class II histone deacetylase genes is associated with poor prognosis in lung cancer patients. *Int. J. Cancer* **112,** 26–32.
Pacholec, M., Bleasdale, J. E., Chrunyk, B., Cunningham, D., Flynn, D., Garofalo, R. S., Griffith, D., Griffor, M., Loulakis, P., Pabst, B., Qiu, X., Stockman, B., *et al.* (2010). SRT1720, SRT2183, SRT1460, and resveratrol are not direct activators of SIRT1. *J. Biol. Chem.* **285,** 8340–8351.
Rahman, I., and Adcock, I. M. (2006). Oxidative stress and redox regulation of lung inflammation in COPD. *Eur. Respir. J.* **28,** 219–242.
Rahman, I., and Biswas, S. K. (2004). Non-invasive biomarkers of oxidative stress: Reproducibility and methodological issues. *Redox Rep.* **9,** 125–143.
Rahman, I., Gilmour, P. S., Jimenez, L. A., and MacNee, W. (2002). Oxidative stress and TNF-alpha induce histone acetylation and NF-kappaB/AP-1 activation in alveolar epithelial cells: Potential mechanism in gene transcription in lung inflammation. *Mol. Cell. Biochem.* **234–235,** 239–248.
Rajendrasozhan, S., Yang, S. R., Kinnula, V. L., and Rahman, I. (2008). SIRT1, an antiinflammatory and antiaging protein, is decreased in lungs of patients with chronic obstructive pulmonary disease. *Am. J. Respir. Crit. Care Med.* **177,** 861–870.
Reznick, A. Z., Cross, C. E., Hu, M. L., Suzuki, Y. J., Khwaja, S., Safadi, A., Motchnik, P. A., Packer, L., and Halliwell, B. (1992). Modification of plasma proteins by cigarette smoke as measured by protein carbonyl formation. *Biochem. J.* **286,** 607–611.
Roth, S. Y., Denu, J. M., and Allis, C. D. (2001). Histone acetyltransferases. *Annu. Rev. Biochem.* **70,** 81–120.
Sasaki, H., Moriyama, S., Nakashima, Y., Kobayashi, Y., Kiriyama, M., Fukai, I., Yamakawa, Y., and Fujii, Y. (2004). Histone deacetylase 1 mRNA expression in lung cancer. *Lung Cancer* **46,** 171–178.
Sethuraman, M., McComb, M. E., Huang, H., Huang, S. Q., Heibeck, T., Costello, C. E., and Cohen, R. A. (2004). Isotope-coded affinity tag (ICAT) approach to redox proteomics: Identification and quantitation of oxidant-sensitive cysteine thiols in complex mixturs. *J. Proteome Res.* **3,** 1228–1233.
Sheehan, D. (2006). Detection of redox-based modification in two-dimensional electrophoresis proteomic separations. *Biochem. Biophys. Res. Commun.* **349,** 455–462.
Spange, S., Wagner, T., Heinzel, T., and Kramer, O. H. (2009). Acetylation of non-histone proteins modulates cellular signalling at multiple levels. *Int. J. Biochem. Cell Biol.* **41,** 185–198.
Stadtman, E. R., and Berlett, B. S. (1997). Reactive oxygen-mediated protein oxidation in aging and disease. *Chem. Res. Toxicol.* **10,** 485–494.

Sun, J. M., Chen, H. Y., Moniwa, M., Litchfield, D. W., Seto, E., and Davie, J. R. (2002). The transcriptional repressor Sp3 is associated with CK2-phosphorylated histone deacetylase 2. *J. Biol. Chem.* **277,** 35783–35786.

Sun, J. M., Spencer, V. A., Chen, H. Y., Li, L., and Davie, J. R. (2003). Measurement of histone acetyltransferase and histone deacetylase activities and kinetics of histone acetylation. *Methods* **31,** 12–23.

Taplick, J., Kurtev, V., Lagger, G., and Seiser, C. (1998). Histone H4 acetylation during interleukin-2 stimulation of mouse T cells. *FEBS Lett.* **436,** 349–352.

Taplick, J., Kurtev, V., Kroboth, K., Posch, M., Lechner, T., and Seiser, C. (2001). Homo-oligomerisation and nuclear localisation of mouse histone deacetylase 1. *J. Mol. Biol.* **308,** 27–38.

Tomita, K., Barnes, P. J., and Adcock, I. M. (2003). The effect of oxidative stress on histone acetylation and IL-8 release. *Biochem. Biophys. Res. Commun.* **301,** 572–577.

Toyokuni, S., Miyake, N., Hiai, H., Hagiwara, M., Kawakishi, S., Osawa, T., and Uchida, K. (1995). The monoclonal antibody specific for the 4-hydroxy-2-nonenal histidine adduct. *FEBS Lett.* **359,** 189–191.

Umstead, T. M., Freeman, W. M., Chinchilli, V. M., and Phelps, D. S. (2009). Age-related changes in the expression and oxidation of bronchoalveolar lavage proteins in the rat. *Am. J. Physiol. Lung Cell Mol. Physiol.* **296,** L14–L29.

Wang, H., Cao, R., Xia, L., Erdjument-Bromage, H., Borchers, C., Tempst, P., and Zhang, Y. (2001). Purification and functional characterization of a histone H3-lysine 4-specific methyltransferase. *Mol. Cell* **8,** 1207–1217.

Weinmann, A. S. (2004). Novel ChIP-based strategies to uncover transcription factor target genes in the immune system. *Nat. Rev. Immunol.* **4,** 381–386.

Wolffe, A. P. (1997). Transcriptional control. Sinful repression. *Nature* **387,** 16–17.

Woo, H. A., Chae, H. Z., Hwang, S. C., Yang, K. S., Kang, S. W., Kim, K., and Rhee, S. G. (2003). Reversing the inactivation of peroxiredoxins caused by cysteine sulfinic acid formation. *Science* **300,** 653–656.

Workman, J. L., and Kingston, R. E. (1998). Alteration of nucleosome structure as a mechanism of transcriptional regulation. *Annu. Rev. Biochem.* **67,** 545–579.

Xu, W. S., Parmigiani, R. B., and Marks, P. A. (2007). Histone deacetylase inhibitors: Molecular mechanisms of action. *Oncogene* **26,** 5541–5552.

Yamada, S., Kumazawa, S., Ishii, T., Nakayama, T., Itakura, K., Shibata, N., Kobayashi, M., Sakai, K., Osawa, T., and Uchida, K. (2001). Immunochemical detection of a lipofuscin-like fluorophore derived from malondialdehyde and lysine. *J. Lipid Res.* **42,** 1187–1196.

Yan, Y., Chen, H., and Costa, M. (2004). Chromatin immunoprecipitation assays. *Methods Mol. Biol.* **287,** 9–19.

Yang, S. R., Chida, A. S., Bauter, M. R., Shafiq, N., Seweryniak, K., Maggirwar, S. B., Kilty, I., and Rahman, I. (2006). Cigarette smoke induces proinflammatory cytokine release by activation of NF-kappaB and posttranslational modifications of histone deacetylase in macrophages. *Am. J. Physiol. Lung Cell Mol. Physiol.* **291,** L46–L57.

Yang, S. R., Wright, J., Bauter, M., Seweryniak, K., Kode, A., and Rahman, I. (2007). Sirtuin regulates cigarette smoke-induced proinflammatory mediator release via RelA/p65 NF-kappaB in macrophages *in vitro* and in rat lungs *in vivo*: Implications for chronic inflammation and aging. *Am. J. Physiol. Lung Cell Mol. Physiol.* **292,** L567–L576.

Yang, W., Steen, H., and Freeman, M. R., (2008). Proteomic approaches to the analysis of multiprotein signaling complexes. *Proteomics* **8,** 832–851.

Yang, S. R., Valvo, S., Yao, H., Kode, A., Rajendrasozhan, S., Edirisinghe, I., Caito, S., Adenuga, D., Henry, R., Fromm, G., Maggirwar, S., Li, D. J., *et al.* (2008). IKK alpha causes chromatin modification on pro-inflammatory genes by cigarette smoke in mouse lung. *Am. J. Respir. Cell Mol. Biol.* **38,** 689–698.

Yang, S. R., Yao, H., Rajendrasozhan, S., Chung, S., Edirisinghe, I., Valvo, S., Fromm, G., McCabe, M. J., Jr., Sime, P. J., Phipps, R. P., Li, J. D., Bulger, M., *et al.* (2009). RelB is differentially regulated by IkappaB Kinase-alpha in B cells and mouse lung by cigarette smoke. *Am. J. Respir. Cell Mol. Biol.* **40,** 147–158.

Yao, H., Edirisinghe, I., Rajendrasozhan, S., Yang, S. R., Caito, S., Adenuga, D., and Rahman, I. (2008). Cigarette smoke-mediated inflammatory and oxidative responses are strain-dependent in mice. *Am. J. Physiol. Lung Cell Mol. Physiol.* **294,** L1174–L1186.

Yao. H., Hwang, J. W., Moscat, J., Diaz-Meco, M. T., Leitges, M., Kishore, N., Li, X., and Rahman, I. (2010). Protein kinase Czeta mediates cigarette smoke/aldehyde- and lipopolysaccharide-induced lung inflammation and histone modification. *J. Biol. Chem.* **285,** 5405–5416.

CHAPTER FOURTEEN

CHARACTERIZATION OF PROTEIN TARGETS OF MAMMALIAN THIOREDOXIN REDUCTASES

Anton A. Turanov,* Dolph L. Hatfield,[†] *and* Vadim N. Gladyshev*

Contents

1. Introduction	246
2. Preparation of TR-immobilized Affinity Resins	247
2.1. Materials	249
2.2. Method	249
3. Identification of Targets of Mammalian TRs in Cell Lysates	250
3.1. Materials	250
3.2. Method	252
4. Concluding Remarks and Future Perspectives	252
Acknowledgments	253
References	253

Abstract

Mammalian thioredoxin reductases (TRs) are members of the pyridine nucleotide-disulfide oxidoreductase family. The main function of these enzymes is to maintain thioredoxins (Trxs) in the reduced state. The accessibility and high reactivity of selenocysteine in the C-terminal tetrapeptide allows mammalian TRs to couple to a range of substrates from proteins, such as Trx, to small molecules, such as selenite and hydroperoxides. However, identification of physiological substrates of TRs remains a challenge, with new targets identified primarily by testing random candidates in *in vitro* assays. The reaction mechanism of TRs supports a procedure that could trap substrates in a covalent nonproductive complex with TRs. Accordingly, attachment of TRs to affinity matrices allows isolation and identification of these targets. Application of this method revealed that Trxs are the major targets of TRs and supported efficient isolation of Trx substrates on TR affinity columns. We suggest that this procedure may be used as a general method of affinity isolation of Trxs and other TR substrates.

* Division of Genetics, Department of Medicine, Brigham & Women's Hospital and Harvard Medical School, Boston, Massachusetts, USA
[†] Molecular Biology of Selenium Section, Laboratory of Cancer Prevention, Center for Cancer Research, National Cancer Institute, National Institutes of Health, Bethesda, Maryland, USA

1. Introduction

Mammalian thioredoxin reductases (TRs) are members of the pyridine nucleotide-disulfide oxidoreductase family. The main function of these enzymes is to maintain thioredoxins (Trxs) in the reduced state. Three TRs have been identified in mammals: TR1 (also known as TrxR1 or Txnrd1) localized in the cytosol/nucleus (Gladyshev et al., 1996; Holmgren and Bjornstedt, 1995), TR3 (also known as TrxR2 or Txnrd2) targeted to mitochondria (Gasdaska et al., 1999; Lee et al., 1999), and thioredoxin/glutathione reductase TGR (also known as Txnrd3 or TR2) that is primarily expressed in testes (Sun et al., 2001). Mammalian TRs are homodimeric enzymes with a head to tail arrangement of subunits. The N-terminal redox-active dithiol in one subunit and the C-terminal selenolthiol site of the other subunit form a redox-active center (Biterova et al., 2005; Sandalova et al., 2001). The proposed mechanism of mammalian TR involves the reduction of the N-terminal active site disulfide by NADPH via enzyme-bound FAD, the electron transfer from the N-terminal dithiol to the C-terminal selenenylsulfide, and finally reduction of the substrate by the C-terminal selenolthiol (Cheng et al., 2009; Zhong et al., 2000). TGR differs from other TRs in that it also has an N-terminal glutaredoxin domain, which can be reduced by the C-terminal selenolthiol as well as by glutathione. In addition, TRs occur in a variety of forms that differ in their N-terminal regions.

The accessibility and high-reactivity selenocysteine (Sec) in the C-terminal tetrapeptide allows mammalian TRs to be active *in vitro* with a wide range of substrates from small molecules, such as selenite and lipid hydroperoxides, to proteins such as Trx, protein disulfide isomerase (PDI), and glutathione peroxidases (GPxs) (Arner, 2009; Gromer et al., 2004). The major intracellular substrates of TRs are thought to be Trxs, which in turn deliver reducing equivalents to many cellular proteins. One example is ribonucleotide reductase, which is essential for DNA synthesis and converts ribonucleotides to deoxyribonucleotides. Trxs also reduce methionine sulfoxide reductases and peroxiredoxins, and therefore are involved in the repair of oxidized proteins and in redox signaling via hydrogen peroxide (Arner and Holmgren, 2000; Holmgren, 1985). In addition, the Trx system participates in many cellular pathways by controlling the redox state of transcription factors via cysteines critical for DNA binding; these transcription factors include NF-κB, AP-1, and p53 (Lillig and Holmgren, 2007). Therefore, TRs are involved in the control of cellular growth, proliferation, viability, and apoptosis through the control of Trx activity and cellular redox state. Another major cellular redox system is the glutathione system, which is also powered by NADPH. Trx and glutathione systems work in parallel, but also may overlap in function.

Identification of substrates of TRs has been a slow process, with new targets identified mostly on a one-to-one basis. However, the reaction mechanism of TRs supports a method that could trap TR substrates in a covalent nonproductive complex with TRs. Accordingly, attachment of TRs to affinity matrices allows isolation and identification of these targets. This method is described in this chapter with focus on targets for TR1 and TR3.

2. Preparation of TR-immobilized Affinity Resins

Preparation of TR-immobilized resins is based on an established procedure that employs cyanogen bromide (CNBr)-activated Sepharose 4B. CNBr-activated matrices, such as Sepharose- or agarose-based, are commonly used materials in proteins analysis. They are widely used for preparation of protein affinity resins to study protein–protein interactions, antibody isolation, and protein purification. Such affinity resins were previously used to characterize targets of plant and animal Trxs, wherein the immobilized Trxs were used in which the resolving Cys in the active site was mutated to Ser or Ala (Motohashi *et al.*, 2001; Schwertassek *et al.*, 2007). Removal of the resolving Cys allowed trapping of Trx-interacting proteins due to formation of stable mixed disulfides between monothiol Trx and its substrate. Subsequently, Trx targets could be eluted from the Trx resin by reducing the mixed disulfide with a reducing agent such as dithiothreitol (DTT).

We adapted a similar catalytic mechanism-based method to identify targets of TRs. The following recombinant mammalian TR forms were prepared: (i) wild-type TR containing a C-terminal GCUG tetrapeptide (this sample was a 1:1 mixture of the Sec-containing form and the form truncated at the Sec UGA codon due inefficiency of Sec insertion into recombinant TRs), (ii) a mutant in which Cys in the C-terminal tetrapeptide was replaced with Ser (GSUG) (this form also was a 1:1 mixture of the Sec-containing form and the form truncated at Ser), (iii) a mutant in which Sec was replaced with Cys (GCCG), (iv) a mutant in which Sec in the C-terminal tetrapeptide was replaced with Cys and Cys was replaced with Ser (GSCG), (v) a mutant in which both Cys and Sec were replaced with Ser residues (GSSG), and (vi) a truncated mutant in which the Sec codon functioned as a stop signal (GC-stop). Each TR form was then linked to a CNBr-activated Sepharose to prepare TR-immobilized affinity resins. It was previously found that Sec is essential for catalysis by mammalian

TRs (Gromer et al., 2003; Zhong and Holmgren, 2000). This is thought to act as the attacking group that reduced the active site disulfide in oxidized Trxs, forming an intermediate selenenylsulfide bond between TR and Trx. The Cys adjacent to Sec in the sequence is thought to act as a resolving residue, which reduces the intermediate selenenylsulfide (Arner, 2009). Alternatively, this Cys might be the attacking residue and Sec may be the resolving residue. In this case, one would expect to find differences in the ability of mutant TR forms to bind target proteins. In particular, GSUG, GC, and GSCG mutants would be expected to stabilize the mixed selenenylsulfide or disulfide bonds between TR and Trx (or others targets), whereas GCUG, GCCG, and especially GSSG mutants would be expected to either complete the reaction or not interact with the substrate. Thus, the target proteins present in cell lysates could be enriched through formation of mixed disulfide or selenenylsulfide intermediates with some immobilized TR, whereas other mutant TR forms could serve as controls. Schematic representation of our method is shown in Fig. 14.1.

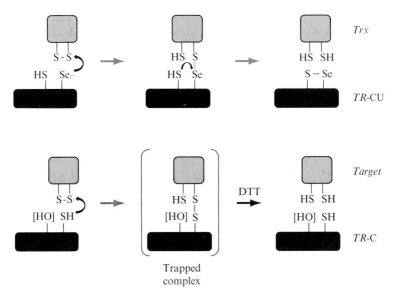

Figure 14.1 Schematic representation of the mechanism-based method for identification of TR targets. *Upper panel*: putative mechanism of Trx (shown in gray) reduction by TR (shown in black), in which the reduced selenolthiol group in TR provides, in a two-step process, reducing equivalents to the active site disulfide in Trx. TR-CU designates the wild-type form of protein containing Cys and Sec in the C-terminal tetrapeptide. *Lower panel*: mechanism-based procedure for identification of TR targets. A C-truncated mutant of an immobilized TR (([OH] indicates a non-redox residue in the TR active site) designated as TR-C) allows trapping of the target via a disulfide bond. Subsequent addition of DTT elutes the target, which is then subjected to tryptic digest and a MS/MS sequencing procedure for protein identification.

2.1. Materials

Recombinant mouse or human TRs; various forms differing in their C-terminal regions (see examples of mutants in the section above)
HisBind HiTrap metal-affinity column (GE HealthCare)
2′,5′-ADP-Sepharose (GE HealthCare)
CNBr-activated Sepharose 4B (Sigma)
Buffer A: 0.1 M NaHCO$_3$, 0.5 M NaCl, pH 8.5
Buffer B: 0.2 M glycine, pH 8.0

2.2. Method

1. The recombinant wild-type and C-terminal mutant form of human or mouse TR are prepared as described previously (Turanov *et al.*, 2006). Briefly, BL21(DE) *Escherichia coli* cells transformed with pET-28-TR constructs are grown in LB medium and protein expression is induced with 0.5 mM IPTG when cells reach $\sim$0.6 $A_{600\text{nm}}$. After the induction, cells are grown overnight at 30 °C, harvested by centrifugation and stored at -80 °C until used.
2. *E. coli* cells are resuspended at 4 °C in 50 mM sodium phosphate, 500 mM NaCl, 20 mM imidazole, pH 8.0, containing Protease Inhibitor Cocktail EDTA-free (Roche) and sonicated. Following centrifugation for 30 min at 14,000 rpm at 4 °C, the supernatant is loaded onto a 5-ml HisBind column (GE HealthCare). Recombinant proteins containing N-terminal His-tag are eluted with a linear gradient of 20–250 mM imidazole in loading buffer. Fractions containing TR are pooled, dialyzed against phosphate-buffered saline (PBS), pH 7.4, and applied to a 5-ml 2′,5′-ADP-Sepharose (GE HealthCare) column equilibrated with PBS. The column is washed extensively with PBS and the bound proteins are eluted with 1 M NaCl in PBS, then fractions containing pure proteins are dialyzed against PBS and stored at -80 °C.
3. Different TR forms (10–20 mg) in 100 mM sodium carbonate buffer, pH 8.5, containing 0.5 M NaCl (coupling buffer A) are incubated with freshly activated CNBr-activated Sepharose 4B according to the manufacturer's protocol. The resin is gently mixed with a end-over-end mixer for 2 h at room temperature or overnight at 4 °C.
4. Unbound protein ligand is washed from resin with buffer A and unreacted groups are blocked with 0.2 M glycine, pH 8.0 (buffer B) for 2 h at room temperature or overnight at 4 °C.
5. After several washing steps with buffer A, the TR-affinity resin is equilibrated with PBS or another buffer suitable for following steps.
6. The bright yellow color of the resin and colorless protein solution after protein binding will indicate successful preparation of the TR-affinity

resin. The immobilized TR is quantified based on the difference between the amounts of TR initially used and remaining in solution after the coupling reaction. Typically, more than 90% of protein could be immobilized.

3. Identification of Targets of Mammalian TRs in Cell Lysates

To identify cellular targets, the immobilized TR forms are reduced with NADPH, briefly washed with a buffer to remove the excess reductant, and mouse or rat liver mitochondrial or cytosolic lysates are added to each resin. Following incubation and washing, the target proteins are then eluted by adding a DTT-containing buffer. Eluted proteins are concentrated and analyzed by sodium dodecyl sulfate–polyacrylamide gel electrophoresis (SDS–PAGE). Resulting protein gels are stained by Coomassie Blue or Silver staining. Analysis of stained gels may reveal several protein bands, which are present independent of the TR form used. The binding of these proteins is either nonspecific or does not involve the C-terminal tetrapeptide, and these proteins are deemed as nonspecific TR targets, or at least they do not associate due to the redox chemistry of TR. Next, specific protein bands can be cut from the gel and subjected to protein identification by mass spectrometry analysis, LC–MS/MS. Identification of TR targets is summarized in Scheme 14.1.

Using TR-affinity columns, we identified cytosolic and mitochondrial Trxs as major TR targets in mouse and rat liver (Turanov *et al.*, 2006). We found that that the TR-C form (truncated TR with only one Cys-residue in C-terminal active site) was particularly efficient in binding Trx, whereas the TR-SS form (double Ser mutant form, used as negative control) did not show significant Trx binding, as expected (Fig. 14.2).

3.1. Materials

Mouse and rat tissues (liver)
TR-affinity resins
Phosphate-buffered saline
Buffer A: 20 mM HEPES, pH 7.5, 250 mM sucrose, 1 mM EDTA, 1 mM EGTA, 10 mM KCl
Buffer B: 50 mM Tris–HCl, pH 8.0, 200 mM NaCl
Other reagents: NADPH, DTT (Sigma), Complete Protease Inhibitor Cocktail (Roche Applied Science)

Characterization of Protein Targets of Mammalian Thioredoxin Reductases 251

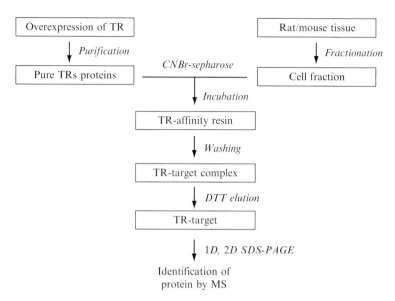

Scheme 14.1 Identification of targets of mammalian TRs in cell lysates.

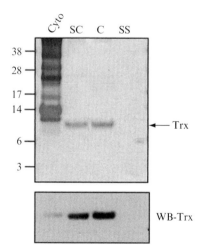

Figure 14.2 Identification of Trx1 as the major target of TR1. Rat liver cytosolic fractions were incubated with different TR1-immobilized resins. Bound proteins were eluted with DTT and separated by SDS–PAGE. SC, C, and SS indicate immobilized TR mutant forms. The gel was stained with Coomassie Blue. Molecular weights of protein standards (in kDa) are shown on the *right*. Arrow shows migration of Trx1. Lower panel shows Western blotting (WB) with anti-Trx1 antibodies.

3.2. Method

1. Preparation of cellular fractions is carried out by differential centrifugation. Rat livers (10–20 g) or mouse livers (2–5 g) are washed twice in ice-cold PBS and lysed on ice with 60 Dounce strokes with a tight fitting pestle in buffered sucrose (buffer A) containing 20 mM HEPES, pH 7.5, 250 mM sucrose, 1 mM EDTA, 1 mM EGTA, 10 mM KCl, and a tablet/10 ml of Complete Protease Inhibitor Cocktail (Roche Applied Science). After two centrifugation runs at 1000×g for 30 min to discard nuclei, mitochondria were pelleted at 10,000×g for 30 min, washed once, resuspended in buffer A and stored at −80 °C. The cytosolic fraction was obtained by ultracentrifugation of the postmitochondrial supernatant at 100,000×g for 1 h. To obtain mitochondrial lysate, an aliquot of the mitochondrial fraction is thawed and resuspended in an appropriate buffer or PBS, sonicated, and the mitochondrial lysate is collected by centrifugation at 10,000×g. Protein content is determined by the Bradford assay (Bio-Rad).
2. TR-immobilized affinity resins are prepared as described above using CNBr-Sepharose. The immobilized TR is initially reduced with 0.2 mM NADPH for 30 min in buffer B. In order to search for TR targets, rat or mouse liver mitochondrial or cytosolic lysates in buffer B or PBS (10–50 mg of total protein) are incubated with 1–2 ml of TR-immobilized resin containing 5–10 mg of TR at room temperature for 1 h under gentle stirring. The resins are washed with 50 mM Tris–HCl, pH 8.0, and 200 mM NaCl (buffer B) to remove nonspecifically bound proteins. The washing is repeated until the absorbance of the wash fraction at 280 nm is negligible. Finally, the resin is suspended in 50 mM Tris–HCl, pH 8.0, 200 mM NaCl, and 10 mM DTT and the mixture is incubated for 30 min at room temperature. Eluted proteins are separated from the resin by centrifugation. The proteins are concentrated and separated on SDS–PAGE gels, visualized by Western blot analyses or subjected to staining with Coomassie or Silver Staining Kit (Bio-Rad) and to identification of proteins by LC–MS/MS.

4. Concluding Remarks and Future Perspectives

The described method can be used to search for TR targets in different tissues and cellular compartments, such as the ER, cytosol, and mitochondria. Lysates prepared from other organisms as well as biological fluids containing Trxs should also be amenable to this method. We further suggest that this mechanism-based affinity method can be used to isolate TR targets such as Trx from a variety of biological samples for further proteomic analyses, for example, posttranslational modifications and binding partners. Figure 14.2 shows a typical result with mouse TR1 and illustrates specificity of this method. Trx substrates could also be efficiently isolated on the TR3 affinity columns. Therefore, we suggest that this should be a general method of affinity isolation of Trxs and other TR substrates. Ultimately, the proteomic method including the efficient 2D SDS–PAGE and LC–MS/MS analysis can be developed for high-throughput analysis of TR-interacting proteins. The concept behind this method can also be transferred to certain other pyridine nucleotide-disulfide oxidoreductases for identification of their respective targets.

ACKNOWLEDGMENTS

This work was supported by NIH grants to V. N. G. and the Intramural Research Program of the National Institutes of Health, NCI, Center for Cancer Research to D. L. H.

REFERENCES

Arner, E. S. (2009). Focus on mammalian thioredoxin reductases—Important selenoproteins with versatile function. *Biochim. Biophys. Acta* **1790**, 495–526.
Arner, E. S., and Holmgren, A. (2000). Physiological functions of thioredoxin and thioredoxin reductase. *Eur. J. Biochem.* **267**, 6102–6109.
Biterova, E. I., Turanov, A. A., Gladyshev, V. N., and Barycki, J. J. (2005). Crystal structures of oxidized and reduced mitochondrial thioredoxin reductase provide molecular details of the reaction mechanism. *Proc. Natl. Acad. Sci. USA* **102**, 15018–15023.
Cheng, Q., Sandalova, T., Lindqvist, Y., and Arner, E. S. (2009). Crystal structure and catalysis of the selenocysteine thioredoxin reductase 1. *J. Biol. Chem.* **284**, 3998–4008.
Gasdaska, P. Y., Berggeren, M. M., Berry, M. J., and Powis, G. (1999). *FEBS Lett.* **442**, 105–111.
Gladyshev, V. N., Jeang, K. T., and Stadtman, T. C. (1996). Selenocysteine, identified as the penultimate C-terminal residue in human T-cell Thioredoxin reductase, corresponds to TGA in the human placental gene. *Proc. Natl. Acad. Sci. USA* **93**, 6146–6151.
Gromer, S., Johansson, L., Bauer, H., Arscott, L. D., Rauch, S., Ballou, D. P., Williams, C. H. Jr., Schirmer, R. H., and Arner, E. S. (2003). Active sites of thioredoxin reductases: Why selenoproteins? *Proc. Natl. Acad. Sci. USA* **100**, 12618–12623.
Gromer, S., Urig, S., and Becker, K. (2004). The thioredoxin system—From science to clinic. *Med. Res. Rev.* **24**, 40–89.
Holmgren, A. (1985). Thioredoxin. *Annu. Rev. Biochem.* **54**, 237–271.
Holmgren, A., and Bjornstedt, M. (1995). Thioredoxin and thioredoxin reductase. *Methods Enzymol.* **252**, 199–208.
Lee, S. R., Kim, J. R., Kwon, K. S., Yoon, H. W., Levine, R. L., Ginsburg, A., and Rhee, S. G. (1999). Molecular cloning and characterization of a mitochondrial selenocysteine-containing thioredoxin reductase from rat liver. *J. Biol. Chem.* **274**, 4722–4734.
Lillig, C. H., and Holmgren, A. (2007). Thioredoxin and related molecules—From biology to health and disease. *Antioxid. Redox Signal.* **9**, 25–47.

Motohashi, K., Kondoh, A., Stumpp, M. T., and Hisabori, T. (2001). Comprehensive survey of proteins targeted by chloroplast thioredoxin. *Proc. Natl. Acad. Sci. USA* **98,** 11224–11229.

Sandalova, T., Zhong, L., Lindqvist, Y., Holmgren, A., and Schneider, G. (2001). Three-dimensional structure of a mammalian thioredoxin reductase: Implications for mechanism and evolution of a selenocysteine-dependent enzyme. *Proc. Natl. Acad. Sci. USA* **98,** 9533–9538.

Schwertassek, U., Balmer, Y., Gutsher, M., Weingarten, L., Preuss, M., Engelhard, J., Winkler, M., and Dick, T. P. (2007). Selective redox regulation of cytokine receptor signaling by extracellular thioredoxin-1. *EMBO J.* **26,** 3086–3097.

Sun, Q. A., Kirnarsky, L., Sherman, S., and Gladyshev, V. N. (2001). Selenoprotein oxidoreductase with specificity for thioredoxin and glutathione systems. *Proc. Natl. Acad. Sci. USA* **98,** 3673–3678.

Turanov, A. A., Su, D., and Gladyshev, V. N. (2006). Mouse mitochondrial thioredoxin reductase: Characterization of alternative cytosolic forms and cellular targets. *J. Biol. Chem.* **281,** 22953–22963.

Zhong, L., and Holmgren, A. (2000). Essential role of selenium in the catalytic activities of mammalian thioredoxin reductase revealed by characterization of recombinant enzymes with selenocysteine mutations. *J. Biol. Chem.* **275,** 18121–18128.

Zhong, L., Arner, E. S., and Holmgren, A. (2000). Structure and mechanism of mammalian thioredoxin reductase: The active site is a redox-active selenolthiol/selenenylsulfide formed from the conserved cysteine-selenocysteine sequence. *Proc. Natl. Acad. Sci. USA* **97,** 5854–5859.

CHAPTER FIFTEEN

ALTERATION OF THIOREDOXIN REDUCTASE 1 LEVELS IN ELUCIDATING CANCER ETIOLOGY

Min-Hyuk Yoo,* Bradley A. Carlson,* Petra Tsuji,*,†,‡ Robert Irons,* Vadim N. Gladyshev,§ *and* Dolph L. Hatfield*

Contents

1. Introduction	256
2. Materials and Methods	257
2.1. Mammalian cell lines and mice	257
2.2. Targeted removal and reexpression of TR1	258
2.3. Monitoring TR1 expression	259
2.4. Tumor formation activity of TR1 knockdown cells assessed by *in vitro* and *in vivo* assays	261
3. Results and Discussion	263
3.1. Knockdown of TR1	264
3.2. Phenotypic changes in siTR1 transfected cells	264
3.3. Tumorigenesis of TR1-deficient cells	266
3.4. *In vitro* and *in vivo* metastatic analysis of TR1-deficient cells	268
3.5. Reexpression of TR1 in TR1-deficient cells	269
4. Conclusions and Future Perspectives	271
Acknowledgments	272
References	272

Abstract

Thioredoxin reductase 1 (TR1) is a major antioxidant and redox regulator in mammalian cells and appears to function as a double-edged sword in that it has roles in preventing and promoting/sustaining cancer. TR1 is overexpressed in many cancer cells and targeting its removal often leads to a reversal in numerous malignant characteristics which has marked this selenoenzyme as a prime target for cancer therapy. Since alterations in TR1 activity may lead to a better

* Molecular Biology of Selenium Section, Laboratory of Cancer Prevention, Center for Cancer Research, National Cancer Institute, National Institutes of Health, Bethesda, Maryland, USA
† Cancer Prevention Fellowship Program, National Cancer Institute, National Institutes of Health, Bethesda, Maryland, USA
‡ Nutritional Science Research Group, Division of Cancer Prevention, National Cancer Institute, National Institutes of Health, Bethesda, Maryland, USA
§ Division of Genetics, Department of Medicine, Brigham and Women's Hospital and Harvard Medical School, Boston, Massachusetts, USA

Methods in Enzymology, Volume 474
ISSN 0076-6879, DOI: 10.1016/S0076-6879(10)74015-5

understanding of the etiology of cancer and new avenues for providing better therapeutic procedures, we have described herein techniques for removing and reexpressing TR1 employing RNAi technology and for assessing the catalytic activity of this enzyme.

1. Introduction

Selenium has long been known to have a role in decreasing the incidence of certain forms of cancer (reviewed in Lu *et al.*, 2009; Selenius *et al.*, 2010; Zeng and Combs, 2008). More recently, it has been shown that selenium-containing proteins (selenoproteins) play a role in preventing colon (Irons *et al.*, 2006) and prostate cancers (Diwadkar-Navsariwala *et al.*, 2006), and have roles in protecting against other forms of cancer (Hatfield and Gladyshev, 2009; Rayman *et al.*, 2009; Selenius *et al.*, 2010; Steinbrenner and Sies, 2009). It has also become apparent that selenium has a role in promoting and/or sustaining malignancy. For example, thioredoxin reductase 1 (TR1) is a selenium-containing enzyme that appears to be a double-edged sword in having roles in preventing as well as promoting/sustaining cancer (e.g., see Arnér, 2009; Hatfield, 2007; Hatfield *et al.*, 2009; Yoo *et al.*, 2006, 2007a). This essential selenoenzyme has an important role in embryogenesis (Jakupoglu *et al.*, 2005), is expressed in all cell types and organs, and is one of the major antioxidant and redox regulators in mammalian cells (Gromer *et al.*, 2005; Holmgren, 2006). It also has significant roles in transcription, DNA repair, cell proliferation, and angiogenesis (see Arnér, 2009 for review; and Arnér and Holmgren, 2006; Biaglow and Miller, 2005; Fujino *et al.*, 2006; Rundlöf and Arnér, 2004 for earlier studies). TR1 maintains thioredoxin1 (Trx1) in the reduced state and therefore has an essential role in the Trx system (Arnér and Holmgren, 2000). The selenium-containing moiety, selenocysteine (Sec), is the C-terminal penultimate amino acid in TR1 (Gladyshev *et al.*, 1996) and this Sec residue is required for TR1's catalytic activity (Nordberg *et al.*, 1998; Zhong and Holmgren, 2000; Zhong *et al.*, 1998). TR1 has also been reported to have a role in activating the p53 tumor suppressor and other tumor suppressor activities (Merrill *et al.*, 1999), and is targeted by carcinogenic electrophilic compounds (Moos *et al.*, 2003). These observations—the fact that one of the foremost characteristics of most, if not all, cancer cells is that they suffer from oxidative stress (Arnér, 2009; Arnér and Holmgren, 2006; Biaglow and Miller, 2005; Fujino *et al.*, 2006; Rundlöf and Arnér, 2004; Steinbrenner and Sies, 2009), and the findings that TR1 acts by reducing Trx1 and other redox regulators in cells (see Arnér, 2009 for review; and Arnér and Holmgren, 2000; Gromer *et al.*, 2005; Holmgren, 2006 for earlier studies)—provide strong evidence that this selenoenzyme plays a major role in cancer prevention.

There is also strong evidence that TR1 has an important role in cancer promotion and/or cancer maintenance. For example, TR1 is overexpressed in many cancers and cancer cell lines (Arnér, 2009; Arnér and Holmgren, 2006; Biaglow and Miller, 2005; Fujino et al., 2006; Gundimeda et al., 2009; Hedstrom et al., 2009; Rundlöf et al., 2004; Yoo et al., 2007a), and this selenoenzyme may have similar roles in malignant and normal cells in preventing or reducing oxidative stress. Numerous inhibitors of TR1 and other potent anticancer drugs that target TR1 activity and alter cancer-related properties also suggest that this protein has a major role in promoting and/or sustaining cancer (see Arnér, 2009; Chew et al., 2010; Gandin et al., 2010; Honeggar et al., 2009; Lam et al., 2009; Li et al., 2009; Selenius et al., 2010; Urig and Becker, 2006; Yan et al., 2009; Yoo et al., 2007a and references therein). Furthermore, comparing the expression of TR1 to that of other selenoproteins in a variety of cancer cell lines showed that TR1 was uniquely overexpressed (Yoo et al., 2007a). In addition, knockdown of TR1 in a mouse lung carcinoma cell line using RNAi technology reversed several malignant characteristics of these cells including manifesting a dramatic reduction in tumor progression and metastasis in mice injected with the TR1-deficient cell line compared with mice injected with the corresponding cell line expressing normal levels of TR1 (Yoo et al., 2006). The targeted removal of TR1 in a mouse cell line driven by oncogenic *k-ras* (DT) resulted in this mouse line losing self-sufficiency of growth, exhibiting a defective progression in the S-phase and a decreased expression of DNA polymerase α (Yoo et al., 2007a). Each of the above studies demonstrating that the impediment of TR1 activity in cancer cells upsets the malignancy process strongly suggests that this selenoenzyme is an excellent target for cancer therapy.

Since alteration of TR1 activity and function may provide insights into the etiology of cancer, the techniques used for targeting the knockdown of TR1 and its reexpression in cancer cells, the assays used for monitoring TR1 expression and the means of assessing reduction of tumor formation and metastasis related to TR1 deficiency should provide important avenues for exploring the underlying mechanisms of this disease. We, therefore, describe herein these techniques.

2. Materials and Methods

2.1. Mammalian cell lines and mice

Mouse Lewis lung carcinoma cells (LLC1) (obtained from ATCC) were used for targeting TR1 removal and examining the resulting effects of TR1 deficiency on various cancer-related properties including tumor formation and metastasis. LLC1 cells were selected for these studies as they are immunologically compatible with C57BL/6 mice and are known to form solid tumors when injected into the flank of these mice and to metastasize to lungs when injected into their tail vein. Mouse kidney (TCMK-1 obtained from ATCC) cells were used to compare the efficiency of TR1 knockdown constructs by transient transfection and the relative effects of the four siTR1 targeting constructs on the efficiency of removal of this selenoprotein. DT cells (provided by Dr. Yoon Sang Cho-Chung), which had been transformed with k-ras and are malignant k-ras expressing cells, were derived from NIH 3T3 cells (Noda et al., 1983) and used for stable transfection with a siTR1/siGPx1 double knockdown vector (Xu et al., 2009) as described below. C57BL/6 mice were used for a tumor formation assay by injecting malignant cells into their flanks and for a metastasis assay by injecting malignant cells into their tail veins and analyzing tumor formation in the lungs.

2.2. Targeted removal and reexpression of TR1

The knockdown vector used for this study, pU6-m3, was modified from pSilencer 2.1-U6 Hygro (Ambion, Inc.) by introducing the following changes: (1) the G and C bases at positions 468 and 469 were changed to a single A that made the U6 promoter closer in sequence to the corresponding wild-type gene; (2) the EcoRI site at position 4110 was deleted; and (3) a new XhoI site was inserted at position 384. Introduction of the two cloning sites resulted in a vector encoding multiple siRNA target sequences. Candidate siRNA target regions were chosen in the 3′-UTR of mouse TR1 and glutathione peroxidase 1 (GPx1) mRNA (accession number: NM_015762, NM_008160) by using the online siDESIGN program (Dharmacon, Inc.). Four of each sense–antisense oligonucleotides for TR1 and GPx1 knockdown candidate sequences were cloned into the pU6-m3 knockdown construct by using the BamH1–HindIII cloning site (Xu et al., 2009; Yoo et al., 2007b). TR1 candidate target regions are shown in Fig. 15.1 along with their targeting sequences. Each candidate knockdown construct was confirmed by sequencing. Knockdown efficiencies were examined by analyzing the levels of TR1 or GPx1 protein expression by ^{75}Se-labeling after transient transfection into mouse kidney (TCMK-1) cells. The mouse TR1 and GPx1 genes, including the 5′- and 3′-UTRs, were amplified by PCR using cDNA from TCMK-1 cells and cloned into a pcDNA3.1 expression vector (Invitrogen). To circumvent the knockdown site for knock-in (reexpression) of TR1 and GPx1, the genes were modified by introducing mutations in the siRNA target region by PCR using mutant primers, siTR1-3m, 5′-gtcttagtctca**aggtaccta**tgtctaatgtc-3′, and GPx1-3m, 5′-**gc**ga**g**agatg**gg**ttcaata-3′, where the bold letters indicate mutated regions. TR1 has an AU-rich mRNA instability element (designated ARE) located in the 3′-UTR that affects the stability of the mRNA. To increase TR1 mRNA stability, the ARE was deleted (bases 2407–3310) using BlpI and

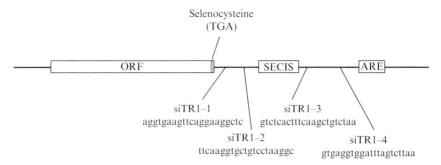

Figure 15.1 Relative positions of the siTR1-targeted regions within the *TR1* gene. The overlapping reading frame (ORF) of *TR1* is shown along with the TGA Sec codon that occurs as the penultimate codon in the gene. The relative positions of the SECIS element, the ARE region, and the four siTR1 RNAs, which are designated 1–4 (see text), along with their targeting sequences, are shown.

*Xho*I restriction sites. Each of the resulting constructs generated for knock-in of TR1 and GPx1 were designated as TR1 ki and GPx1 ki as discussed below and in Fig. 15.6.

Knockdown of TR1 or double knockdown of TR1 and GPx1 was carried out by transfecting LLC1 or DT cells as follows: 3×10^5 cells/well were seeded in a 6-well plate, transfected with 4 µg of each knockdown construct using Lipofectamine 2000 (Invitrogen) and incubated overnight. Cells were grown in the presence of 500 µg/ml of hygromycin B (Mediatech, Inc.) until the untransfected cells died and the transfected cells formed colonies. The resulting TR1 and TR1/GPx1 knockdown cells were confirmed by ^{75}Se-labeling, Western and Northern blot analyses and catalytic activity assays.

2.3. Monitoring TR1 expression

2.3.1. Northern blot analysis

Northern blot analysis of TR1 was carried out by isolating total RNA using TRIzol (Invitrogen) according to the manufacturer's instructions. The resulting RNA preparation was electrophoresed in 1× MOPS buffer (3-(*N*-morpholino)propanesulfonic acid) on a 1% agarose gel containing 1× MOPS buffer and 6% formaldehyde and transblotted in 10× SSC (saline-sodium citrate) to a Hybond N+ nylon membrane (GE Healthcare). Membranes were hybridized with a ^{32}P-labeled TR1 cDNA probe prepared by PCR of the entire coding region of the *TR1* gene from cDNA of TCMK-1 cells, washed twice with 2× SSC, 0.1% SDS and once with 0.1× SSC, 0.1% SDS, exposed to a PhosphorImager (GE Healthcare), and quantified using ImageQuant software (GE Healthcare). Ribosomal 18S and 28S RNA were used as loading controls.

2.3.2. ^{75}Se-labeling

To examine the expression of TR1 and other selenoproteins, cells were seeded onto a 6-well plate (3×10^5 cells/well) and incubated for 18 h. 40 μCi of ^{75}Se (20 nM; specific activity 1000 Ci/mmol; University of Missouri Research Reactor) were added to each well and the cells incubated for 24 h. Labeled cells were washed with ice-cold PBS twice, lysed with lysis buffer (20 mM Tris–Cl, 150 mM NaCl, 1% Triton X-100, 0.5% sodium deoxycholate, 10 mM NaF, 5 mM EDTA, and protease inhibitor cocktail), harvested using a cell scraper, incubated on ice with intermittent vortexing and prepared by centrifugation at 13,000$\times g$ for 10 min. Protein concentrations of lysates were measured using BCA Protein Assay Reagent (Thermo Fisher Scientific Inc.). Thirty micrograms of each protein sample were separated by electrophoresis on a 4–12% Bis-Tris NuPAGE gel. The gel, following electrophoresis, was stained with Coomassie Blue, dried, exposed to a PhosphorImager, and quantified using ImageQuant software.

2.3.3. Western blot analysis

Cells were washed with ice-cold PBS and whole cell lysates prepared as above. Thirty micrograms of each protein sample were electrophoresed on 4–12% Bis-Tris NuPAGE gels, the separated proteins transferred to a PVDF membrane, and then incubated initially with primary antibody according to the manufacturer's recommendation and finally with HRP-conjugated secondary antibody (Cell Signaling Technology). Membranes were reacted with SuperSignal West Dura Extended Duration Substrate (Pierce) and exposed to X-ray film.

2.3.4. Thioredoxin reductase activity assay

Activity of TR was determined spectrophotometrically by the method of Holmgren and Bjornstedt (1995) and Hill *et al.* (1997) as modified by Hintze *et al.* (2003) and Smith and Levander (2002). The assay is based on the reduction of 5,5′-dithio-bis(2-nitrobenzoic acid (DTNB) by TR1. The time-dependent increase in maximal absorbance was measured at 412 nm (extinction coefficient 27.2 or 13,600 M^{-1} cm^{-1} for TNB) using a microplate reader (Molecular Devices, Spectramax Plus 384 with Softmax Pro software). Since the assay is not completely specific for TR activity, specific activity was determined by subtracting the measured activity in the presence of the thioredoxin TR inhibitor aurothioglucose (ATG; Sigma) from the total activity. A unit of activity was defined as 1.0 μmol 5-thio-2-nitrobenzoic acid formed per min/mg protein. Protein concentrations were measured using the BCA reagent.

- *Materials*
  Phosphate buffer: 0.25 M Na$_2$HPO$_4$
  Inhibition buffer: 0.25 mM ATG in cold phosphate buffer (protected from light at 4 °C)

Reaction mix (for six microwells (one reaction ± inhibitor in triplicate)):
75 μl 0.035 M β-NADPH
75 μl 0.2 M EDTA
600 μl 0.25 M phosphate buffer
117 μl 0.063 M DTNB (in ethanol)
15 μl bovine serum albumin (BSA, 20 mg/ml)
685 μl dH$_2$O

1. Cells were harvested with cold phosphate buffer (300 μl/well), sonicated on ice, centrifuged, and the supernatant was kept on ice until measurement. The reaction mix minus the NADPH was kept at 37 °C with the NADPH added immediately prior to adding the sample at measurement.
2. Blank controls consisted of (a) a buffer-blank in triplicate, containing 70 μl phosphate buffer and 220 μl complete reaction mix; and (b) a buffer + ATG-blank in triplicate, containing 50 μl phosphate buffer, 220 μl complete reaction mix and 20 μl inhibition buffer.
3. For each sample, two reactions were set up in triplicate in a 96-well plate.
   - *Reaction 1 (sample)*: 25 μl cell lysate; 45 μl phosphate buffer; 220 μl complete reaction mix (amount of cell lysate may be doubled, adjusting the amount of phosphate buffer in the reaction mix).
   - *Reaction 2 (sample + ATG)*: 25 μl cell lysate; 25 μl phosphate buffer; 220 μl complete reaction mix; 20 μl inhibition buffer.
4. Once the complete reaction mix (37 °C) was added, the plate was measured immediately with a Spectramax Plus384 microplate reader. Automix was initiated before first read. Using the kinetic assay setup with path length check, and wavelength set at 412 nm, absorption was measured for 2 min in four 15-s intervals. The appropriate negative controls were subtracted from the results obtained for the samples.
5. To calculate the specific TR activity in the samples, the change in absorbance/min (slope) from samples with ATG was subtracted from the slope calculated for samples without ATG. The resulting value was divided by the extinction coefficient 27.2, adjusted for the amount of protein per sample, and subsequently expressed as μmol TNB/min/mg protein.

2.4. Tumor formation activity of TR1 knockdown cells assessed by *in vitro* and *in vivo* assays

2.4.1. Anchorage-independent growth assay

To assess the anchorage-independent growth of cells, 1000 control or TR1 knockdown (siTR1) cells were suspended in noble agar (0.35%) in growth medium (3.5 ml) containing 10% fetal bovine serum (FBS) and the cells were evenly spread onto 60 mm plates that had been previously covered with a 4 ml basal layer of noble agar (0.7%) containing Dulbecco's modified Eagle's medium without FBS. Plates were placed into a humidified CO_2 incubator at 37 °C for 14 days and 500 μl of fresh growth medium was added onto agar plates every 5 days to replenish media and prevent the plates from drying out. At the end of the incubation period, colonies were visualized by staining the plates overnight with 500 μl of p-iodonitrotetrazolium violet (0.5 mg/ml, INT) at 37 °C in the above incubator and the plates photographed.

2.4.2. Tumor formation assays

To assess tumor formation or metastatic capabilities of TR1-deficient (siTR1) and control cell lines, 2×10^5 control (pU6-m3) or siTR1 cells that had been maintained in a linear growth phase were either subcutaneously injected into the flanks of mice or into the tail veins of mice, respectively. Mice that were injected subcutaneously into their flanks were monitored every 2 days for 2 weeks, and after the 2-week period, mice were euthanized, tumor tissue excised, tumors fixed with 10% neutral-buffered formalin and stored at 4 °C for further analysis as described below. Mice that were injected into tail veins were euthanized after 4 weeks, autopsied for examining the extent of tumor formation in all tissues and the lungs removed, photographed, fixed with 10% neutral-buffered formalin and stored at 4 °C for further analysis. All mice used for assessing tumor formation and metastatic capabilities of the two cell lines were 5-week-old C57BL/6 females. Animal care was in accordance with the National Institutes of Health institutional guidelines under the expert direction of Dr. John Dennis (NCI, NIH, Bethesda, MD).

2.4.3. *In vitro* chemotaxis and invasion assays

To further examine the effect of TR1 knockdown in cancer cell lines, 4×10^4 of either control or TR1 knockdown cells were seeded in the upper chamber of a 24-well transwell migration and invasion assay plate (BD Biosciences) containing a matrigel-coated membrane (8.0 μm pore size) and 500 μl of serum-deficient media to trigger serum gradient-induced chemotaxis. The cells were incubated in the upper chamber of the plate for 18 h to allow cells to migrate through the membrane to the serum-containing media (750 μl) in the lower well. The membrane was washed with PBS twice and fixed with 100% methanol. Cells were visualized by hematoxylin staining and those cells that did not move across the membrane were removed with swab. Cells that migrated through the membrane were photographed.

3. Results and Discussion

We targeted the knockdown of TR1 in various malignant cells and examined the resulting effects of TR1 deficiency on cancer development (Yoo et al., 2006, 2007a). The 3'-UTRs in eukaryotic selenoprotein mRNAs contain the selenocysteine insertion sequence (SECIS) element that governs the incorporation of Sec in response to UGA and therefore is essential for the expression of this class of proteins (Low et al., 2000). This information afforded us an opportunity to target the removal of selenoproteins and their subsequent reexpression without disrupting the protein internal coding sequence of the mRNA (Yoo et al., 2007b). Other investigators have also used RNAi technology to target the removal of TR1 in various normal or cancer cells (e.g., see Eriksson et al., 2009; Honeggar et al., 2009; Liu and Shen, 2009; Watson et al., 2008) and the reader is referred to these studies to see additional approaches of targeting TR1 knockdown that vary from those described herein. One advantage of using mouse cancer cell lines in RNAi technology studies is that the resulting protein-deficient cells can be injected into mice that are not immunocompromised to further examine the effect of the loss of the targeted protein on, for example, tumorigenicity and metastasis. It should also be noted that RNAi technology has some pitfalls that must be taken into consideration to fully assess the effects of altering the expression of the targeted protein. For example, off-targeting is one of the major concerns to rule out in order to ensure that any observed phenotypic or functional changes are due totally to alteration in activity of the specifically intended protein and not to an unintended target. The subject of off-targeting has been addressed by us in considerable detail elsewhere (Xu et al., 2009), and the simplest and easiest means of overcoming this potential problem is to initially target more than one site in the intended protein and determine the effects of each siRNA to ensure that those siRNAs acting efficiently to remove the target give similar phenotypic results (see Section 3.1). It should also be noted that the level of mRNA knockdown may be quite substantial in some cases and may not coincide with the level of the loss of the corresponding protein wherein the protein expression may remain quite high. The level of protein, therefore, should be analyzed following the knockdown of the corresponding mRNA target by Western blot analysis and, in case of selenoproteins, also by ^{75}Se-labeling. In addition, in assessing levels of TR1, another potential problem is that the Sec moiety occurs at the C-terminal penultimate position and the UGA codon can serve as a stop codon resulting in a truncated protein which migrates at the same place on gels as the normal protein following electrophoresis. Although ^{75}Se-labeling can be used to address these latter potential problems in accurately assessing

TR1 levels, it is often best to additionally measure the catalytic activity of TR1 that explicitly demonstrates the active level of this selenoenzyme relative to that found in the corresponding control cell line or tissue.

3.1. Knockdown of TR1

The positions of the four siTR1 targeting sites, designated siTR1-1, siTR1-2, siTR1-3, and siTR1-4, relative to each other and to the SECIS element and the ARE within the 3′-UTR of TR1 are shown in Fig. 15.1. The corresponding constructs were generated and TCMK-1 cells were transiently transfected with each construct and the control vector, pU6-m3, as described in Section 2. The transfected cell lines were labeled with ^{75}Se and the resulting labeled selenoproteins from each are shown in Fig. 15.2A. TCMK-1 cells that were not transfected were also labeled and the resulting selenoproteins examined as a control along with the other control cells (pU6-m3 transfected cells). Cells transfected with pU6-m3, siTR1-1, and siTR1-2, and the untransfected cells expressed similar selenoprotein levels including the level of TR1. Since siTR1-1 and siTR1-2 did not target TR1 reduction, they were not further examined. Cells transiently transfected with the siTR1-3 and siTR1-4, however, manifested dramatically reduced levels of TR1. LLC1 cells were then stably transfected with these two constructs and the control construct. The knockdown of TR1 relative to untransfected LLC1 cells and pU6-m3 transfected control cells was examined following ^{75}Se-labeling, Western blotting (Fig. 15.2B) and Northern blotting (Fig. 15.2C). Both sites that were targeted in TR1 mRNA resulted in a dramatic reduction in the expression of this selenoprotein relative to that observed with the control cells as assessed by these three criteria. However, the knockdown of TR1 appeared to be slightly more efficient with siTR1-3 than with siTR1-4. We therefore examined the catalytic activity of this selenoenzyme with only siTR1-3 which also demonstrated the virtual loss of activity (Fig. 15.2D).

3.2. Phenotypic changes in siTR1 transfected cells

It was important to determine whether the two TR1 targeting constructs acted similarly in causing phenotypic changes in transfected cells to rule out the effect of off-targeting (see Xu et al., 2009 and references therein). A characteristic of many cancer cells, unlike most normal cells, is that they can grow unanchored in soft agar. The ability of the two cell lines stably transfected with siTR1-3 and siTR1-4 to grow in soft agar was compared to the two control cell lines, LLC1 and LLC/pU6-m3 (Fig. 15.3). The anchorage-independent growth of siTR1-3 and siTR1-4 transfected cells was inhibited, suggesting that reduction in TR1 expression altered this malignant characteristic. siTR1-4 appeared to grow slightly better than

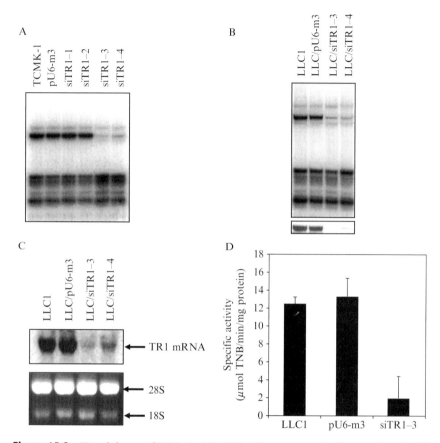

Figure 15.2 Knockdown of TR1. In (A), DT cells were transiently transfected with the control construct, pU6-m3, or with the siRNA constructs, siTR1-1, siTR1-2, siTR1-3, or siTR1-4. The untransfected cells (lane 1) and the resulting transfected cells (lanes 2–5 as indicated) were labeled with ^{75}Se, proteins extracted, 30 μg of each sample applied to a gel, the gel electrophoresed and exposed to a PhosphorImager as described in Section 2. In (B)–(D), LLC1 cells were stably transfected with pU6-m3, siTR-1-3, or siTR1-4, the untransfected and transfected cells examined by (B) labeling each cell line with ^{75}Se as indicated in the figure, proteins samples prepared, electrophoresed and exposed a PhosphorImager as given in (A); (C) preparing each cell line for Northern blot analysis and blotting using 28S and 18S ribosomal RNAs as loading controls; and (D) preparing only the untransfected cells, pU6-m3, and siTR1-3 cell lines for catalytic analysis and analyzing TR activity as described in Section 2. (The figures in panels B and D were reproduced from Yoo et al., 2006.)

siTR1-3 which is consistent with the observations above showing that the former targeting construct was not as efficient in removing TR1 expression by Western blotting and TR1 mRNA expression by Northern blotting. However, it should also be noted that the LLC/pU6-m3 transfected cells

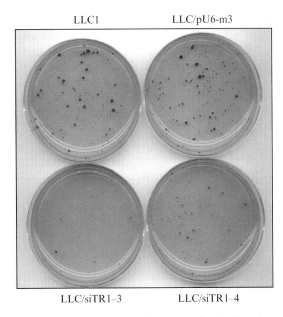

Figure 15.3 Phenotypic changes of TR1-deficient cells. LLC1 cells stably transfected with the pU6-m3 control, siTR1-3, or siTR1-4 constructs were suspended in noble agar/growth medium and evenly spread on plates covered with 0.7% noble agar and grown for 14 days as described in Section 2. At the end of the growth period, colonies were stained overnight with ρ-iodonitrotetrazolium violet, visualized, and photographed.

appeared to grow slightly more efficiently than the corresponding untransfected cells. The LLC/siTR1-3 transfected cells manifested other phenotypic changes that included morphology, a slight reduction in growth rate and a dramatic reduction in the levels of two cancer-related proteins, Hgf and Opn1 (data not shown; see Yoo et al., 2006), suggesting that additional cancer-related properties are altered making the TR1-deficient cells more like normal cells.

3.3. Tumorigenesis of TR1-deficient cells

Further studies were carried out only with the siTR1-3 construct. Mice were injected in the flank with LLC1 cells stably transfected with the siTR1-3 construct or the pU6-m3 construct to assess the ability of these two cell lines to form a solid tumor (Fig. 15.4A). Progression of tumor formation was determined after 2 weeks by euthanizing the mice, and removing and examining the tumors. Mice injected with the control, pU6-m3 cells had much larger tumors with an average weight of 0.341 g compared to the mice injected with the siTR1-3 transfected cells that had

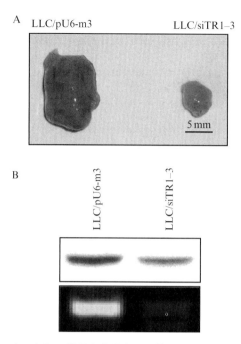

Figure 15.4 Tumorigenicity of TR1-deficient cells. In (A) Mice were injected in the flank with LLC1 cells transfected with the pU6-m3 or siTR1-3 constructs, tumor formation monitored every 2 days, and after 2 weeks, mice were euthanized, tumors removed, weighed and the weights averaged from three separate mice (average weight 0.341 g in mice injected with the pU6-m3 control construct and 0.063 g in mice injected with the siTR1-3 construct) and photographed. In (B), TR1 Western blot analysis of tumor extracts (shown in the upper panel) and PCR analysis of genomic DNA extracted from tumors of the pU6-m3 and siTR1-3 constructs (shown in the lower panel). (The figures were reproduced from Yoo et al., 2006.)

tumors with an average weight of 0.063 g (see Fig. 15.4A and figure legend). The tumors that formed in mice injected with the TR1-deficient cells were due to the loss of the siTR1-3 targeting vector as shown by Western blotting and PCR analysis (Fig. 15.4B). The reason that the siTR1-3 vector was lost is that all the constructs described herein were retained in stably transfected cells with hygromycin B, whereas mice injected with cells encoding the construct could not be treated with this drug. PCR analysis of genomic DNA from both the pU6-m3 and siTR1 tumors demonstrated that the siTR1-3 construct was lost in the tumor cells (Fig. 15.4B, lower panel). Western blot analysis showed higher levels of TR1 in tumors developed from siTR1-3 transfected cells than siTR1-3 cells, suggesting that TR1 was required for tumor growth and that the tumors arose due to reversal of the TR1 knockdown in this cancer model

(Fig. 15.4B, upper panel). The Western blot analysis study is consistent with the finding in the PCR experiment.

3.4. *In vitro* and *in vivo* metastatic analysis of TR1-deficient cells

The metastatic potential of malignant cells can be examined *in vitro* prior to an *in vivo* analysis by assessing their invasiveness and chemostatic potential. The effect of TR1 deficiency on the metastatic potential of malignant cells was examined by using an *in vitro* invasive and chemotactic assay and comparing the efficiencies over an 18 h period of pU6-m3 control and siTR1-3 transfected LLC1 cells to migrate through the serum gradient environment and the basement membrane (Fig. 15.5A). The TR1-deficient cells manifested a significantly decreased migration compared to the control cells, suggesting that this selenoenzyme is involved with the process of migration in malignant cells (Fig. 15.5A). The involvement of TR1 in chemotaxis such

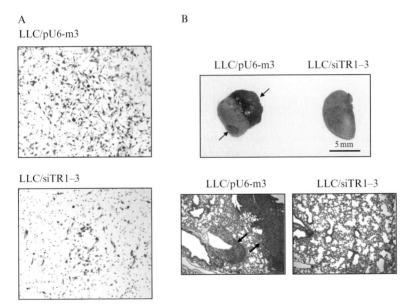

Figure 15.5 Effects of TR1 knockdown on chemotaxis and metastasis. In (A), LLC/pU6-m3 or siTR1-3 transfected cells (4×10^4 cells/chamber) were incubated in a transwell migration and invasion assay plate containing a matrigel-coated membrane and migrated cells were stained with hematoxylin and photographed as described in Section 2. In (B), metastasis was assessed by injecting LLC/pU6-m3 or siTR1-3 transfected cells into the tail veins of mice, and after 4 weeks, lungs were removed photographed (upper panel) and the slides shown in the lower panel were prepared from lung samples from both animals, H&E stained and photographed as described in Section 2. (The figures shown in panel B were reproduced from Yoo *et al.*, 2006.)

as a role in sensing the serum gradient, invasion of the TR1-deficient cells through the extracellular matrix, and the intracellular migratory pathway in the TR1-deficient cells need to be further examined to elucidate the molecular mechanism(s) of TR1 and its probable substrate(s) in the metastatic potential of malignant cells. However, the metastatic potential of the LLC/siTR1-3 cells as assessed by an *in vitro* assay was severely altered suggesting that metastasis in an *in vivo* assay would also likely be altered.

To assess the *in vivo* metastatic ability of pU6-m3 control and siTR1-3 cells, lungs of mice injected in the tail vein with either of these cell lines were examined 4 weeks postinjection for the presence of tumors (Fig. 15.5B). No tumors were found in the mice injected with the TR1 knockdown cells, whereas tumors were present in the lungs of mice injected with the pU6-m3 control cells. Lung tissue from mice with apparent normal lungs that had been injected with the siTR1-deficient cells and from mice with cancerous lungs that had been injected with the pU6-m3 control cells were examined for pathological changes (Fig. 15.5B, lower panel). Extensive malignancy was observed in mice injected with the pU6-m3 control cells, but only normal tissue was found in mice injected with the siTR1 knockdown cells.

3.5. Reexpression of TR1 in TR1-deficient cells

In assessing the role of TR1 in the etiology of cancer by means of using RNAi technology, it is important to have a means of reexpressing the knocked down protein to further examine its role in the malignancy process. For example, the reintroduction of a protein can be done wherein the exogenous protein contains a His-tag and is expressed in the absence of the corresponding endogenous protein for examining alterations that may occur in the targeted protein that might otherwise be undetected. The His-tagged protein can then be isolated and purified in the absence of endogenous protein for further study. Since we have also developed techniques for simultaneous removal of multiple selenoproteins and their subsequent reexpression, either individually or simultaneously (Xu *et al.*, 2009; Yoo *et al.*, 2007b), and such procedures can be used to further explore the role of selenoproteins in cancer etiology, the technology of multiple knockdown and knock-in is further considered. DT cells were used for stable transfection with an siTR1/siGPx1 double knockdown construct for their simultaneous removal and subsequent knocking back-in by transiently transfecting with the construct that either individually or simultaneously circumvented the targeting vector resulting in the reexpression of one or both of the removed selenoproteins. Transfected cells were labeled with ^{75}Se to monitor the expression of selenoproteins in transfected DT cells (Fig. 15.6). Two of the control cell lines stably transfected with either pU6-m3 or the double siTR1/siGPx1 knockdown vector are shown, respectively, in lanes 1 and 2 of the figure. The double targeting vector effectively

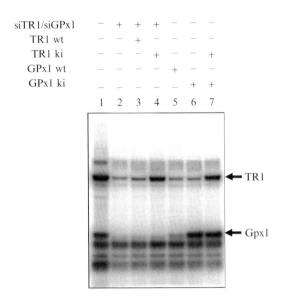

Figure 15.6 Simultaneous knockdown of TR1 and GPx1 and their individual or simultaneous reexpression. DT cells were stably transfected initially with either the pU6-m3 control construct or the siTR1/siGPx1 double knockdown construct. The resulting stably transfected siTR1/siGPx1 double knockdown cells were subsequently transiently transfected with either the control pcDNA3.1 expression vector or with one of the following expression vectors encoding either the TR1 wild-type gene (designated TR1 wt in the figure), TR1 knock-in gene (designated TR1 ki), GPx1 wild-type gene (designated GPx1 wt), GPx1 knock-in gene (designated GPx1 ki), or TR1 and GPx1 knock-in genes (designated TR1 ki and GPx1 ki). All cell lines were labeled with ^{75}Se, cell extracts prepared and electrophoresed. In lane 1, the cells that were stably transfected with the pU6-m3 control construct were transiently transfected with pcDNA3 expression control vector, and in lanes 2–7, the cells that were stably transfected with the double siTR1-siGPx1 construct were transiently transfected as follows: lane 2, the pcDNA3.1 expression control vector; lane 3, TR1 wt; lane 4, TR1 ki; lane 5, GPx1 wt; lane 6, GPx1 ki; and lane 7, TR1 ki-GPx1 ki. (The figure was reproduced from Xu *et al.*, 2009.)

removed TR1 and GPx1 expression. Replacement of the TR1 and GPx1 wild-type genes did not result in expression of the corresponding selenoproteins due to the presence of the siTR1 and siGPx1 RNAs generated from the stably transfected vector (Fig. 15.6, lanes 3 and 5, respectively). Circumventing the targeting regions with transiently transfected constructs (designated in the figure as "ki" for knock-in) that either individually or collectively bypassed the siRNAs by encoding *TR1* and/or *GPx1* genes with mutations corresponding to the siRNAs resulted in reexpression of the corresponding selenoprotein or selenoproteins (Fig. 15.6, lanes 4, 6, and 7).

The only variation in the above procedure in order to isolate the reintroduced exogenous protein is to prepare the circumvention construct with an N-terminal or a C-terminal His-tag on one or the other of the reintroduced proteins.

4. Conclusions and Future Perspectives

Reducing the expression of TR1 in cancer cells using RNAi technology has provided new insights into the underlying roles of this selenoenzyme in the malignant process. For example, TR1 deficiency in LLC1 cells resulted in a reversal of their ability to grow unanchored in soft agar and altered their tumorigenicity and metastatic properties making each of these characteristics more similar to those of normal cells. In fact, it appeared that TR1 activity was required in tumor formation following injection of the TR1 knockdown cells as the resulting tumors that formed had lost their TR1 knockdown construct (Fig. 15.4B).

The means by which levels of TR1 are measured in cells and tissues is also an important factor to consider as partially or fully inactive TR1 can also arise. For example, inactive TR1 can arise by the cessation of translation at the UGA codon and partially active TR1 can result from replacement of Sec with Cys. Labeling cells in culture or tumors (i.e., by labeling the mouse) and examining the resulting levels of TR1 following electrophoresis on gels and/or measuring catalytic activity in cells and tissues can overcome such possible problems.

Reexpression of TR1 in TR1 knockdown cells provides a means of expressing exogenous protein in the absence of the corresponding endogenous protein. By inserting a His-tag at the N- or C-terminus for isolating the protein, this technique can open new opportunities of exploring the function of TR1 in malignant and normal cells.

Other investigators have compared the effects of reducing TR1 expression by siRNA knockdown and specific inhibitors and have found that TR1 and Trx1 may act independently of each other in some situations in the malignancy process (Eriksson *et al.*, 2009; Watson *et al.*, 2008). These studies also provide insights into other possible pathways for exploring the underlying roles of TR1 and Trx1 in malignancy.

It should also be noted that TR1-deficient DT cells lost their self-sufficiency in growth and were found to have a defective progression in their S-phase and a reduced expression of DNA polymerase α (Yoo *et al.*, 2007a). These studies suggested that TR1 is essential for self-sufficiency in growth and provide further evidence that this selenoenzyme acts as a procancer protein in DT cells and a prime target in cancer therapy. Interestingly, self-sufficiency in growth is one of the six hallmarks of cancer as defined by Hanahan and Weinberg (2000). Knockdown of TR1 in LLC1 cells demonstrated that TR1 has a role in tissue invasion and metastasis (Yoo *et al.*, 2006) that is another of the Hanahan and Weinstein cancer hallmarks. Thus, RNAi technology has been used to show that TR1 plays a major role in two of the six cancer hallmarks and can likely be used to demonstrate whether this selenoenzyme has roles in the other four cancer hallmarks which are insensitive to antigrowth signals, evading apoptosis, sustained angiogenesis, and limitless replicative potential (Hanahan and Weinberg, 2000). In fact, we have observed that TR1 deficiency in a breast cancer cell line, EMT6, altered the ability of these cells to evade apoptosis and we are examining the role of TR1 in the other cancer hallmarks (M.-H. Yoo, B. A. Carlson, V. N. Gladyshev, D. L. Hatfield, unpublished data).

Unquestionably, RNAi technology has proven to be a powerful tool in the study of protein function, and now, in the study of selenoprotein function in cancer. The various approaches employing this technology by others (Eriksson et al., 2009; Honeggar et al., 2009; Liu and Shen, 2009; Watson et al., 2008) and us (Xu et al., 2009; Yoo et al., 2006, 2007a,b) have yielded many new insights into the underlying roles of TR1 in the malignancy process, and undoubtedly, will shed much more light on the molecular mechanisms driving cancer.

ACKNOWLEDGMENTS

This work was supported by the Intramural Research Program of the National Institutes of Health, NCI, Center for Cancer Research to D. L. H. and NIH grants to V. N. G.

REFERENCES

Arnér, E. S. (2009). Focus on mammalian thioredoxin reductases—Important selenoproteins with versatile functions. *Biochim. Biophys. Acta* **1790**, 495–526.

Arnér, E. S. J., and Holmgren, A. (2000). Physiological functions of thioredoxin and thioredoxin reductase. *Eur. J. Biochem.* **267**, 6102–6109.

Arnér, E. S. J., and Holmgren, A. (2006). The thioredoxin system in cancer. *Semin. Cancer Biol.* **16**, 420–426.

Biaglow, J. E., and Miller, R. A. (2005). The thioredoxin reductase/thioredoxin system: Novel redox targets for cancer therapy. *Cancer Biol. Ther.* **4**, 6–13.

Chew, E. H., Nagle, A. A., Zhang, Y., Scarmagnani, S., Palaniappan, P., Bradshaw, T. D., Holmgren, A., and Westwell, A. D. (2010). Cinnamaldehydes inhibit thioredoxin reductase and induce Nrf2: Potential candidates for cancer therapy and chemoprevention. *Free Radic. Biol. Med.* **48**, 98–111.

Diwadkar-Navsariwala, V., Prins, G. S., Swanson, S. M., Birch, L. A., Ray, V. H., Hedayat, S., Lantvit, D. L., and Diamond, A. M. (2006). Selenoprotein deficiency accelerates prostate carcinogenesis in a transgenic model. *Proc. Natl. Acad. Sci. USA* **103**, 8179–8184.

Eriksson, S. E., Prast-Nielsen, S., Flaberg, E., Szekely, L., and Arnér, E. S. (2009). High levels of thioredoxin reductase 1 modulate drug-specific cytotoxic efficacy. *Free Radic. Biol. Med.* **47**, 1661–1671.

Fujino, G., Noguchi, T., Takeda, K., and Ichijo, H. (2006). Thioredoxin and protein kinases in redox signaling. *Semin. Cancer Biol.* **16**, 427–435.

Gandin, V., Fernandes, A. P., Rigobello, M. P., Dani, B., Sorrentino, F., Tisato, F., Bjornstedt, M., Bindoli, A., Sturaro, A., Rella, R., and Marzano, C. (2010). Cancer cell death induced by phosphine gold(I) compounds targeting thioredoxin reductase. *Biochem. Pharmacol.* **79,** 90–101.

Gladyshev, V. N., Jeang, K. T., and Stadtman, T. C. (1996). Selenocysteine, identified as the penultimate C-terminal residue in human T-cell thioredoxin reductase, corresponds to TGA in the human placental gene. *Proc. Natl. Acad. Sci. USA* **93,** 6146–6151.

Gromer, S., Eubel, J. K., Lee, B. L., and Jacob, J. (2005). Human selenoproteins at a glance. *Cell. Mol. Life Sci.* **62,** 2414–2437.

Gundimeda, U., Schiffman, J. E., Gottlieb, S. N., Roth, B. I., and Gopalakrishna, R. (2009). Negation of the cancer-preventive actions of selenium by over-expression of protein kinase Cepsilon and selenoprotein thioredoxin reductase. *Carcinogenesis* **30,** 1553–1561.

Hanahan, D., and Weinberg, R. A. (2000). The hallmarks of cancer. *Cell* **100,** 57–70.

Hatfield, D. L. (2007). Thioredoxin reductase 1: A double-edged sword in cancer prevention and promotion. *CCR Frontiers Sci.* **6,** 8–10.

Hatfield, D. L., and Gladyshev, V. N. (2009). The outcome of Selenium and Vitamin E Cancer Prevention Trial (SELECT) reveals the need for better understanding of selenium biology. *Mol. Interv.* **9,** 18–21.

Hatfield, D. L., Yoo, M. H., Carlson, B. A., and Gladyshev, V. N. (2009). Selenoproteins that function in cancer prevention and promotion. *Biochim. Biophys. Acta* **1790,** 1541–1545.

Hedstrom, E., Eriksson, S., Zawacka-Pankau, J., Arnér, E. S., and Selivanova, G. (2009). p53-dependent inhibition of TrxR1 contributes to the tumor-specific induction of apoptosis by RITA. *Cell Cycle* **8,** 3576–3583.

Hill, K. E., McCollum, G. W., and Burk, R. F. (1997). Determination of thioredoxin reductase activity in rat liver supernatant. *Anal. Biochem.* **253,** 123–125.

Hintze, K. J., Wald, K. A., Zeng, H., Jeffery, E. H., and Finley, J. W. (2003). Thioredoxin reductase in human hepatoma cells is transcriptionally regulated by sulforaphane and other electrophiles via an antioxidant response element. *J. Nutr.* **133,** 2721–2727.

Holmgren, A. (2006). In "Selenium: Its molecular biology and role in human health," 2nd edn. (D. L. Hatfield, M. J. Berry, and V. N. Gladyshev, eds.), pp. 183–194. Springer Science+Business Media, New york.

Holmgren, A., and Bjornstedt, M. (1995). Thioredoxin and thioredoxin reductase. *Methods Enzymol.* **252,** 199–208.

Honeggar, M., Beck, R., and Moos, P. J. (2009). Thioredoxin reductase 1 ablation sensitizes colon cancer cells to methylseleninate-mediated cytotoxicity. *Toxicol. Appl. Pharmacol.* **241,** 348–355.

Irons, R., Carlson, B. A., Hatfield, D. L., and Davis, C. D. (2006). Both selenoproteins and low molecular weight selenocompounds reduce colon cancer risk in mice with genetically impaired selenoprotein expression. *J. Nutr.* **136,** 1311–1317.

Jakupoglu, C., Przemeck, G. K., Schneider, M., Moreno, S. G., Mayr, N., Hatzopoulos, A. K., de Angelis, M. H., Wurst, W., Bornkamm, G. W., Brielmeier, M., and Conrad, M. (2005). Cytoplasmic thioredoxin reductase is essential for embryogenesis but dispensable for cardiac development. *Mol. Cell. Biol.* **25,** 1980–1988.

Lam, J. B., Chow, K. H., Xu, A., Lam, K. S., Liu, J., Wong, N. S., Moon, R. T., Shepherd, P. R., Cooper, G. J., and Wang, Y. (2009). Adiponectin haploinsufficiency promotes mammary tumor development in MMTV-PyVT mice by modulation of phosphatase and tensin homolog activities. *PLoS One* **4,** e4968.

Li, S., Zhang, J., Li, J., Chen, D., Matteucci, M., Curd, J., and Duan, J. X. (2009). Inhibition of both thioredoxin reductase and glutathione reductase may contribute to the anticancer mechanism of TH-302. *Biol. Trace Elem. Res.* (in press).

Liu, Z. B., and Shen, X. (2009). Thioredoxin reductase 1 upregulates MCP-1 release in human endothelial cells. *Biochem. Biophys. Res. Commun.* **386,** 703–708.

Low, S. C., Grundner-Culemann, E., Harney, J. W., and Berry, M. J. (2000). SECIS–SBP2 interactions dictate selenocysteine incorporation efficiency and selenoprotein hierarchy. *EMBO J.* **19,** 6882–6890.

Lu, J., Berndt, C., and Holmgren, A. (2009). Metabolism of selenium compounds catalyzed by the mammalian selenoprotein thioredoxin reductase. *Biochim. Biophys. Acta* **1790,** 1513–1519.

Merrill, G. F., Dowell, P., and Pearson, G. D. (1999). The human p53 negative regulatory domain mediates inhibition of reporter gene transactivation in yeast lacking thioredoxin reductase. *Cancer Res.* **59,** 3175–3179.

Moos, P. J., Edes, K., Cassidy, P., Massuda, E., and Fitzpatrick, F. A. (2003). Electrophilic prostaglandins and lipid aldehydes repress redox-sensitive transcription factors p53 and hypoxia-inducible factor by impairing the selenoprotein thioredoxin reductase. *J. Biol. Chem.* **278,** 745–750.

Noda, M., Selinger, Z., Scolnick, E. M., and Bassin, R. H. (1983). Flat revertants isolated from Kirsten sarcoma virus-transformed cells are resistant to the action of specific oncogenes. *Proc. Natl. Acad. Sci. USA* **80,** 5602–5606.

Nordberg, J., Zhong, L., Holmgren, A., and Arnér, E. S. (1998). Mammalian thioredoxin reductase is irreversibly inhibited by dinitrohalobenzenes by alkylation of both the redox active selenocysteine and its neighboring cysteine residue. *J. Biol. Chem.* **273,** 10835–10842.

Rayman, M. P., Combs, G. F., Jr., and Waters, D. J. (2009). Selenium and vitamin E supplementation for cancer prevention. *JAMA* **301,** 1876.

Rundlöf, A. K., and Arnér, E. S. J. (2004). Regulation of the mammalian selenoprotein thioredoxin reductase 1 in relation to cellular phenotype, growth, and signaling events. *Antioxid. Redox Signal.* **6,** 41–52.

Rundlöf, A. K., Janard, M., Miranda-Vizuete, A., and Arnér, E. S. (2004). Evidence for intriguingly complex transcription of human thioredoxin reductase 1. *Free Radic. Biol. Med.* **36,** 641–656.

Selenius, M., Rundlof, A. K., Olm, E., Fernandes, A. P., and Bjornstedt, M. (2010). Selenium and selenoprotein thioredoxin reductase in the prevention, treatment and diagnostics of cancer. *Antioxid. Redox Signal.* **12,** 867–880.

Smith, A. D., and Levander, O. A. (2002). High-throughput 96-well microplate assays for determining specific activities of glutathione peroxidase and thioredoxin reductase. *Methods Enzymol.* **347,** 113–121.

Steinbrenner, H., and Sies, H. (2009). Protection against reactive oxygen species by selenoproteins. *Biochim. Biophys. Acta* **1790,** 1478–1485.

Urig, S., and Becker, K. (2006). On the potential of thioredoxin reductase inhibitors for cancer therapy. *Semin. Cancer Biol.* **16,** 452–465.

Watson, W. H., Heilman, J. M., Hughes, L. L., and Spielberger, J. C. (2008). Thioredoxin reductase-1 knock down does not result in thioredoxin-1 oxidation. *Biochem. Biophys. Res. Commun.* **368,** 832–836.

Xu, X. M., Yoo, M. H., Carlson, B. A., Gladyshev, V. N., and Hatfield, D. L. (2009). Simultaneous knockdown of the expression of two genes using multiple shRNAs and subsequent knock-in of their expression. *Nat. Protoc.* **4,** 1338–1348.

Yan, C., Shieh, B., Reigan, P., Zhang, Z., Colucci, M. A., Chilloux, A., Newsome, J. J., Siegel, D., Chan, D., Moody, C. J., and Ross, D. (2009). Potent activity of indolequinones against human pancreatic cancer: Identification of thioredoxin reductase as a potential target. *Mol. Pharmacol.* **76,** 163–172.

Yoo, M. H., Xu, X. M., Carlson, B. A., Gladyshev, V. N., and Hatfield, D. L. (2006). Thioredoxin reductase 1 deficiency reverses tumor phenotype and tumorigenicity of lung carcinoma cells. *J. Biol. Chem.* **281,** 13005–13008.

Yoo, M. H., Xu, X. M., Carlson, B. A., Patterson, A. D., Gladyshev, V. N., and Hatfield, D. L. (2007a). Targeting thioredoxin reductase 1 reduction in cancer cells inhibits self-sufficient growth and DNA replication. *PLoS ONE* **2,** e1112.

Yoo, M. H., Xu, X. M., Turanov, A. A., Carlson, B. A., Gladyshev, V. N., and Hatfield, D. L. (2007b). A new strategy for assessing selenoprotein function: siRNA knockdown/knock-in targeting the 3'-UTR. *RNA* **13,** 921–929.

Zeng, H., and Combs, G. F., Jr. (2008). Selenium as an anticancer nutrient: Roles in cell proliferation and tumor cell invasion. *J. Nutr. Biochem.* **19,** 1–7.

Zhong, L., and Holmgren, A. (2000). Essential role of selenium in the catalytic activities of mammalian thioredoxin reductase revealed by characterization of recombinant enzymes with selenocysteine mutations. *J. Biol. Chem.* **275,** 18121–18128.

Zhong, L., Arnér, E. S., Ljung, J., Aslund, F., and Holmgren, A. (1998). Rat and calf thioredoxin reductase are homologous to glutathione reductase with a carboxyl-terminal elongation containing a conserved catalytically active penultimate selenocysteine residue. *J. Biol. Chem.* **273,** 8581–8591.

CHAPTER SIXTEEN

REGULATION OF APOPTOSIS SIGNAL-REGULATING KINASE 1 IN REDOX SIGNALING

Kazumi Katagiri, Atsushi Matsuzawa, *and* Hidenori Ichijo

Contents

1. Overview	278
2. Materials	281
2.1. Cell culture	281
2.2. Reagents	282
2.3. Plasmids	282
2.4. Antibodies	282
2.5. Solutions	282
3. Methods	283
3.1. *In vitro* binding assay for ASK1 and Trx	283
3.2. Detection of endogenous ASK1 activation by phospho- ASK antibody	285
3.3. *In vitro* kinase assay	286
4. Comment	286
References	287

Abstract

Apoptosis signal-regulating kinase 1 (ASK1) is a member of the mitogen-activated protein kinase (MAPK) kinase kinase family and elicits a wide variety of cellular responses to various types of stress through activation of the JNK and p38 MAPK pathways. ASK1 is preferentially activated in response to oxidative stress, but this regulatory mechanism is still not completely understood. In our previous report, thioredoxin (Trx), which is an antioxidant protein and plays pivotal roles in maintaining intracellular redox balance, inhibited ASK1 kinase activity by direct binding to ASK1 under normal conditions. Under oxidative conditions, ASK1 is dissociated from Trx and therefore fully activated. The active site of Trx contains two cysteine residues that undergo

Laboratory of Cell Signaling, Graduate School of Pharmaceutical Sciences, The University of Tokyo, Hongo, Bunkyo-ku, Tokyo, Japan reversible oxidation to form a disulfide bond with each other, so that the conformation of Trx is changed by intracellular redox conditions. Thus, the oxidative stress-induced conformational change of Trx is particularly important for interaction with and regulation of ASK1, and elucidation of the regulatory mechanisms of ASK1 by Trx is critical to understanding the intracellular redox signaling. In this chapter, we review the regulatory mechanisms of ASK1 activity by Trx, and describe a method for monitoring *in vitro* binding between Trx and ASK1 under various redox conditions. In addition, we present methods to detect the oxidative stress-induced activation of ASK1 in the cells by Western blot analysis and *in vitro* kinase assay. The techniques presented in this chapter will be useful for a range of investigations into intracellular redox signaling.

1. Overview

Living organisms are constantly exposed to reactive oxygen species (ROS). While ROS have an important function as signal mediators, their overproduction in cells induces oxidative stress, leading to apoptosis (Genestra, 2007; Janssen-Heininger *et al.*, 2008; Orrenius *et al.*, 2007). Therefore, the cells must have defense mechanisms against excessive production of ROS. The thioredoxin (Trx) system is one of the major intracellular defense mechanisms against oxidative stress (Powis and Montfort, 2001). Trx is an antioxidant protein and contains two cysteine residues in the active site. Under reduced conditions, the two cysteines in the active site of Trx have two thiols. Under oxidized conditions, these thiols undergo reversible oxidation to form an intramolecular disulfide bond with each other.

Previously, we showed that Trx inhibited oxidative stress-induced apoptosis signaling (Saitoh *et al.*, 1998). In unstimulated cells, Trx bound directly to apoptosis signal-regulating kinase 1 (ASK1) and inhibited ASK1 kinase activity. ASK1 is a member of the mitogen-activated protein kinase (MAPK) kinase kinase family and activates the MAPK kinase 4 (MKK4)/MKK7-JNK and MKK3/6-p38 pathways in response to various stresses, including oxidative stress, tumor necrosis factor-α (TNF-α), lipopolysaccharide (LPS), extracellular ATP, endoplasmic reticulum stress, and calcium overload (Ichijo *et al.*, 1997; Matsuzawa *et al.*, 2005; Nishitoh *et al.*, 1998, 2002; Noguchi *et al.*, 2008; Takeda *et al.*, 2004, 2008; Tobiume *et al.*, 2001). TNF-α, LPS, and ATP activate ASK1 through production of ROS. Our recent studies have suggested that ROS-dependent activation of ASK1 is required for the oxidative stress-induced death of various cells (Noguchi *et al.*, 2008; Tobiume *et al.*, 2001). ASK1 is constitutively oligomerized through its C-terminal coiled-coil (CCC) domain, which is necessary for basal activity (Gotoh and Cooper, 1998; Tobiume *et al.*, 2002). In unstimulated cells, Trx inhibits the homophilic interaction through the N-terminal coiled-coil (NCC) domain of ASK1, which is required for ROS-induced full activation of ASK1 (Fujino et al., 2007). Under oxidative stress, Trx is dissociated from ASK1, and autophosphorylation-dependent full activation of ASK1 is induced by tight oligomerization of ASK1 through its NCC domain. Thus, Trx has pivotal roles as a safety lock in apoptosis signaling (Fig. 16.1). Furthermore, an *in vitro* binding assay revealed that ASK1 interacts with the reduced form of Trx, but not with the oxidized form of Trx or with a Trx mutant in which both active site cysteines are replaced with serine residues, resulting in the loss of the reducing activity (Saitoh et al., 1998). This result indicates that the binding between Trx and ASK1 depends on the redox state of Trx. Our methods for the *in vitro* binding assay of ASK1 and Trx are presented below as Methods 1–3, and the results of the assay are presented in Fig. 16.2.

Next, we will introduce a method to measure ASK1 kinase activity in the cells. In our previous report, ASK1 activity was shown to be tightly regulated by phosphorylation at the activation segment of the kinase domain of ASK1 (Tobiume et al., 2002). Thr 838 and Thr 845 are located within the activation segment of human and mouse ASK1, respectively,

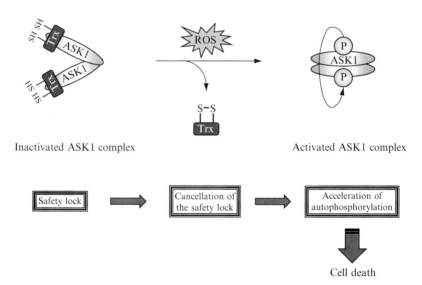

Figure 16.1 Schematic model of ROS-induced activation of ASK1 regulated by Trx. ASK1 is constitutively oligomerized through its CCC domain, and its kinase activity is inhibited by direct binding with Trx. The production of ROS induces oxidization of Trx and dissociation of Trx from ASK1. Following tight oligomerization of ASK1 through the NCC domain, autophosphorylation and activation of ASK1 are promoted.

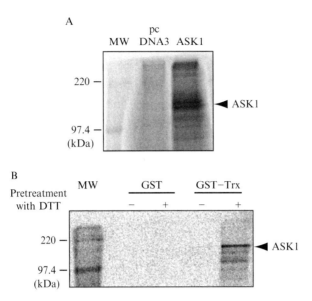

Figure 16.2 The interaction between ASK1 and Trx depends on the redox state of Trx. (A) Expression of *in vitro*-translated ^{35}S-ASK1 by using a TNT T7 Quick Coupled Transcription/Translation System. The expression plasmid pcDNA3 for 3′-HA-tagged ASK1 was used, and the intact pcDNA3 vector was used as a negative control. (B) Binding of *in vitro*-translated ^{35}S-ASK1 to the reduced form of GST-Trx protein. Purified GST-fused Trx or GST alone as a negative control was immobilized on the beads, then treated with 1 mM DTT or left untreated on ice for 30 min, washed with buffer A, and subjected to an *in vitro* binding assay as described in Method 3. MW; molecular weight.

and the phosphorylation of these threonine residues is essential for the activation of ASK1 (Fig. 16.3). The phospho-ASK antibody, which recognizes the phosphorylation of these threonine residues, can be used to monitor the activation status of ASK1 in response to a diverse array of stresses, including oxidative stress. Below, we present our method for using this antibody in a Western blot analysis to detect the oxidative stress-induced activation of ASK1 (Method 4), and the results of the analysis are presented in Fig. 16.4.

The kinase activity of ASK1 can be also detected by *in vitro* kinase assay. ASK1, a member of the MAPK kinase kinase family, phosphorylates MAPK kinases such as MKK4/MKK7 and MKK3/MKK6 as substrates. Therefore, we also describe a method for *in vitro* kinase assay using recombinant kinase-dead MKK6, GST-MKK6KD, as a substrate (Method 5). This method can be used to evaluate ASK1 kinase activity and can also be applied to other kinases.

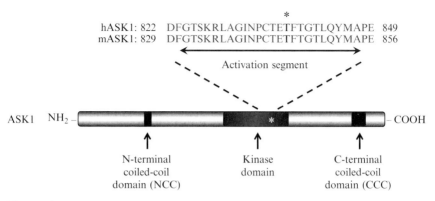

Figure 16.3 Schematic representation of the domain structure of ASK1. Two coiled-coil domains (NCC and CCC) play pivotal roles in the homophilic interaction and activation of ASK1. The upper part of this figure shows the amino acid sequences of the activation segments in the ASK1 kinase domain. The numbers indicate amino acid positions of both ends of the activation segments. A phosphorylation site is essential for the activation of ASK1 (Thr 838 and Thr 845 of human and mouse ASK1, respectively), and is indicated by an asterisk (*). The phospho-ASK antibody recognizes the phosphorylation of these threonine residues.

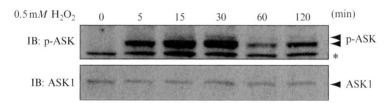

Figure 16.4 H_2O_2-induced activation of endogenous ASK1. Confluent HEK293A cells were treated with 0.5 mM H_2O_2 for the indicated periods. The cell extracts were subjected to immunoblotting with anti-total ASK1 and anti-phospho-ASK antibodies. The asterisk (*) indicates a nonspecific reactive band.

2. Materials

2.1. Cell culture

HEK293A cells were grown in Dulbecco's modified Eagle's medium (DMEM; Sigma, St. Louis, MO) containing 10% heat-inactivated fetal bovine serum (FBS; Bio West, Nuaille, France), 4.5 g/l glucose, and 100 units/ml penicillin-G (Meiji Seika, Tokyo, Japan) in 5% CO_2 at 37 °C.

2.2. Reagents

Transfection was performed with Tfx-50 (Promega, Madison, WI) or FuGENE6 (Roche, Mannheim, Germany) according to the manufacturer's instructions. Immunoblot analysis was performed with an ECL system (GE Healthcare, Buckinghamshire, UK). H_2O_2 was purchased from Wako (Tokyo, Japan). Protein G-Sepharose beads, protein A-Sepharose beads, and glutathione-Sepharose beads were purchased from GE Healthcare. LB medium was purchased from Sigma. *In vitro*-translated ASK1 was prepared with a TNT T7 Quick Coupled Transcription/Translation System (Promega).

2.3. Plasmids

HA-tagged human ASK1 expression plasmids in pcDNA3 vector (Invitrogen, Carlsbad, CA) were generated as described previously (Ichijo *et al.*, 1997; Saitoh *et al.*, 1998). The plasmids of GST-tagged human Trx and GST-tagged human MKK6KD for the bacterial fusion protein were constructed in pGEX-4T-1 vector (GE Healthcare) (Saitoh *et al.*, 1998).

2.4. Antibodies

Antibodies to HA (12CA5 and 3F10; Roche), GST (GE Healthcare), and Trx (FL-105; Santa Cruz Biotechnology, Santa Cruz, CA) were purchased from commercial sources. Rabbit polyclonal antibody to phospho-ASK was described previously (Tobiume *et al.*, 2002). Rat monoclonal antibody to ASK1 (TC003; Wako) was established using immunized rats with a GST-tagged C-terminal fragment of mouse ASK1 corresponding to amino acids 948–1380 as described previously (Takeda *et al.*, 2007).

2.5. Solutions

TBS-T: 20 mM Tris–HCl, pH 8.0, 137 mM NaCl, and 0.1% (v/v) Tween 20
Buffer A: 20 mM Tris–HCl, pH 7.5, 150 mM NaCl, 10 mM EDTA, 1% (v/v) Triton X-100, and 1% (w/v) deoxycholate
Lysis buffer: 20 mM Tris–HCl, pH 7.5, 150 mM NaCl, 10 mM EDTA, 1% (v/v) Triton X-100, 1% (w/v) deoxycholate, 1 mM phenylmethylsulfonyl fluoride (PMSF), and 5 μg/ml leupeptin or 150 units/ml aprotinin
IVK lysis buffer: 20 mM Tris–HCl, pH 7.5, 150 mM NaCl, 5 mM EGTA, 1% (v/v) Triton X-100, 1% (w/v) deoxycholate, 12 mM β-glycerophosphate, 10 mM NaF, 1 mM sodium orthovanadate, 3 mM dithiothreitol (DTT), 1 mM PMSF, and 5 μg/ml leupeptin or 150 units/ml aprotinin

Wash buffer: 20 mM Tris–HCl, pH 7.5, 150 mM NaCl, 5 mM EGTA, 2 mM DTT, and 1 mM PMSF

IVK buffer: 20 mM Tris–HCl, pH 7.5, 20 mM $MgCl_2$, 100 μM ATP, and 5 μCi of [γ-^{32}P] ATP (Perkin-Elmer, Norwalk, CT)

$2 \times$ *SDS-sample buffer*: 80 mM Tris–HCl, pH 8.8, 80 mg/l bromophenol blue, 28.8% (v/v) glycerol, and 4% (w/v) sodium dodecyl sulfate (SDS)

Fixative 1: 7% (v/v) acetic acid and 30% (v/v) methanol

Fixative 2: 4% (v/v) glycerol and 25% (v/v) methanol

$10 \times PBS$: 8% (w/v) NaCl, 0.2% (w/v) KCl, 2.9% (w/v) $Na_2HPO_4 \cdot 12H_2O$, and 0.2% (w/v) KH_2PO_4

3. Methods

3.1. *In vitro* binding assay for ASK1 and Trx

3.1.1. Method 1: Preparation of glutathione *S*-transferase-tagged thioredoxin (GST-Trx)

(1) To purify GST-Trx, *Escherichia coli* BL21 cells transformed with pGEX-4T-1 vector for expression of GST-Trx are cultured in 50 ml LB medium containing 20 mg ampicillin/l.

(2) When the cells are grown to confluence, 50 ml of culture medium is all added to 500 ml LB medium containing 20 mg ampicillin/l and placed in a shaking incubator (37 °C, 150 rpm) for 2 h.

(3) After incubation, the expression of GST-Trx is induced by further incubation with isopropyl-β-D-thiogalactopyranoside (IPTG; 0.1 mM final concentration) in a shaking incubator (30 °C, 150 rpm) for 2 h.

(4) After induction, the culture medium is centrifuged at 5,000 rpm for 20 min to collect *E. coli* cells. Following the removal of supernatant, the pelleted cells are resuspended in 30 ml PBS containing 10 mM EDTA. This suspension is centrifuged at 3,000 rpm for 30 min and the supernatant is removed.

(5) The pelleted cells are resuspended in 30 ml PBS containing 10 mM EDTA and 1% (v/v) Triton X-100, and the suspension is sonicated to destroy the cell bodies. To prevent increased temperature, the suspension is sometimes placed on ice.

(6) The sonicated suspension is centrifuged at 18,500 rpm for 30 min at 4 °C, and the supernatant fluid is separated.

(7) The supernatant is incubated at 4 °C on a rotator with about 1 ml of glutathione-Sepharose beads that have been washed three times with PBS containing 10 mM EDTA.

(8) After overnight incubation, to purify GST-Trx immobilized on glutathione-Sepharose beads and remove Triton X-100, the beads are sufficiently washed with PBS containing 10 mM EDTA.
(9) To check the purification of GST-Trx, the proteins immobilized on the beads are boiled at 98 °C for 5 min with the same volume of SDS-sample buffer containing 1 mM DTT, and each sample is separated by sodium dodecyl sulfate–polyacrylamide gel electrophoresis (SDS–PAGE).
(10) The separated proteins are fixed with Fixative 1 and Fixative 2 solution.
(11) The gel is stained with Coomassie brilliant blue.

3.1.2. Method 2: Preparation of ASK1 protein

It is difficult to purify a full-length ASK1 protein from *E. coli*, because ASK1 may be partitioned into the insoluble fraction. Therefore, we recommend a TNT T7 Quick Coupled Transcription/Translation System (Promega). The T7 promoter is the most efficient for the transcription and translation in this system. Thus, in this procedure, the 3′-HA-tagged ASK1 expression plasmid in pcDNA3 is used, and the intact vector is used as a negative control.

The experiment is performed according to the procedure recommended by Promega, except that twice the amount of plasmid is added to the reaction mixture. Because the binding between ASK1 and Trx is dependent on the expression levels of radiolabeled and epitope-tagged ASK1, these expression levels should be determined by detection of radiolabel incorporation and immunoblotting analysis. Our methods for these procedures are presented below.

(I) Detection of radiolabel incorporation
  (1) After *in vitro* translation, ^{35}S-labeled ASK1 protein is separated by the SDS–PAGE method used in Method 1-(9) and -(10).
  (2) The fixed gel is dried, and the labeled proteins are detected by an image analyzer. The results are indicated in Fig. 16.2 A.
(II) Detection of epitope-tagged ASK1
  (1) Because the reaction mixture is crude, it is difficult to detect 3′-HA-tagged ASK1 by Western blot analysis. To remove unnecessary proteins, immunoprecipitation is performed. The reaction mixture is diluted about 10-fold with buffer A and incubated with anti-HA antibody (12CA5; Roche) at 4 °C on a rotator.
  (2) After overnight incubation, the solution is incubated with protein G-Sepharose beads for 1 h at 4 °C on a rotator.
  (3) The beads are washed three times with buffer A.

(4) To assess the expression of 3′-HA-tagged ASK1, the proteins immobilized on the beads are resolved by SDS–PAGE and transferred to a polyvinylidene difluoride membrane.
(5) After blocking with 5% skim milk in TBS-T, the membrane is probed with anti-HA antibody (3F10; Roche). An ECL system (GE Healthcare) is used for the detection.

3.1.3. Method 3: *In vitro* binding assay

(1) To prepare the reduced form of Trx, GST-Trx immobilized on the glutathione-Sepharose beads is incubated with buffer A containing 1 mM DTT at 4 °C for 30 min on a rotator.
(2) The beads are washed three times with buffer A for the removal of DTT.
(3) *In vitro*-translated ASK1 is incubated with DTT-treated or -untreated beads in buffer A at 4 °C for 30 min. For easy detection, we recommend that more than 50 μl of reaction mixture be incubated with the beads in each sample. If the volume of beads is too large, nonspecific binding may be more detectable. Therefore, we often use 15 μl of beads in each sample.
(4) After incubation, the beads are washed five times with buffer A.
(5) Each sample is analyzed by SDS–PAGE and Western blotting. In this assay, the amount of GST-Trx should be equal in each sample. To evaluate the amount of GST-Trx, the lower part of the SDS–PAGE gel is cut out, transferred, and immunoblotted with anti-GST antibody (GE Healthcare) or anti-Trx antibody (FL-105; Santa Cruz Biotechnology). To detect ^{35}S-labeled ASK1, the upper part of the SDS–PAGE gel is analyzed in the same manner as in Method 2-(I). The results of this experiment are presented in Fig. 16.2 B. In this experiment, GST protein was used as a negative control.

3.2. Detection of endogenous ASK1 activation by phospho- ASK antibody

3.2.1. Method 4

(1) HEK293A cells are grown in DMEM supplemented with 10% heat-inactivated FBS and 100 units/ml of penicillin-G in 5% CO_2 at 37 °C.
(2) Confluent HEK293A cells in a 24-well plate are stimulated with 0.5 mM H_2O_2.
(3) After the indicated periods, the cells are lysed by 120 μl of lysis buffer supplemented with PhosSTOP (Roche) at 4 °C.
(4) 40 μl of the cell extract is subjected to immunoblotting with anti-total ASK1 and anti-phospho-ASK antibodies. The immunoblotting is

performed as described above in Method 2-(II)-(4) and -(5). The results are shown in Fig. 16.4.

Antibodies against activated ASK1 are commercially available from several companies, including Cell Signaling Technology and Wako.

3.3. *In vitro* kinase assay

3.3.1. Method 5

(1) Confluent HEK293A cells expressing 3′-HA-tagged ASK1 in a 6-well plate are treated with 1 mM H_2O_2 or left untreated. Transfection is performed using Tfx-50 (Promega) or FuGENE6 (Roche) according to the manufacturer's instruction.
(2) After stimulation, the cells are lysed by 800 μl of IVK lysis buffer at 4 °C.
(3) To determine the expression level of 3′-HA-tagged ASK1, 50 μl of the cell extract is subjected to immunoblotting with anti-HA antibody (3F10) by the above-described method.
(4) 750 μl of the cell extract is incubated with anti-HA antibody (12CA5) for 1 h on ice, and then 20 μl of protein-A Sepharose is added and the extract is incubated for an additional 20 min on ice.
(5) The beads are washed twice with wash buffer and incubated with 0.2 g of GST-MKK6KD in IVK buffer at room temperature for 10 min (final volume 30 μl).
(6) The reaction is terminated by addition of SDS-sample buffer and the supernatant is subjected to SDS–PAGE.
(7) The fixed and dried gel or transferred membrane is exposed to a BAS imaging plate (Fuji, Tokyo, Japan). ASK1 activity is obtained as an image of phosphorylated GST-MKK6KD. This method was previously performed in a study by Tobiume *et al.* (2002).

4. Comment

Many previous reports have demonstrated that ASK1 is regulated by various binding partners (Takeda *et al.*, 2008). We identified Trx as one of the key components in the regulation of oxidative stress-induced activation of ASK1 (Saitoh *et al.*, 1998). Trx binds directly to ASK1 and inhibits its kinase activity in the cells. Although the precise inhibitory mechanism of Trx in ASK1 activation is not yet completely understood, in this chapter we showed by *in vitro* binding assay that the binding of Trx to ASK1 can be monitored by controlling the redox state of Trx. The advantage of this method is that the redox state of recombinant proteins can be directly controlled *in vitro*. It is difficult to regulate the redox state of proteins in cell lines. Although GST-Trx protein immobilized on the beads can be stored at 4 °C for about 1 month, recombinant proteins should be used as soon as possible. Furthermore, because Trx could be oxidized in air, pretreatment with DTT (as described in Method 3-(1)) should be performed just before incubation with *in vitro*-translated ASK1.

Phospho-ASK antibodies, such as that used in the present study, have been shown to be a useful and accurate tool for detecting and quantifying the amount of activated ASK1 (Tobiume *et al.*, 2002). In recent years, several other antibodies against activated ASK1 have also become commercially available. The use of such antibodies to monitor the amount of activated ASK1 under various conditions will be crucial to improving our understanding of the physiological and pathophysiological roles of ASK1.

In this chapter, we focused on redox signaling mediated by the ASK1–Trx system. Recently, many reports have suggested that ROS are produced not only as by-products of aerobic metabolism but also as signaling mediators controlling cell proliferation, migration, and differentiation (Janssen-Heininger *et al.*, 2008; Rhee *et al.*, 2000). Redox-regulating and -regulated proteins, including Trx and ASK1, may also play important roles in these physiological events. We hope that this volume will engender new biochemical and biomedical investigations into the roles of ROS.

REFERENCES

Fujino, G., Noguchi, T., Matsuzawa, A., Yamauchi, S., Saitoh, M., Takeda, K., and Ichijo, H. (2007). Thioredoxin and TRAF family proteins regulate reactive oxygen species-dependent activation of ASK1 through reciprocal modulation of the N-terminal homophilic interaction of ASK1. *Mol. Cell. Biol.* **27,** 8152–8163.

Genestra, M. (2007). Oxyl radicals, redox-sensitive signalling cascades and antioxidants. *Cell. Signal.* **19,** 1807–1819.

Gotoh, Y., and Cooper, J. A. (1998). Reactive oxygen species- and dimerization-induced activation of apoptosis signal-regulating kinase 1 in tumor necrosis factor-alpha signal transduction. *J. Biol. Chem.* **273,** 17477–17482.

Ichijo, H., Nishida, E., Irie, K., ten Dijke, P., Saitoh, M., Moriguchi, T., Takagi, M., Matsumoto, K., Miyazono, K., and Gotoh, Y. (1997). Induction of apoptosis by ASK1, a mammalian MAPKKK that activates SAPK/JNK and p38 signaling pathways. *Science* **275,** 90–94.

Janssen-Heininger, Y. M., Mossman, B. T., Heintz, N. H., Forman, H. J., Kalyanaraman, B., Finkel, T., Stamler, J. S., Rhee, S. G., and van der Vliet, A. (2008). Redox-based regulation of signal transduction: principles, pitfalls, and promises. *Free Radic. Biol. Med.* **45,** 1–17.

Matsuzawa, A., Saegusa, K., Noguchi, T., Sadamitsu, C., Nishitoh, H., Nagai, S., Koyasu, S., Matsumoto, K., Takeda, K., and Ichijo, H. (2005). ROS-dependent activation of the TRAF6-ASK1-p38 pathway is selectively required for TLR4-mediated innate immunity. *Nat. Immunol.* **6,** 587–592.

Nishitoh, H., Saitoh, M., Mochida, Y., Takeda, K., Nakano, H., Rothe, M., Miyazono, K., and Ichijo, H. (1998). ASK1 is essential for JNK/SAPK activation by TRAF2. *Mol. Cell* **2,** 389–395.

Nishitoh, H., Matsuzawa, A., Tobiume, K., Saegusa, K., Takeda, K., Inoue, K., Hori, S., Kakizuka, A., and Ichijo, H. (2002). ASK1 is essential for endoplasmic reticulum stress-induced neuronal cell death triggered by expanded polyglutamine repeats. *Genes Dev.* **16,** 1345–1355.

Noguchi, T., Ishii, K., Fukutomi, H., Naguro, I., Matsuzawa, A., Takeda, K., and Ichijo, H. (2008). Requirement of reactive oxygen species-dependent activation of ASK1-p38 MAPK pathway for extracellular ATP-induced apoptosis in macrophage. *J. Biol. Chem.* **283,** 7657–7665.

Orrenius, S., Gogvadze, V., and Zhivotovsky, B. (2007). Mitochondrial oxidative stress: Implications for cell death. *Annu. Rev. Pharmacol. Toxicol.* **47,** 143–183.

Powis, G., and Montfort, W. R. (2001). Properties and biological activities of thioredoxins. *Annu. Rev. Biophys. Biomol. Struct.* **30,** 421–455.

Rhee, S. G., Bae, Y. S., Lee, S. R., and Kwon, J. (2000). Hydrogen peroxide: A key messenger that modulates protein phosphorylation through cysteine oxidation. *Sci. STKE* **2000,** pe1.

Saitoh, M., Nishitoh, H., Fujii, M., Takeda, K., Tobiume, K., Sawada, Y., Kawabata, M., Miyazono, K., and Ichijo, H. (1998). Mammalian thioredoxin is a direct inhibitor of apoptosis signal-regulating kinase (ASK) 1. *EMBO J.* **17,** 2596–2606.

Takeda, K., Matsuzawa, A., Nishitoh, H., Tobiume, K., Kishida, S., Ninomiya-Tsuji, J., Matsumoto, K., and Ichijo, H. (2004). Involvement of ASK1 in Ca^{2+}-induced p38 MAP kinase activation. *EMBO Rep.* **5,** 161–166.

Takeda, K., Shimozono, R., Noguchi, T., Umeda, T., Morimoto, Y., Naguro, I., Tobiume, K., Saitoh, M., Matsuzawa, A., and Ichijo, H. (2007). Apoptosis signal-regulating kinase (ASK) 2 functions as a mitogen-activated protein kinase in a heteromeric complex with ASK1. *J. Biol. Chem.* **282,** 7522–7531.

Takeda, K., Noguchi, T., Naguro, I., and Ichijo, H. (2008). Apoptosis signal-regulating kinase 1 in stress and immune response. *Annu. Rev. Pharmacol. Toxicol.* **48,** 199–225.

Tobiume, K., Matsuzawa, A., Takahashi, T., Nishitoh, H., Morita, K., Takeda, K., Minowa, O., Miyazono, K., Noda, T., and Ichijo, H. (2001). ASK1 is required for sustained activations of JNK/p38 MAP kinases and apoptosis. *EMBO Rep.* **2,** 222–228.

Tobiume, K., Saitoh, M., and Ichijo, H. (2002). Activation of apoptosis signal-regulating kinase 1 by the stress-induced activating phosphorylation of pre-formed oligomer. *J. Cell. Physiol.* **191,** 95–104.

CHAPTER SEVENTEEN

Protocols for the Detection of S-Glutathionylated and S-Nitrosylated Proteins In Situ

Scott W. Aesif,[*] Yvonne M. W. Janssen-Heininger,[*] and Niki L. Reynaert[†]

Contents

1. Introduction	290
2. Protein S-Glutathionylation	290
2.1. In situ detection of S-glutathionylated proteins	291
3. Protein S-Nitrosylation	293
3.1. In situ detection of S-nitrosylated proteins	294
4. Summary	295
References	295

Abstract

The oxidation of protein cysteine residues represents significant posttranslational modifications that impact a wide variety of signal transduction cascades and diverse biological processes. Oxidation of cysteines occurs through reactions with reactive oxygen as well as nitrogen species. These oxidative events can lead to irreversible modifications, such as the formation of sulfonic acids, or manifest as reversible modifications such as the conjugation of glutathione with the cysteine moiety, a process termed S-glutathionylation (also referred to as S-glutathiolation, or protein mixed disulfides). Similarly, S-nitrosothiols can also react with the thiol group in a process known as S-nitrosylation (or S-nitrosation). It is the latter two events that have recently come to the forefront of cellular biology through their ability to reversibly impact numerous cellular processes. Herein we describe two protocols for the detection of S-glutathionylated or S-nitrosylated proteins in situ. The protocol for the detection of S-glutathionylated proteins relies on the catalytic specificity of glutaredoxin-1 for the reduction of S-glutathionylated proteins. The protocol for the detection of S-nitrosylated proteins represents a modification of the previously described biotin switch protocol, which relies on ascorbate in the presence of chelators to

[*] Department of Pathology, University of Vermont College of Medicine, Burlington, Vermont, USA
[†] Department of Respiratory Medicine, Maastricht University, Maastricht, The Netherlands

Methods in Enzymology, Volume 474 © 2010 Elsevier Inc.
ISSN 0076-6879, DOI: 10.1016/S0076-6879(10)74017-9 All rights reserved.

decompose *S*-nitrosylated proteins. These techniques can be applied *in situ* to elucidate which compartments in tissues are affected in diseased states whose underlying pathologies are thought to represent a redox imbalance.

1. Introduction

Cysteines with low pK_a values ranging from 4 to 5 as dictated by the charges and the electrophilic nature of surrounding amino acids are described as being "reactive" with regard to their susceptibility to undergo oxidative modifications. The pK_a of most cysteines within a protein's structure is 8.5 and not considered to be highly susceptible to oxidative modification and are thus considered "nonreactive" (Janssen-Heininger *et al.*, 2008; Meng *et al.*, 2002). In nature, free cysteines are found to be conserved within the primary sequence in numerous classes of proteins throughout species, suggesting that there may be other roles intended for cysteine residues beyond metal coordination and structural disulfide formation. Notably, there are numerous redox-based modifications applicable to cysteine residues, including hydroxylation (SOH), nitrosylation (SNO), glutathionylation (SSG), and the formation of inter/intramolecular disulfides (S–S), with each modification having the potential to modify not only protein structure but also function in a widely diverse fashion (Giles *et al.*, 2003; Jacob *et al.*, 2003; Ji *et al.*, 1999; Kim *et al.*, 2002; Mallis and Thomas, 2000).

2. Protein *S*-Glutathionylation

Protein *S*-glutathionylation represents the conjugation of the low molecular weight antioxidant molecule glutathione to cysteines within a protein. As described, during oxidative stress, cysteines are among the most vulnerable with regard to oxidative modifications. As an antioxidant molecule, GSH is present within cells at millimolar concentrations (1–10 mM) and it is believed that conjugation of GSH to oxidized cysteines serves a protective mechanism in the prevention of overoxidation. More recently, however, protein *S*-glutathionylation has been compared to O-phosphorylation with regard to its impact on protein structure and function, and its ability to reversibly impact signaling pathways (Dalle-Donne *et al.*, 2007; Ghezzi, 2005; Holmgren *et al.*, 2005). This comparison is further bolstered by the observation that protein *S*-glutathionylation occurs not only in response to overt oxidative injury but also in pathophysiological states, and in settings where ratios of GSH to oxidized GSH (GSSG) are low (i.e., 100:1 vs. 3:1). Further, the heterogeneity of cysteine reactivity with oxidants imparts a unique specificity toward protein *S*-glutathionylation formation. Finally, there exist mechanisms by which S-glutathionylation formation and resolution are exquisitely regulated (Dalle-Donne et al., 2007, 2008; Gallogly and Mieyal, 2007; Ghezzi, 2005; Shelton et al., 2005).

Glutaredoxins (Grx) are low molecular weight (9–14 kDa) enzymes of the oxidoreductase class of enzymes, and under physiological conditions play an important role in the reduction of S-glutathionylated proteins. It is worthy of mention that under conditions of oxidative stress when GSSG concentrations are increased, Grx causes increases in protein S-glutathionylation, instead of decreases. Of the mammalian isoforms, Grx1 is the most extensively described and characterized. Localized primarily in the cytoplasm, Grx1 compared to other thiol–disulfide oxidoreductases (i.e., Grx2 and thioredoxin (Trx)) is significantly more efficient in catalyzing the deglutathionylation of proteins. Despite its lower concentration within cells (1 μM compared to 10 μM for Trx), Grx1 has a 10-fold lower K_m value for ribonucleotide reductase and a 5000-fold higher k_{cat}/K_m for S-glutathionylated cysteines in vitro as compared to Trx (Chrestensen et al., 2000; Holmgren, 1976; Holmgren et al., 1978). Under physiological conditions where GSH/GSSG ratios are high, Grx1 functions through the monothiol mechanism in the reduction of mixed disulfides. With regard to cellular processes, Grx1 has been shown to be important in cellular differentiation, regulation of transcription factor activation, and apoptosis (Anathy et al., 2009; Pineda-Molina et al., 2001; Takashima et al., 1999). We have utilized the catalytic specificity of Grx1 to successfully detect S-glutathionylated proteins in paraffin embedded tissues in situ. The protocol is schematically depicted in Fig. 17.1, and described in the following section.

2.1. In situ detection of S-glutathionylated proteins

1. After dewaxing tissue samples in three changes of xylene, tissue is rehydrated in 100%, 95%, and 75% ethanol, and washed in one change of Tris-buffered saline (TBS). Free thiol groups are blocked with 40 mM

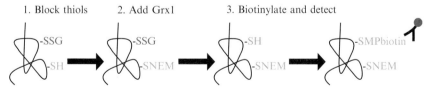

Figure 17.1 Schematic representation of the Grx1-catalyzed cysteine derivatization protocol. Free thiols are blocked by alkylation with NEM (1), followed by Grx1-catalyzed reduction of S-glutathionylated proteins (2), and finally newly reduced cysteines are labeled using MPB and detected using streptavidin-conjugated fluorophores (3).

N-ethylmaleimide (NEM) in buffer that contains 25 mM HEPES, pH 7.4, 0.1 mM EDTA, pH 8.0, 0.01 mM neocuproine, and 1% Triton for 30 min.
2. After three washes with TBS, S-glutathionylated cysteine groups are reduced by incubation with a reaction mix that contains recombinant Grx1 and the components necessary for its activity as follows: 13.5 μg/ml recombinant human Grx1, 35 μg/ml GSSG reductase, 1 mM GSH, 1 mM NADPH, 18 μmol EDTA, and 137 mM Tris–HCl, pH 8.0 for 30 min.
3. After three washes with TBS, newly reduced cysteine residues are labeled with 1 mM N-(3-maleimidylpropionyl) biocytin (MPB) dissolved in 25 mM HEPES, pH 7.4, 0.1 mM EDTA, pH 8.0, 0.01 mM neocuproine for 30 min.
4. Excess MPB is removed by three washes with TBS.
5. Tissue samples are incubated with 0.5 μg/ml streptavidin-conjugated fluorophores for 30 min. Nuclei are counter stained using appropriate concentrations of dye (Fig. 17.2 A and B).

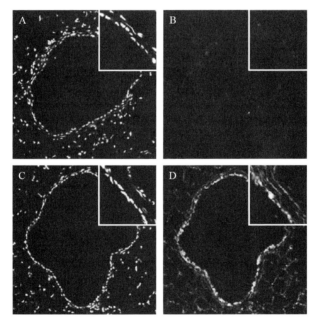

Figure 17.2 Representative image of untreated paraffin embedded murine lungs stained for S-glutathionylated proteins using the Grx1-catalyzed derivatization protocol and visualization via confocal laser scanning microscopy (B and D). Panels A and C represent nuclear counter stains for the tissues evaluated in (B) and (D), respectively. Panel D: Following the protocol outlined herein, reactivity is seen primarily within the epithelial cells of the conducting airways (inset) with some parenchymal reactivity. Panel B: As a negative control, GSH was omitted from the reaction mix, resulting in a loss of Grx-1-catalyzed labeling. Magnification = 200×. Insets: zoom = 4×.

Prior to derivatization of tissues, all buffers are prepared fresh, and buffers containing MPB protected from direct exposure to light. The reaction mix containing Grx1 is prepared immediately prior to application. All steps are conducted at room temperature.

Following Grx1-catalyzed cysteine derivatization, samples are routinely analyzed by confocal microscopy. Use of confocal imaging is critical for optimal detection of S-glutathionylated proteins over potential background signals, and to elucidate tissue or cellular compartments that harbor S-glutathionylated proteins which would be impossible using wide field fluorescence. The specificity of Grx1 for S-glutathionylated proteins is highlighted by the competitive inhibition of detectable signal by incubating samples with 2.5 mg/ml S-glutathionylated bovine serum albumin (BSA), as compared to fully reduced BSA. Omission of Grx1 from the reaction mix should be an important reagent control as it highlights the requirement of the presence of Grx1 in the observed labeling reaction. Furthermore, omission of free GSH from the reaction mix completely abrogates detectable signal owing to its necessity in reducing catalytically active Grx1 (Fig. 17.2). As an additional negative control, samples should be incubated with dithiothreitol (DTT) prior to incubation with NEM, in order to assess the efficiency of blocking free cysteine residues. As a positive control, tissue sections can be incubated with 400 μM diamide and 1 mM GSH for 10 min prior to application of NEM (Aesif et al., 2009; Reynaert et al., 2006).

3. Protein S-Nitrosylation

Nitric oxide (NO) is an important signaling molecule that exerts many of its effects through the posttranslational modification S-nitrosylation. S-Nitrosylation is the oxidation of cysteine thiols by NO-derived species such as N_2O_3, but also by transnitrosylation from low molecular weight nitrosothiols such as S-nitrosylated glutathione (GSNO) or nitrosylated proteins (Stamler et al., 2001). To date, a large number of targets for S-nitrosylation have been identified and linked to functional consequences. Importantly, protein regulation by S-nitrosylation has been coupled to physiological stimuli that involve both receptor-mediated activation of nitric oxide synthases (NOS), as well as stimulus-induced denitrosylation (Benhar et al., 2009; Gow et al., 2002; Hess et al., 2005). Multiple mechanisms for denitrosylation exist, including SNO decomposition by various nonenzymatic compounds, ascorbate, and two enzymatic pathways, namely Trx1 and GSNO reductase (Benhar et al., 2009).

Sample preparation and storage for the determination of nitrosothiol content and protein nitrosylation requires great care since these moieties are readily decomposed by a myriad of factors. Furthermore, only recently have technological advances been made to facilitate research into this redox-

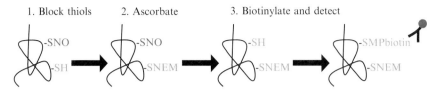

Figure 17.3 Schematic representation of the protocol for detecting S-nitrosylated proteins *in situ*. Free thiols are blocked by alkylation with NEM (1), S-nitrosylated cysteines are reduced using ascorbate (2), and finally newly reduced cysteines are labeled using MPB and detected using streptavidin-conjugated fluorophores (3).

dependent posttranslational modification. For instance, an antibody was developed against nitrosocysteine that has been used among others to visualize protein S-nitrosylation in various tissues (Gow *et al.*, 2002). S-Nitrosylation of specific proteins can be achieved by measuring the release of NO by these proteins using chemiluminescence, or by the biotin-switch protocol as described by Jaffrey and Snyder (2001) and subsequently published variations thereof. Finally, we have successfully adapted this biotin derivatization procedure to detect protein S-nitrosylation *in situ* (Ckless *et al.*, 2004). The protocol is schematically depicted in Fig. 17.3 and described in the following section.

3.1. *In situ* detection of *S*-nitrosylated proteins

1. After dewaxing tissue sections in three changes of xylene, tissue is rehydrated in 100%, 95%, and 75% ethanol, and washed three times with phosphate-buffered saline (PBS) containing 0.4 mM EDTA and 0.04 mM neocuproine. Next, free thiol groups are blocked with 40 mM NEM in PBS containing 0.4 mM EDTA, 0.04 mM neocuproine, and 2.5% SDS for 30 min.
2. After removal of blocking solution and three washes with PBS containing 0.4 mM EDTA and 0.04 mM neocuproine, sections are incubated with 1 mM sodium ascorbate in PBS for 15 min to reduce S-nitrosylated proteins.
3. Newly reduced cysteine residues are then labeled with 0.1 mM MPB in PBS for 30 min.
4. Excess MPB is removed by three washes with PBS containing 0.4 mM EDTA and 0.04 mM neocuproine.
5. Tissues are next incubated for 30 min with 0.5 µg/ml of a fluorophore-labeled streptavidin in order to visualize biotinylated proteins and nuclei are counterstained with an appropriate dye.

All procedures prior to labeling with MPB must be performed under protection from direct light and all reagents must be dissolved freshly before their use. All incubation steps in the protocol occur at room temperature.

Following the biotin switch protocol to label protein S-nitrosylation, tissues are analyzed by confocal microscopy. Use of confocal imaging is critical to optimal detection of S-nitrosylated proteins over potential background signal, and to elucidate tissue or cellular compartments that harbor S-nitrosylated proteins, which would be impossible to achieve using wide field fluorescence. As a negative control, omission of the ascorbate decomposition step should be performed, which, in the case of optimal blocking of reduced cysteines, should result in the absence of labeling. Alternatively, S-nitrosothiols can be decomposed with 1 mM ascorbate, or subjected UV-mediated prephotolysis at 335 nm (Forrester $et\ al.$, 2009) prior to blocking. These steps should result in the absence of labeling. As a positive control, tissue sections can be incubated with a nitrosothiol such as GSNO or L-CysNO (100–500 μM) prior to the application of NEM.

4. Summary

The protocols presented herein represent robust avenues for the *in situ* detection of S-glutathionylated and S-nitrosylated proteins in tissues. They are, however, not without potential and long debatable pitfalls, as neither protocol absolutely excludes the possibility of false positivity due to the reduction of intermolecular disulfides, or other oxidative modifications. The labeling and detection of newly generated thiols depend on the specificity of ascorbate and human Grx1 toward S-nitrosylated and S-glutathionylated proteins, respectively. As screening tools, however, these assays and adaptations thereof that undoubtedly will be developed in the future will provide valuable insights into which tissue or organ compartments are affected by reversible cysteine oxidations.

REFERENCES

Aesif, S. W., et al. (2009). In situ analysis of protein S-glutathionylation in lung tissue using glutaredoxin-1-catalyzed cysteine derivatization. *Am. J. Pathol.* **175,** 36–45.

Anathy, V., et al. (2009). Redox amplification of apoptosis by caspase-dependent cleavage of glutaredoxin 1 and S-glutathionylation of Fas. *J. Cell Biol.* **184,** 241–252.

Benhar, M., et al. (2009). Protein denitrosylation: Enzymatic mechanisms and cellular functions. *Nat. Rev. Mol. Cell Biol.* **10,** 721–732.

Chrestensen, C. A., et al. (2000). Acute cadmium exposure inactivates thioltransferase (Glutaredoxin), inhibits intracellular reduction of protein-glutathionyl-mixed disulfides, and initiates apoptosis. *J. Biol. Chem.* **275,** 26556–26565.

Ckless, K., et al. (2004). In situ detection and visualization of S-nitrosylated proteins following chemical derivatization: Identification of Ran GTPase as a target for S-nitrosylation. *Nitric Oxide* **11,** 216–227.

Dalle-Donne, I., et al. (2007). S-glutathionylation in protein redox regulation. *Free Radic. Biol. Med.* **43,** 883–898.

Dalle-Donne, I., et al. (2008). Molecular mechanisms and potential clinical significance of S-glutathionylation. *Antioxid. Redox Signal.* **10,** 445–473.

Forrester, M. T., et al. (2009). Detection of protein S-nitrosylation with the biotin-switch technique. *Free Radic. Biol. Med.* **46,** 119–126.

Gallogly, M. M., and Mieyal, J. J. (2007). Mechanisms of reversible protein glutathionylation in redox signaling and oxidative stress. *Curr. Opin. Pharmacol.* **7,** 381–391.

Ghezzi, P. (2005). Regulation of protein function by glutathionylation. *Free Radic. Res.* **39,** 573–580.

Giles, G. I., et al. (2003). Evaluation of sulfur, selenium and tellurium catalysts with antioxidant potential. *Org. Biomol. Chem.* **1,** 4317–4322.

Gow, A. J., et al. (2002). Basal and stimulated protein S-nitrosylation in multiple cell types and tissues. *J. Biol. Chem.* **277,** 9637–9640.

Hess, D. T., et al. (2005). Protein S-nitrosylation: Purview and parameters. *Nat. Rev. Mol. Cell Biol.* **6,** 150–166.

Holmgren, A. (1976). Hydrogen donor system for *Escherichia coli* ribonucleoside-diphosphate reductase dependent upon glutathione. *Proc. Natl. Acad. Sci. USA* **73,** 2275–2279.

Holmgren, A., et al. (1978). Thioredoxin from *Escherichia coli*. Radioimmunological and enzymatic determinations in wild type cells and mutants defective in phage T7 DNA replication. *J. Biol. Chem.* **253,** 430–436.

Holmgren, A., et al. (2005). Thiol redox control via thioredoxin and glutaredoxin systems. *Biochem. Soc. Trans.* **33,** 1375–1377.

Jacob, C., et al. (2003). Sulfur and selenium: the role of oxidation state in protein structure and function. *Angew. Chem. Int. Ed. Engl.* **42,** 4742–4758.

Jaffrey, S. R., and Snyder, S. H. (2001). The biotin switch method for the detection of S-nitrosylated proteins. *Sci. STKE* **2001,** pl1.

Janssen-Heininger, Y. M., et al. (2008). Redox-based regulation of signal transduction: Principles, pitfalls, and promises. *Free Radic. Biol. Med.* **45,** 1–17.

Ji, Y., et al. (1999). S-nitrosylation and S-glutathiolation of protein sulfhydryls by S-nitroso glutathione. *Arch. Biochem. Biophys.* **362,** 67–78.

Kim, S. O., et al. (2002). OxyR: A molecular code for redox-related signaling. *Cell* **109,** 383–396.

Mallis, R. J., and Thomas, J. A. (2000). Effect of S-nitrosothiols on cellular glutathione and reactive protein sulfhydryls. *Arch. Biochem. Biophys.* **383,** 60–69.

Meng, T. C., et al. (2002). Reversible oxidation and inactivation of protein tyrosine phosphatases in vivo. *Mol. Cell* **9,** 387–399.

Pineda-Molina, E., et al. (2001). Glutathionylation of the p50 subunit of NF-kappaB: A mechanism for redox-induced inhibition of DNA binding. *Biochemistry* **40,** 14134–14142.

Reynaert, N. L., et al. (2006). In situ detection of S-glutathionylated proteins following glutaredoxin-1 catalyzed cysteine derivatization. *Biochim. Biophys. Acta* **1760,** 380–387.

Shelton, M. D., et al. (2005). Glutaredoxin: Role in reversible protein S-glutathionylation and regulation of redox signal transduction and protein translocation. *Antioxid. Redox Signal.* **7,** 348–366.

Stamler, J. S., et al. (2001). Nitrosylation. The prototypic redox-based signaling mechanism. *Cell* **106,** 675–683.

Takashima, Y., et al. (1999). Differential expression of glutaredoxin and thioredoxin during monocytic differentiation. *Immunol. Lett.* **68,** 397–401.

CHAPTER EIGHTEEN

Synthesis, Quantification, Characterization, and Signaling Properties of Glutathionyl Conjugates of Enals

Sanjay Srivastava,* Kota V. Ramana,[†] Aruni Bhatnagar,* and Satish K. Srivastava[†]

Contents

1. Introduction	298
2. Synthesis, Quantification, and Characterization of Reagent Glutathionyl Conjugates of HNE	300
2.1. GS-HNE	300
2.2. Synthesis, purification, and characterization of reagent GS-DHN	303
2.3. Synthesis and characterization of esterified glutathionyl conjugates	303
3. Metabolism of HNE	305
3.1. Cellular metabolism of HNE	305
3.2. Systemic metabolism of HNE	306
3.3. Systemic metabolism of acrolein	307
4. Signaling Properties of Glutathionyl Conjugates of HNE	309
4.1. Effect of glutathionyl conjugates of HNE on acute peritonitis	309
5. Conclusions	310
Acknowledgments	311
References	311

Abstract

Oxidation of lipids generates large quantities of highly reactive α,β-unsaturated aldehydes (enals). Enals and their protein adducts accumulate in the tissues of several pathologies. *In vitro*, low concentrations of enals such as HNE (4-hydroxy *trans*-2-nonenal) affect cell signaling whereas high concentrations

---

* Diabetes and Obesity Center, University of Louisville, Louisville, Kentucky, USA
[†] Department of Biochemistry and Molecular Biology, University of Texas Medical Branch, Galveston, Texas, USA

of enals are cytotoxic. Direct conjugation of the C2–C3 double bond of enals with the sulfhydryl group of GSH is a major route for the metabolism and detoxification of enals. Recently, we found that glutathionyl conjugate of HNE (GS-HNE) enhances the peritoneal leukocyte infiltration and stimulates the formation of proinflammatory lipid mediators. Moreover, the reduced form of the glutathione conjugate of HNE (GS-DHN) elicits strong mitogenic signaling in smooth muscle cells. In this chapter we discuss the methods to study the metabolism of enals and the redox signaling properties of glutathionyl conjugates of HNE.

1. Introduction

Oxidation of lipids generates a variety of bioactive molecules including lipid radicals, peroxides, isoprostanes, and epoxides. Of these, aldehydes, generated from the degradation of lipid hydroperoxides, are the major end products (Esterbauer et al., 1991). Among the free aldehydes generated from lipid peroxidation, the α,β-unsaturated aldehydes (enals) are highly reactive. The α,β-unsaturation in aldehydes generates two reactive centers in the molecule: a polarized carbonyl that could participate in the formation of Schiff bases and an electrophilic δ carbon that undergoes Michael addition reactions (Esterbauer et al., 1991). Consequently, enals are several orders of magnitude more reactive than corresponding saturated aldehydes or the aromatic aldehydes. Among the enals, the C9 unsaturated aldehyde 4-hydroxy *trans*-2-nonenal (HNE) is the most abundant. HNE is generated primarily from the peroxidation of ω-6 polyunsaturated fatty acids, and under some conditions, accounts for >95% of the enals produced (Benedetti et al., 1980). Normal HNE concentrations in tissue and plasma range from 0.8 to 2.8 μM and the concentration of HNE in oxLDL has been estimated to be 150 mM (Esterbauer et al., 1991). Apart from HNE, lipid peroxidation also generates highly reactive C3 enal, acrolein; C4 enal, crotonaldehyde; and the C6 enal, hexenal. Acrolein is also generated from myeloperoxidase, metabolism of anticancer drug cyclophosphamide and allyl amine, a dye extensively used in the dye industry (Bhatnagar, 2006; Uchida et al., 1998). High concentrations of acrolein and crotonaldehyde are also present in automobile exhaust and cigarette smoke (Bhatnagar, 2006; Uchida et al., 1998). Hexenal is one of the most abundant aldehydes present in fruits and vegetables and is also extensively used as a flavoring agent (Feron et al., 1991).

Enals and their protein adducts accumulate in several diseased tissues such as atherosclerotic lesions (Srivastava et al., 2009; Uchida et al., 1998), inflamed arteries (Rittner et al., 1999), ischemic hearts (Eaton et al., 1999), failing hearts (Srivastava *et al.*, 2002), and neuronal lesions associated with Parkinson (Yoritaka *et al.*, 1996) and Alzheimer (Montine *et al.*, 1997) disease. The accumulation of enals and their protein adducts is indicative of oxidative stress. *In vitro*, enals such as HNE affect multiple signaling pathways (Dianzani, 1998, 2003; Robino *et al.*, 2001), including membrane conductances and signal transduction cascades (Dianzani, 1998, 2003; Robino *et al.*, 2001). In several different cell types, exposure to HNE leads to the activation of JNK (Dianzani, 1998, 2003; Robino *et al.*, 2001) and in stellate cells, HNE has been shown to covalently modify the p46 and p54 isoforms of JNK (Parola *et al.*, 1998). Data from our laboratory (Ruef *et al.*, 2000) showed that HNE is a potent mitogen for vascular smooth muscle cells (VSMC) and Ishii *et al.* (2004) found that HNE induces the expression of scavenger receptors in VSMC. However, the mechanisms by which enals such as HNE exert cell signaling are not quite clear.

Our studies show that enals such as HNE are metabolized either by direct conjugation with glutathione or by oxidative or reductive transformations (Hill *et al.*, 2009; Srivastava *et al.*, 1998b, 2001). The conjugation of unsaturated aldehydes to glutathione is catalyzed by the glutathione S-transferase (GSTs) isoforms, whereas oxidation of HNE to 4-hydroxynonanoic acid (HNA) is catalyzed by aldehyde dehydrogenases. Both free and glutathiolated HNE are efficient substrates of aldose reductase (AR), and reduction via AR appears to be a significant fate of enals and their glutathione conjugates *in vivo*.

Although enals such as HNE readily form glutathione conjugates, the formation of GS-HNE may not by itself be sufficient for detoxification. We found that in VSMC, GS-HNE activates PKC and stimulates NF-κB- and AP-1-dependent gene transcription and smooth muscle cell proliferation (Ramana *et al.*, 2006). Our recent studies show that GS-HNE stimulates the formation of superoxide radical and induces the expression of CD11b in polymorphonuclear leukocytes (PMNs) *in vitro*, and intraperitoneal injection with GS-HNE significantly increases the peritoneal leukocyte infiltration and stimulates the formation of proinflammatory lipid mediators (Spite *et al.*, 2009). Similarly, the glutathionyl conjugate of acrolein has been shown to increase superoxide formation (Adams and Klaidman, 1993) and glutathione conjugates of enals have been shown to be markedly toxic and can induce DNA damage or stimulate the production of reactive oxygen species (Dittberner *et al.*, 1995; Horvath *et al.*, 1992). Initially, it was thought that the reduction of aldehyde–glutathione conjugate by AR would annul the reactivity and therefore the toxicity of the conjugate. However, on contrary to this expectation, we found that the reduced form of the glutathione conjugate of HNE (GS-DHN) elicits strong mitogenic signaling in smooth muscle cells (Ramana *et al.*, 2006). Incubation of cell permeable esters of GS-HNE and GS-DHN with VSMC stimulated protein kinase C, NF-κB, AP-1, and promoted cell growth (Ramana *et al.*, 2006). Inhibition of AR, the protein which catalyzes the reduction of GS-HNE to GS-DHN, prevented the GS-HNE-induced protein kinase C, NF-κB, AP-1 activation and cell proliferation, but had no effect on GS-DHN-induced cell signaling and growth. Pretreatment of the cells with antibodies raised against the transporters of glutathione conjugates—RLIP (Ral-binding protein) or multidrug resistance protein-2, promotes the mitogenic effects of GS-DHN. Similarly, inhibition of RLIP by siRNA, enhances the cell growth in response to glutathionyl conjugates of HNE, whereas overexpression of RLIP diminishes the glutathionyl conjugates-induced cell proliferation. Moreover, RLIP-knockout mice show higher levels of GS-HNE in the liver (Warnke *et al.*, 2008). Here, we briefly describe the methods for the formation, characterization, and quantification of glutathione conjugates of enals and their signaling properties.

2. Synthesis, Quantification, and Characterization of Reagent Glutathionyl Conjugates of HNE

2.1. GS-HNE

HNE and other α,β-unsaturated aldehydes are strong electrophiles that react spontaneously with reduced glutathione. The reaction is stiochiometric, and is accelerated by inorganic phosphate (Esterbauer *et al.*, 1991). Other nucleophiles such as histidine, lysine, and carnosine also react with these aldehydes but the rate of conjugate formation with these nucleophiles is usually an order of magnitude lower than with GSH. The GS-DHN is stable at acidic pH and the conjugate, once prepared, could be stored either in an acidic or an aprotic (acetonitrile) medium.

2.1.1. Synthesis
Reagent HNE and 4-(^{3}H)-HNE (specific activity of 175 mCi/mmol) were synthesized as described before (Srivastava *et al.*, 1998b). Because of their high reactivity, several unsaturated aldehydes are synthesized and stored as esters. Before the experiments, the free aldehyde is released by the acid hydrolysis of the ester (Srivastava *et al.*, 1998b).

Described below briefly is the method for the synthesis of GS-HNE:

1. Dissolve 100 nmol of nonradioactive HNE and 100 nmol of radiolabeled HNE containing 100,000 cpm 4-(^{3}H)-HNE in 990 μl of 0.1 *M* potassium phosphate, pH 7.0 in a quartz cuvette.
2. Add 10 μl of 50 m*M* GSH, prepared in ice-cold water. Incubate the mixture at 37 °C in a temperature-controlled water bath.

3. Monitor the reaction by following the consumption of HNE at 224 nm (molar extinction coefficient = 13,750). Formation of the glutathione conjugates results in a decrease in the absorbance at 224 due to saturation of the conjugated diene chromophore. Typically, under these experimental conditions, >99% HNE is consumed in 30 min.

A similar procedure could be used to synthesize conjugates of other enals. Short chain enals react faster with the sulfhydryl group of GSH than the longer chain enals. The rate of reaction of acrolein with GSH is 25-fold faster than crotonaldehyde and the rate of reaction of enals decreases further with an increase in hydrocarbon chain length. Presence of a hydroxyl group at the fourth carbon of medium chain enals (C5–C10) increases the rate of reaction of enals with the sulfhydryl group of GSH by four- to eightfold, whereas introduction of an additional double bond between the C4 and C5 of hydrocarbon chain completely abolishes their reactivity with GSH (Srivastava et al., 1999).

2.1.2. Purification and quantification

For the synthesis of GS-HNE, HNE is incubated with an excess of GSH. To remove unreacted GSH, the conjugate has to be purified by HPLC. This is best accomplished by using reverse-phase (RP) HPLC. In our laboratory, we typically use a Varian RP ODS C_{18} column preequilibrated with 0.1% aqueous trifluoroacetic acid. The column is eluted using a gradient consisting of solvent A (0.1% aqueous trifluoroacetic acid) and solvent B (100% acetonitrile) at a flow rate of 1 ml min^{-1}. The gradient is established such that solvent B reaches 24% in 20 min, 26% in 30 min and is held at this value for 10 min. In the next 10 min, solvent B reaches 60%, and in an additional 5 min it reaches 100% and is held at this value for 10 min. Samples are collected every 1 min and 50 μl aliquots are used to count the radioactivity on a scintillation counter. As shown in Fig. 18.1A, under these conditions, GS-HNE elutes with a retention time of 22 min. Typically, we recover 86 ± 6% radioactivity applied to the column by this protocol.

2.1.3. Characterization of GS-HNE

The chemical identity of reagent GS-HNE purified by HPLC is established by electrospray ionization mass spectrometry (ESI-MS). We use a Micro-Mass ZMD 2000 mass spectrometer (Waters-Micromass, Milford, MA). The ESI^{+}-MS operating parameters are as follows: capillary voltage, 3.0 kV; cone voltage, 13 V; extractor voltage, 9 V; source block temperature, 100 °C; and dissolvation temperature, 200 °C. Nitrogen at 3 p.s.i. is used as nebulizer gas. Samples (50 μl) are mixed 1:1 with 50 μl of acetonitrile/water/acetic acid (100/100/0.2) (v/v/v) and applied to the mass spectrophotometer using a Harvard syringe pump at a rate of 5 μl/min. Spectra are acquired at the rate of 200 a.m.u./s over the range of 20–2000 a.m.u.. As

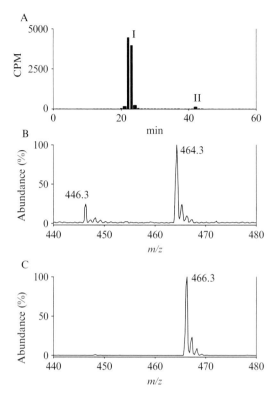

Figure 18.1 Purification and characterization of radiolabeled GS-HNE and GS-DHN. GS-HNE is purified by reverse-phase HPLC on an ODS C_{18} column. One milliliter fractions are collected and the radioactivity is quantified on a scintillation counter (A). The HPLC-purified peak I is analyzed by ESI^+-MS (B). Ions with m/z values of 464.3 and 446.3 represent, $[M + H]^+$ of GS-HNE and its dehydrated daughter ion, respectively. Panel C shows the ESI^+-MS analysis of reagent GS-DHN. Retention time of the reagent GS-DHN on HPLC is identical to that of reagent GS-HNE. ESI^+-MS of the HPLC peak corresponding to reagent GS-DHN shows the molecular ion with m/z 466.3 corresponds to $[M + H]^+$ of GS-DHN.

shown in Fig. 18.1B, the ESI^+-MS of HPLC peak I, eluting at 22 min, shows a strong molecular ion with an m/z value of 464.3, consistent with $[M + H]^+$ of GS-HNE. Another strong ion is generated with an m/z ratio of 446.3, corresponding to the dehydrated daughter ion $[M - H_2O]$ of GS-HNE. That the m/z 446 ion is derived from GS-HNE could be verified by changing the repeller voltage of the electrospray source. An increase in the repeller voltage should increase fragmentation and dehydration leading to an increase in the m/z 446 ion. For quantification, the intensity of the 446 and the 464 m/z ions is added together.

2.2. Synthesis, purification, and characterization of reagent GS-DHN

The GS-HNE conjugate could be reduced to generate GS-1,4-dihydroxynonanol (GS-DHN). Our studies show that the AR catalyzes the reduction of the conjugate (Srivastava et al., 1998a) and that the reduced conjugate is biologically active (Ramana et al., 2006). The GS-HNE conjugate could be reduced enzymatically, using recombinant AR (Abanova, Walnut, CA). Alternatively, the conjugate could be reduced by sodium borohydride.

1. Incubate 10 μg of human recombinant AR in 940 μl of 0.1 M potassium phosphate, pH 6.0, for 5 min at 37 °C in a cuvette.
2. Add 10 μl of 7.5 mM NADPH prepared in ice-cold water.
3. Add 50 μl of GS-[4-(^{3}H)]-HNE, reconstituted in 0.1 M potassium phosphate at a concentration of 200 μM.
4. Monitor the reaction by following the consumption of NADPH at 340 nm. Typically, under these experimental conditions, >99% GS-[4-(^{3}H)]-HNE is reduced in 3 h.
5. Alternatively, to a GS-HNE solution in methanol add 4 M excess of NaBH$_4$. Stir the mixture for 60 min at room temperature and then add 0.1 M HCl dropwise to the reaction mixture to bring the pH to 2.0. Leave the mixture in acidic conditions for 30 min to decompose unreacted NaBH$_4$ and extract the conjugate three times with dichloromethane and purify by HPLC as described above. Under the experimental conditions described, the GS-DHN conjugate coelutes with GS-HNE at the retention time of 22 min.
6. Chemical identity of reagent GS-DHN is established by ESI$^+$-MS as described above. As shown in Fig. 18.1C, upon ESI-MS GS-DHN forms a strong molecular ion with m/z value of 466.3.

2.3. Synthesis and characterization of esterified glutathionyl conjugates

To examine the effect of GS-HNE and GS-DHN on cell signaling, cell permeable esterified forms of their glutathionyl conjugates are used (Ramana et al., 2006). They are synthesized as described below:

1. Incubate 4-(^{3}H)-HNE (100 nmol) with 5 M excess of glutathione monoethyl ester and monitor the consumption of HNE at 224 nm as described above.
2. For the synthesis of the reduced form of the esterified glutathione-HNE conjugate (GS-DHN-ester), incubate aliquots of the GS-HNE-ester with AR and NADPH as described above.

3. Esterified glutathionyl conjugates are purified on HPLC as described above. Unlike the nonesterified conjugates, the GS-DHN-ester and GS-HNE-ester are nicely separated and elute with the retention time of 28 and 31 min, respectively (Fig. 18.2A).
4. Chemical identity of the esterified glutathionyl conjugates is established by ESI^+-MS. ESI^+-MS of HPLC peak I shows a molecular ion with an m/z value of 494.3, consistent with the $[M + H]^+$ of GS-DHN-ester and ESI^+-MS of HPLC peak II shows strong ions with m/z values of 492.2 and 274.2 corresponding to $[M + H]^+$ of GS-HNE-ester and its dehydrated daughter ion, respectively (Fig. 18.2B). The in-source

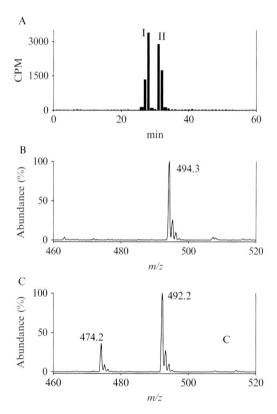

Figure 18.2 Purification and characterization of GS-HNE-ester and GS-DHN-ester. Reagent-esterified glutathionyl conjugates of HNE are separated by reverse-phase HPLC on an ODS C_{18} column (A). Chemical identities of GS-HNE-ester and GS-DHN-ester are established by ESI^+-MS. ESI^+-MS of the HPLC peak I shows strong molecular ion with m/z value of 494.3 corresponding to the $[M + H]^+$ of GS-DHN-ester (B). ESI^+-MS of the HPLC- peak II shows strong ions with m/z values of 492.2 and 474.2; consistent with the $[M + H]^+$ of GS-HNE-ester and its dehydrated daughter ion, respectively (C).

dehydration of GS-HNE-ester could be completely prevented by either decreasing the cone voltage or by lowering the block temperature; however, this results in a 10-fold decrease in signal intensity and thereby decreases the signal-to-noise ratio (data not shown).

3. Metabolism of HNE

3.1. Cellular metabolism of HNE

In most cells, HNE is rapidly metabolized. Cellular metabolism of HNE generates several products that result from the enzymatic reduction or oxidation of the aldehyde and the formation of a glutathione conjugate (GS-HNE), which undergoes reductive metabolism to generate GS-DHN. Glutathionyl conjugates of HNE account for 40–60% of the total metabolites in cardiovascular tissues (Hill *et al.*, 2009; Srivastava *et al.*, 1998b, 2001). Most of these metabolites including the glutathionyl conjugates are extruded out of the cell and <1% of the enal is retained inside the cells or are bound to intracellular proteins. Here, we describe briefly the method for the quantification and characterization of HNE metabolites in PC12 cells. For quantification, we have found it necessary to use radiolabeled [4-(^{3}H)-HNE] HNE.

1. Culture cells in 10 cm^2 dishes to 80% confluency.
2. Prior to the experiment, rinse the cells with prewarmed (37 °C) Hank's balanced salt solution (HBSS).
3. Incubate the cells with 5 μM 4-(^{3}H)-HNE (100,000 cpm) in 2.5 ml HBSS for 30 min.
4. At the end of the incubation, remove the medium and separate the radioactivity by HPLC on an RP C$_{18}$ column (Fig. 18.3A). HPLC conditions are identical to those described for the purification of reagent glutathionyl conjugates. On the basis of the retention time of the synthetic metabolites four major peaks in the eluate are identified to be glutathionyl conjugates of HNE (peak I), 1,4-dihydroxynonene (DHN; peak II), HNA (peak III), and unmetabolized HNE (peak IV). The chemical identity of these peaks is established by ESI$^+$-MS or gas chromatography–mass spectrometry (GC–MS). Two other minor peaks (peaks V and VI) are also observed. Chemical identity of these peaks has yet not been established.
5. As shown in Fig. 18.3B, ESI$^+$-MS of HPLC peak I shows strong molecular ions with m/z value of 466.3 and 464.3, consistent with [M + H]$^+$ GS-DHN and GS-HNE. These data are consistent with our previous studies showing that glutathionyl conjugates of HNE are the major metabolites of HNE in cardiovascular tissues (Hill *et al.*, 2009;

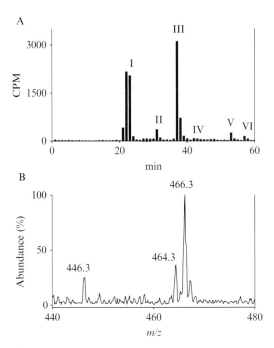

Figure 18.3 Metabolism of HNE in PC12 Cells. PC12 Cells are incubated with 4-(^{3}H)-HNE (5 μM) in 2.5 ml HBSS for 30 min at 37 °C. After the incubation, radioactivity in the incubation medium is separated by HPLC on an ODS C_{18} reverse-phase HPLC column. Panel A shows the HPLC profile of radiolabeled metabolites of [^{3}H]-HNE. Peaks I–IV correspond to the retention time of reagent glutathione conjugates of HNE (GS-HNE and GS-DHN), DHN, HNA, and unmetabolized HNE, respectively. Chemical identities of peaks V and VI have yet not been established. Panel B shows the ESI$^+$-MS chromatogram of HPLC peak I. Molecular ions with m/z value of 464.3 and 466.3 correspond to [M + H]$^+$ of GS-HNE and GS-DHN, respectively. Ion with m/z ratio of 446.3 corresponds to the dehydrated daughter ion of GS-HNE.

Srivastava *et al.*, 1998b, 2001). Sensitivity for the detection of glutathionyl conjugates of enals in tissues could be significantly increased by using ESI$^+$ and select ion monitoring (Warnke *et al.*, 2008).

3.2. Systemic metabolism of HNE

In this method, we describe, how enals such as HNE are metabolized *in vivo*. It is currently believed that glutathionyl conjugates of HNE and related enals are actively extruded into the circulation from the tissues in which they are generated. The conjugates appearing the circulation are taken up by the kidney and further processed and excreted in the urine as mercapturates. Of the total metabolites of HNE recovered in the urine, upon intravenous injection of 4-(^{3}H)-HNE to mouse, N-acetylcysteine-DHN (NAC-DHN), formed from GS-DHN, is the major metabolite and it accounts for 34 ± 4% of radioactivity recovered in the urine. These observations suggest glutathione conjugate formation and its subsequent reduction is the major pathway for the metabolism of HNE *in vivo*. NAC-DHN has also been detected as the physiological component of human and rat urine (Alary *et al.*, 1998). NAC-DHN concentration in the urine could also be used as a biomarker of oxidative stress *in vivo* (Kuiper *et al.*, 2008). Described below is the method for the quantification and characterization of NAC-DHN in the urine of mice exposed to HNE. For identification and quantification, radioactive HNE is required for these experiments.

1. Inject 4-(^{3}H)-HNE (1 mg/kg in PBS; 1.0×10^6 cpm) in the tail vein of C57BL/6 mice (25 ± 3 g). House mice in metabolic cages and provide water and food *ad libitum*.
2. Collect urine for 8 h
3. Aspirate an aliquot of urine (40 μl) in a glass tube and mix it with an equal volume of trichloroacetic acid (20%, w/v) and remove the precipitated proteins by centrifugation.
4. Separate the radioactivity in the supernatant by HPLC (Fig. 18.4A) as described above.
5. Radioactive peak eluting with the retention time of reagent NAC-DHN (26 min), representing 34 ± 4% of the radioactivity recovered from HPLC, is further characterized by ESI$^-$-MS.
6. Samples corresponding to the retention time of 26 min are dried under vacuum, reconstituted in 100 μl of actonitrile:10% ammonium hydroxide (v/v) and applied to the mass spectrophotometer using a Harvard syringe pump at a rate of 5 μl/min. As shown in Fig. 18.4B, ESI$^-$-MS of these peak shows a strong molecular with m/z value of 320.1, consistent with [M − H]$^-$ of NAC-DHN.

3.3. Systemic metabolism of acrolein

Similar to HNE, exposure of mice to the C3 enal acrolein results in the formation of the formation of glutathionyl conjugates which are subsequently metabolized and excreted in the urine as mercapturic acids. Hydroxypropyl mercapturic acid (HPMA) is the major metabolite recovered in the urine of mice exposed to acrolein. The following procedure is used for the quantification and characterization of HPMA in urine. Because acrolein is highly reactive and volatile, it is difficult to synthesize radiolabeled acrolein. Hence, a ^{13}C-standard is required to quantify acrolein-mercapturates by GC–MS. ^{13}C-Acrolein was custom synthesized (Sigma Chemical Co., St. Louis, MO). To study acrolein metabolism:

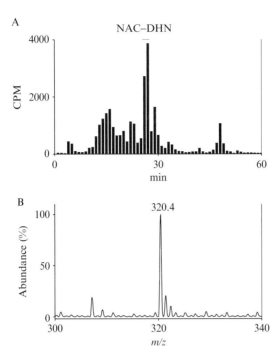

Figure 18.4 Systemic metabolism of HNE. Male C57 are injected with 4-(^{3}H)-HNE (1 mg/kg in PBS) in the tail vein and urine is collect for 8 h. Fifty microliters urine was mixed with an equal volume of trichloroacetic acid (20%, w/v) and the proteins that are precipitated are removed by centrifugation. Radioactivity in the supernatant is separated by HPLC on an ODS C_{18} reverse-phase HPLC column as described in Fig. 18.1. Panel A shows the HPLC profile of radiolabeled metabolites of [^{3}H]-HNE in the urine. Chemical identity of the radiolabeled peak eluting at 26 min, corresponding to the retention time of reagent NAC-DHN, is established by ESI$^-$-MS. As shown in Panel B, ESI$^-$-MS of these peak shows a strong molecular ion with m/z value of 320.1, consistent with [M − H]$^-$ of NAC-DHN.

1. Inject the mice with saline (0.1 ml, i.p.) or acrolein (2 mg/kg) and collect the urine as described above.
2a. Synthesize the internal standard, [$^{13}C_3$]3-HPMA, by incubating [$^{13}C_3$] 3-acrolein with a 10-fold excess of NAC in 0.1 M K$^+$-phosphate, pH 7.4, for 1 h at room temperature. Purify the NAC-propanal generated from this reaction by RP HPLC on an ODS C_{18} column. Reduce the NAC-propanal by incubating with AR or sodium borohydride as described above and purify the product [$^{13}C_3$]3-HPMA by HPLC. Establish the chemical identity by ESI-MS (Conklin et al., 2009).
2b. To quantify HPMA in the urine of control and acrolein exposed mice, mix the internal standard [$^{13}C_3$]3-HPMA (10 nmol) to urine and subject the urine to solid-phase extraction (Conklin et al., 2009).

After conditioning the cartridge, apply the urine on the cartridge, and wash the cartridge with 2% NH$_4$OH (6 ml) followed by methanol (6 ml). Dry the cartridge under N$_2$ and wash with 2% formic acid (6 ml, aq.). Elute the samples with 30% MeOH/2% formic acid and dry them under vacuum. Separate the mercapturates by RP HPLC and collect the peak corresponding to the retention time of reagent HPMA. Derivatize the samples with bistrimethylsilyltrifluoroacetamide for 1 h at 70 °C. Injection samples (1 μl) into an Agilent Technologies (Santa Clara, CA) 6890 N gas chromatograph equipped with an HP-5 capillary column (50 m × 0.2 mm i.d. × 0.5-μm phase thickness) coupled to a 5973 detector operated in the positive chemical ionization mode with ammonia as the reagent gas. Quantify the ion with m/z value of 366, using the ion with m/z value of 369 ($[^{13}C_3]$3-HPMA) as an internal standard (Conklin et al., 2009).

4. Signaling Properties of Glutathionyl Conjugates of HNE

Unsaturated aldehydes are highly reactive. Thus, when generated in cells or delivered exogenously, they induce a variety of stress responses and activate inflammatory signaling. At high concentrations, these aldehydes cause nonspecific toxicity (Esterbauer et al., 1991). In contrast, the glutathionyl conjugates of enals such as HNE, though less reactive than the parent aldehyde, elicit more selective signaling pathways. GS-DHN is a mitogen for VSMC (Ramana et al., 2006) and its formation is required for stimulation of the NF-κB pathway by HNE. GS-HNE, displays strong proinflammatory properties (Spite et al., 2009). Described below are methods to examine the inflammatory properties of GS-HNE in acute peritonitis. Other glutathione conjugates of enals may also have similar properties, but these remain to be studied. Similar studies could be performed to examine the biological effects of other conjugates of HNE and related enals.

4.1. Effect of glutathionyl conjugates of HNE on acute peritonitis

1. Inject saline, GS-HNE, intraperitoneally (each 10 μg) to 6- to 8-week-old male FVB mice.
2. After 4 h, lavage the peritonium with phosphate buffer saline.
3. Count total PMNs by differential analysis using Wright–Giemsa staining and by FACS analysis using PE-conjugated anti-Gr-1 antibodies.

4. For the assessment of vascular permeability, inject Evan's blue dye (1%, i.v.) prior to the intraperitoneal injection of GS-HNE and determine the absorbance (OD_{610} nm) of the peritoneal lavage after 15 min.
5. For the quantification of lipid mediators, process the samples as following:
   (a) Lavage the peritonium with phosphate buffer saline, 4 h after the i.p injection of saline or GS-HNE.
   (b) Mix the exudates with 2 volumes of cold methanol, containing 400 pg of d4-PGE_2 as an internal standard.
   (c) Cool the samples to -20 °C for 30 min to allow protein precipitation.
   (d) Centrifuge the samples at $850 \times g$ for 20 min (4 °C) and perform the solid-phase extraction of the supernatant, using C_{18} SEP-PAK cartridges (Alltech, Deerfield, IL) as per manufacturer's instructions.
   (e) Dry the methyl formate fractions under N_2 and resuspend the samples in methanol and perform the LC/MS/MS analysis for the quantification of proinflammatory lipid mediators LTB_4, PGE_2 and thromboxane B_2 (TXB_2), and monohydroxy fatty acids—5-, 12-, and 15-HETE, and 17-HDHA.
6. For the quantification of cysteinyl leukotriene formation, perform the solid-phase extraction on exudates (5 ml), dry the methanol fractions under N_2 and resuspend the samples in extraction buffer. Quantify the cysteinyl leukotriene formation by ELISA.

5. Conclusions

Our studies show that conjugation of enals with glutathione and subsequent reduction of the glutathionyl conjugates of enals is a major pathway for the cellular metabolism of enals. NAC-DHN and HPMA, the reduced form of the mercapturic acids of HNE and acrolein, respectively, are the major metabolites recovered in the urine of mice exposed to acrolein and DHN. NAC-DHN has also been shown to be present in the urine of healthy human subjects. GS-HNE promotes ROS formation in PMNs and causes acute peritonitis in mice. Glutathionyl conjugates of HNE mediate redox signaling and promote smooth muscle cell proliferation, and the mitogenic effect of glutathionyl conjugates is regulated by glutathione conjugate transporter, RLIP (Fig. 18.5). In summary, glutathionyl conjugates of enals could be critical regulators of redox signaling. Careful employment of methods to synthesize and study the effects of glutathione conjugates could lead to a better understanding of their role in regulating cell physiology and function and the role of oxidative stress in pharmacology and toxicology.

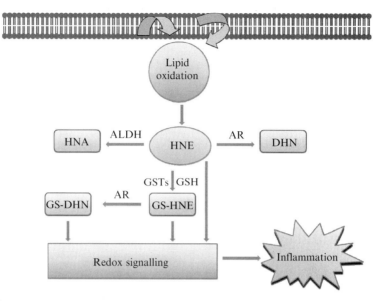

Figure 18.5 Effect of HNE and its glutathionyl conjugates on redox signaling.

ACKNOWLEDGMENTS

This work was supported in parts by NIH grants HL95593, ES17260, ES11594, ES11860, HL59378, RR 24489, GM71036, and DK36118.

REFERENCES

Adams, J. D. Jr., and Klaidman, L. K. (1993). Acrolein-induced oxygen radical formation. *Free Radic. Biol. Med.* **15,** 187–193.

Alary, J., Debrauwer, L., Fernandez, Y., Cravedi, J. P., Rao, D., and Bories, G. (1998). 1,4-Dihydroxynonene mercapturic acid, the major end metabolite of exogenous 4-hydroxy-2-nonenal, is a physiological component of rat and human urine. *Chem. Res. Toxicol.* **11,** 130–135.

Benedetti, A., Comporti, M., and Esterbauer, H. (1980). Identification of 4-hydroxynonenal as a cytotoxic product originating from the peroxidation of liver microsomal lipids. *Biochim. Biophys. Acta* **620,** 281–296.

Bhatnagar, A. (2006). Environmental cardiology: Studying mechanistic links between pollution and heart disease. *Circ. Res.* **99,** 692–705.

Conklin, D. J., Haberzettl, P., Lesgards, J. F., Prough, R. A., Srivastava, S., and Bhatnagar, A. (2009). Increased sensitivity of glutathione S-transferase P-null mice to cyclophosphamide-induced urinary bladder toxicity. *J. Pharmacol. Exp. Ther.* **331,** 456–469.

Dianzani, M. U. (1998). 4-Hydroxynonenal and cell signalling. *Free Radic. Res.* **28,** 553–560.

Dianzani, M. U. (2003). 4-Hydroxynonenal from pathology to physiology. *Mol. Aspects Med.* **24,** 263–272.

Dittberner, U., Eisenbrand, G., and Zankl, H. (1995). Genotoxic effects of the alpha, beta-unsaturated aldehydes 2-trans-butenal, 2-trans-hexenal and 2-trans, 6-cis-nonadienal. *Mutat. Res.* **335**, 259–265.

Eaton, P., Li, J. M., Hearse, D. J., and Shattock, M. J. (1999). Formation of 4-hydroxy-2-nonenal-modified proteins in ischemic rat heart. *Am. J. Physiol.* **276**, H935–H943.

Esterbauer, H., Schaur, R. J., and Zollner, H. (1991). Chemistry and biochemistry of 4-hydroxynonenal, malonaldehyde and related aldehydes. *Free Radic. Biol. Med.* **11**, 81–128.

Feron, V. J., Til, H. P., de, V. F., Woutersen, R. A., Cassee, F. R., and van Bladeren, P. J. (1991). Aldehydes: Occurrence, carcinogenic potential, mechanism of action and risk assessment. *Mutat. Res.* **259**, 363–385.

Hill, B. G., Awe, S. O., Vladykovskaya, E., Ahmed, Y., Liu, S. Q., Bhatnagar, A., and Srivastava, S. (2009). Myocardial ischaemia inhibits mitochondrial metabolism of 4-hydroxy-trans-2-nonenal. *Biochem. J.* **417**, 513–524.

Horvath, J. J., Witmer, C. M., and Witz, G. (1992). Nephrotoxicity of the 1:1 acrolein-glutathione adduct in the rat. *Toxicol. Appl. Pharmacol.* **117**, 200–207.

Ishii, T., Itoh, K., Ruiz, E., Leake, D. S., Unoki, H., Yamamoto, M., and Mann, G. E. (2004). Role of Nrf2 in the regulation of CD36 and stress protein expression in murine macrophages: Activation by oxidatively modified LDL and 4-hydroxynonenal. *Circ. Res.* **94**, 609–616.

Kuiper, H. C., Miranda, C. L., Sowell, J. D., and Stevens, J. F. (2008). Mercapturic acid conjugates of 4-hydroxy-2-nonenal and 4-oxo-2-nonenal metabolites are in vivo markers of oxidative stress. *J. Biol. Chem.* **283**, 17131–17138.

Montine, K. S., Kim, P. J., Olson, S. J., Markesbery, W. R., and Montine, T. J. (1997). 4-Hydroxy-2-nonenal pyrrole adducts in human neurodegenerative disease. *J. Neuropathol. Exp. Neurol.* **56**, 866–871.

Parola, M., Robino, G., Marra, F., Pinzani, M., Bellomo, G., Leonarduzzi, G., Chiarugi, P., Camandola, S., Poli, G., Waeg, G., Gentilini, P., and Dianzani, M. U. (1998). HNE interacts directly with JNK isoforms in human hepatic stellate cells. *J. Clin. Invest.* **102**, 1942–1950.

Ramana, K. V., Bhatnagar, A., Srivastava, S., Yadav, U. C., Awasthi, S., Awasthi, Y. C., and Srivastava, S. K. (2006). Mitogenic responses of vascular smooth muscle cells to lipid peroxidation-derived aldehyde 4-hydroxy-trans-2-nonenal (HNE): Role of aldose reductase-catalyzed reduction of the HNE-glutathione conjugates in regulating cell growth. *J. Biol. Chem.* **281**, 17652–17660.

Rittner, H. L., Hafner, V., Klimiuk, P. A., Szweda, L. I., Goronzy, J. J., and Weyand, C. M. (1999). Aldose reductase functions as a detoxification system for lipid peroxidation products in vasculitis. *J. Clin. Invest.* **103**, 1007–1013.

Robino, G., Zamara, E., Novo, E., Dianzani, M. U., and Parola, M. (2001). 4-Hydroxy-2,3-alkenals as signal molecules modulating proliferative and adaptative cell responses. *Biofactors* **15**, 103–106.

Ruef, J., Liu, S. Q., Bode, C., Tocchi, M., Srivastava, S., Runge, M. S., and Bhatnagar, A. (2000). Involvement of aldose reductase in vascular smooth muscle cell growth and lesion formation after arterial injury. *Arterioscler. Thromb. Vasc. Biol.* **20**, 1745–1752.

Spite, M., Summers, L., Porter, T. F., Srivastava, S., Bhatnagar, A., and Serhan, C. N. (2009). Resolvin D1 controls inflammation initiated by glutathione-lipid conjugates formed during oxidative stress. *Br. J. Pharmacol.* **158**, 1062–1073.

Srivastava, S., Chandra, A., Ansari, N. H., Srivastava, S. K., and Bhatnagar, A. (1998a). Identification of cardiac oxidoreductase(s) involved in the metabolism of the lipid peroxidation-derived aldehyde-4-hydroxynonenal. *Biochem. J.* **329**, 469–475.

Srivastava, S., Chandra, A., Wang, L. F., Seifert, W. E. Jr., Dague, B. B., Ansari, N. H., Srivastava, S. K., and Bhatnagar, A. (1998b). Metabolism of the lipid peroxidation product, 4-hydroxy-trans-2-nonenal, in isolated perfused rat heart. *J. Biol. Chem.* **273**, 10893–10900.

Srivastava, S., Watowich, S. J., Petrash, J. M., Srivastava, S. K., and Bhatnagar, A. (1999). Structural and kinetic determinants of aldehyde reduction by aldose reductase. *Biochemistry* **38**, 42–54.

Srivastava, S., Conklin, D. J., Liu, S. Q., Prakash, N., Boor, P. J., Srivastava, S. K., and Bhatnagar, A. (2001). Identification of biochemical pathways for the metabolism of oxidized low-density lipoprotein derived aldehyde-4-hydroxy trans-2-nonenal in vascular smooth muscle cells. *Atherosclerosis* **158**, 339–350.

Srivastava, S., Chandrasekar, B., Bhatnagar, A., and Prabhu, S. D. (2002). Lipid peroxidation-derived aldehydes and oxidative stress in the failing heart: Role of aldose reductase. *Am. J. Physiol. Heart Circ. Physiol.* **283**, H2612–H2619.

Srivastava, S., Vladykovskaya, E., Barski, O. A., Spite, M., Kaiserova, K., Petrash, J. M., Chung, S. S., Hunt, G., Dawn, B., and Bhatnagar, A. (2009). Aldose reductase protects against early atherosclerotic lesion formation in apolipoprotein E-null mice. *Circ. Res.* **105**, 793–802.

Uchida, K., Kanematsu, M., Morimitsu, Y., Osawa, T., Noguchi, N., and Niki, E. (1998). Acrolein is a product of lipid peroxidation reaction. Formation of free acrolein and its conjugate with lysine residues in oxidized low density lipoproteins. *J. Biol. Chem.* **273**, 16058–16066.

Warnke, M. M., Wanigasekara, E., Singhal, S. S., Singhal, J., Awasthi, S., and Armstrong, D. W. (2008). The determination of glutathione-4-hydroxynonenal (GSHNE), E-4-hydroxynonenal (HNE), and E-1-hydroxynon-2-en-4-one (HNO) in mouse liver tissue by LC-ESI-MS. *Anal. Bioanal. Chem.* **392**, 1325–1333.

Yoritaka, A., Hattori, N., Uchida, K., Tanaka, M., Stadtman, E. R., and Mizuno, Y. (1996). Immunohistochemical detection of 4-hydroxynonenal protein adducts in Parkinson disease. *Proc. Natl. Acad. Sci. USA* **93**, 2696–2701.

CHAPTER NINETEEN

Thioredoxin and Redox Signaling in Vasculature—Studies Using Trx2 Endothelium-Specific Transgenic Mice

Wang Min,[*,†,1] Luyao (Kevin) Xu,[*] Huanjiao (Jenny) Zhou,[†] Qunhua Huang,[*] Haifeng Zhang,[*] Yun He,[*] Xu Zhe,[*] and Yan Luo[†,1]

Contents

1. Introduction	316
2. Methods	318
2.1. Generation and characterization of the EC-specific transgenic mice expressing Trx2	319
2.2. Systemic oxidative stress, total antioxidant, hemodynamic, and echocardiography studies	319
2.3. Serum NO_x, NO release, eNOS phosphorylation, and eNOS enzymatic assay, and vessel function assays	320
2.4. Mouse EC culture and ROS measurement	321
2.5. Localization of Trx2 by confocal immunofluorescence microscopy	322
2.6. Apoptosis assays	322
References	323

Abstract

Increasing evidence supports that reactive oxygen species (ROS) generated from mitochondria in vasculature significantly contribute to human disease. The mitochondrial antioxidant systems, particularly the redox protein thioredoxin-2 (Trx2), provide a primary line of defense against cellular ROS. Using endothelial cell culture and endothelial cell-specific transgenesis of Trx2 gene in mice, we demonstrate the critical roles of Trx2 in regulating endothelium functions. Here, we describe the methods related to generation and characterization of the Trx2

---

* Interdepartmental Program in Vascular Biology and Therapeutics, Yale University School of Medicine, New Haven, Connecticut, USA
† State Key Laboratory of Ophthalmology, Zhongshan Ophthalmic Center, Sun Yat-Sen University, Guangzhou, China
1 Corresponding author: wang.min@yale.edu; luoyan2@mail.sysu.edu.cn

Methods in Enzymology, Volume 474　　　　　　　　　　　　　© 2010 Elsevier Inc.
ISSN 0076-6879, DOI: 10.1016/S0076-6879(10)74019-2　　　　　All rights reserved.

transgenic mice, and the *in vivo* functional assays associated with Trx2 activities. These methods could be applied to functional analyses for other redox genes.

1. INTRODUCTION

It has become clear that changes of cellular/systemic redox state, resulting in increases in inflammation (e.g., TNF) and reactive oxygen species (ROS), represent a common pathogenic mechanism for atherosclerosis. The vascular cell that primarily limits the inflammatory and atherosclerotic process is the endothelial cell (EC). Increasing evidence supports that ROS generated from mitochondria in vasculature significantly contribute to EC dysfunction and atherosclerotic progression. A key system regulating mitochondria redox is the mitochondria-specific thioredoxin (Trx2) system consisting of Trx2, Trx2 reductase (TrxR2), and Trx2-depndent peroxidase (Prx3) (Chang *et al.*, 2004). Trx is a small, multifunctional protein that has a redox-active site (sequence Cys32-Gly-Pro-Cys35). Human Trx1 is a 104-amino-acid protein with a molecular weight of 12 kDa that is localized to the cytosol. Human and other mammalian Trx1s contain two catalytic site Cys residues (Cys32 and Cys35) and three other Cys residues (Cys62, Cys69, and Cys73; all positions are for human Trx1). Trx2 (a 166 amino acid, 18 kDa protein) was identified in mitochondria. Trx2 has a conserved Trx catalytic site and a consensus signal sequence for mitochondrial translocation. Trx2, like Trx1, contains the redox-active site (C90 and C93). It is conceivable that Trx2 in mitochondria undergoes reversible oxidation to the Cys disulfide (Trx-S2), through the transfer of reducing equivalents from the catalytic site Cys residues to a disulfide protein substrate (protein-S2), thus maintaining protein in a reduced state. The reduced form of Trx is then regenerated by Trx reductase (TrxR) at the expense of NAPDH (Holmgren, 2000). Since mitochondria have the most reducing environment among all cellular organelles, a reduced state is critical for maintaining the mitochondria electrochemical potential, generation of ATP, and eliminating ROS (Hansen *et al.*, 2006). Many mitochondria proteins including those involved in the respiratory chain are sensitive to oxidation by ROS, resulting in augmented ROS generation and cell injury (Andreyev *et al.*, 2005). Therefore, reduction of the oxidized protein by Trx2 is critical for maintaining mitochondria in a reduced state and normal function of mitochondria. Genetic knockout of Trx2 or TrxR2 causes embryonic lethality, likely due to increased oxidative stresses. Likewise, cells with deficiency or knockdown of Trx2, TrxR2, or Prx3 accumulate endogenous ROS, and are highly sensitive to TNF (Chang *et al.*, 2004; Conrad *et al.*, 2004; Nonn *et al.*, 2003; Tanaka *et al.*, 2002). Conversely, overexpression of Prx3 in mice improved survival after myocardial infarction (Matsushima *et al.*, 2006). These data suggest that the Trx2 system may provide a primary line of defense against cellular ROS and inflammation.

The protein structure and function as well as the role of Trx1 in vascular system have been extensively studied (Yamawaki et al., 2003). However, little is known for the function and mechanism for the mitochondrial Trx2 antioxidant system in vasculature. Trx2 has been recently shown to have antiapoptotic function by blocking mitochondria-specific apoptosis signaling (Chen et al., 2002; Tanaka et al., 2002). Our data suggest that mitochondrial Trx2 may play critical roles in protecting ROS-induced EC dysfunction. First, we have shown that Trx2 protects EC from ROS-induced mitochondria-dependent apoptosis by inhibiting the activity of proapoptotic protein kinase ASK1 through protein–protein interactions. Trx2 in the mitochondria binds to ASK1 via a specific Cys residue (C30) on ASK1 to regulate mitochondria-dependent apoptotic pathway. Interestingly, a single Cys mutant within the catalytic site of Trx2 (Trx2-C93S which should lack the redox-catalyzing activity), but not the double-mutant (Trx2-C93/90S), retains the ability in ASK1 binding, inhibiting ASK1-induced EC apoptosis. These data suggest that a direct association of Trx2 with ASK1 is critical for the inhibitory effect of Trx2 on ASK1 (Zhang et al., 2004). Second, we have demonstrated a critical role of Trx2 in regulating EC function by increasing NO bioavailability. Specifically, Trx2 expression driven by VE-cadherin promoter in mice increased total antioxidants in plasma and reduced oxidative stress compared to the control littermates. Aortas from Trx2 TG mice increased basal NO bioavailability, resulting in reduced vasoconstriction to phenylephrine (PE) and enhanced vasodilation to acetylcholine. Similarly, isolated mouse aorta EC from Trx2 TG mice markedly increased NO release. It has been shown that ROS not only alters endothelium-dependent vascular relaxation through interaction with NO, but the resultant peroxynitrite also oxidizes eNOS cofactor BH4, causing a deficiency of BH4 and pathogenic uncoupling of eNOS (Landmesser et al., 2003). Our data suggest that Trx2 increases NO bioavailability by scavenging ROS generated from mitochondria. Therefore, Trx2 TG EC or isolated mitochondria from these cells show increased activities in scavenging ROS concomitant with reduced protein nitrotyrosine levels, an indicator of peroxynitrite (Zhang et al., 2007).

For *in vivo* function of Trx2, by using EC-specific transgenesis of mitochondrial form of thioredoxin gene in mice (Trx2 TG), we show the critical roles of Trx2 in regulating endothelium functions. Under physiological condition, Trx2 TG mice have increased oxidative stress and increased NO levels. Consistently, aortas from Trx2 TG mice show reduced vasoconstriction and enhanced vasodilation. EC isolated from Trx2 TG mice show increased capacities in scavenging ROS generated from mitochondria, resulting in increases in NO bioavailability as well as increased resistance to ASK1-mediated apoptosis and increased angiogenic activities (Dai et al., 2009; Zhang et al., 2007). We then examined effect of Trx2 TG in several pathological models. In the ApoE-deficiency mouse model, Trx2 prevents EC dysfunction and reduces

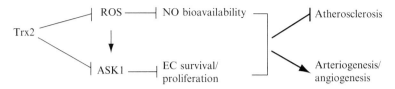

Figure 19.1 A model for Trx2 function in ischemia-mediated angiogenesis. Trx2 maintains EC function by two parallel pathways – scavenging ROS to increase NO bioavailability and inhibiting ASK1 activity to enhance EC survival, leading to enhanced ischemia-mediated arteriogenesis and angiogenesis.

hypercholesterolemia-induced atherosclerotic progression in ApoE-deficiency mice (Zhang *et al.*, 2007). In a variety of angiogenesis models such as VEGF-induced cornea/retina angiogenesis in which NO is an important mediator, Trx2 TG show enhanced angiogenic responses. Finally, we applied Trx2 TG to a femoral artery ligation model in which NO-mediated EC function and ROS-induced EC apoptosis play important roles. Trx2 prevents ischemia-induced ROS production, peroxynitrite formation, ASK1–JNK activation, inflammatory responses, and EC apoptosis. Trx2 TG mice had enhanced arteriogenesis and angiogenesis, enhanced capacity in limb perfusion recovery, and ischemic reserve capacity. To define the relative contributions of Trx2-increased NO and Trx2-reduced ASK1 apoptotic activity to angiogenesis *in vivo*, we examined Trx2 effects on ischemia-induced angiogenesis in eNOS-deficient mice. The eNOS deletion caused severe impairment in the functional flow recovery in response to ischemia. Trx2 expression in eNOS-KO mice still markedly inhibited ischemia-induced ASK1 and EC apoptosis, leading to an enhanced functional flow recovery (Dai *et al.*, 2009).

These *in vivo* and *in vitro* data support that Trx2 maintains EC function by two parallel pathways—scavenging ROS to increase NO bioavailability and inhibiting ASK1 activity to enhance EC survival. Normal EC function is pivotal to maintain antiatherosclerotic state while facilitate vessel repairs upon injury. Therefore, Trx2 TG facilitates ischemia-mediated arteriogenesis and angiogenesis while preventing hypercholesterolemia-induced atherosclerotic progression (Fig. 19.1).

2. Methods

Here, we describe methods related to generation and characterization of the Trx2 transgenic mice and *in vivo* functional assays associated with Trx2 activities. These methods could be applied to functional analyses of other mitochondrial redox genes.

2.1. Generation and characterization of the EC-specific transgenic mice expressing Trx2

The VE-cadherin promoter in TA vector was obtained from Dr. Laura Benjamin (Harvard Medical School). The 3′-UTR (untranslated region) from bovine growth hormone (bGH) was cloned into downstream of the VE-cadherin promoter to generate pVE-pA vector (Zhang et al., 2007). The human Trx2 cDNA with a HA-tag sequence at 3′-end was inserted into the EcoRI and XbaI sites between the VE-cadherin promoter and bGH pA to obtain pVE-Trx2 plasmid. The plasmid was linearized with XhoI digestion and the pronuclear injection was performed at Yale Transgenic Core. The founder transgenics were identified by PCR of tail DNA with a 5′ primer of Trx2 and 3′ primer of HA. Trx2 TG mice were backcrossed with C57BL/6 mice for more than six generations before experiments. All experiments were performed with heterozygous Trx2 TG mice and their non-TG littermates as controls. Transgene expression was visualized by immunofluorescence microscopy. Eight to ten-week-old Trx2 TG or wild-type (WT) littermates were perfused with phosphate-buffered saline (PBS) for 5 min and then with 4% paraformaldehyde for 5 min at physiological temperature and pressure. Blood vessels were then harvested and diffusion-fixed overnight at 4 °C and then dehydrated in 30% sucrose overnight at 4 °C. The vessels were then embedded, and 5-μm sections were stained for rat high-affinity anti-HA antibody (Roche) followed by Alexa Fluor 594 (red)-conjugated anti-rat secondary antibody. Endothelium was visualized by goat anti-CD31 antibody (Santa Cruz) followed by Alexa Fluor 488 (green)-conjugated anti-rat secondary antibody.

2.2. Systemic oxidative stress, total antioxidant, hemodynamic, and echocardiography studies

8-Isoprostane in serum was determined by an ELISA kit from Cayman Chemical. The isoprostanes are a family of eicosanoids of nonenzymatic origin produced by random oxidation of tissue phospholipids by oxygen radicals. Total antioxidant levels in serum were determined by the Total Antioxidant Assay Kit from calbiochem. The assay relies on the ability of antioxidants in the sample to inhibit the oxidation of ABTS (2,2′-azino-di-[3-ethylbenzthiazoline sulfonate]) to $ABTS^+$ by metmyoglobin (a peroxidase). The amount of $ABTS^+$ produced can be monitored by reading the absorbance at 600 nm. NO_x (both nitrite and nitrate) in serum was determined by an assay kit from Cayman Chemical.

For hemodynamic studies, anesthesia was induced by intramuscular ketamine injection. Mice were then intubated, placed on positive pressure ventilation and light anesthesia maintained by inhaled isoflurane. The right common jugular vein was cannulated with polyethylene tubing and a 1.9

French transducer-tipped catheter (Millar Inc., Houston, TX) was advanced into the left ventricle via the right carotid artery. Left ventricular pressures, including high-fidelity positive and negative dP/dt, were measured under basal conditions and during intravenous infusion of graduated doses of dobutamine. Data were recorded by using MacLab software and were analyzed by using the Heartbeat program (UC San Diego). Echocardiograms were obtained on lightly anesthetized mice (isoflurane inhalation via nose cone) by using a 15 M Hz transducer and a Sonos 7500 console. Zoomed 2D images were used to determine a short-axis plane at the level of the papillary muscles and then M-mode was obtained at this level. Measurements were obtained using the 7500 analysis software.

2.3. Serum NO_x, NO release, eNOS phosphorylation, and eNOS enzymatic assay, and vessel function assays

NO levels in serum or EC media were assessed by measuring nitrite levels using a NO-specific chemiluminescence analyzer (Sievers, Boulder, CO). Data are reported as NO_2^- accumulation per mg protein (Fulton *et al.*, 1999). Serum or media were deproteinized and samples containing NO_2^- were refluxed in glacial acetic acid containing sodium iodide. Under these conditions, NO_2^- was quantitatively reduced to NO, which was quantified by a chemiluminescence detector after reaction with ozone in a NO analyzer. In all the experiments, NO_2^- release was inhibitable by a NOS inhibitor. The eNOS activity assay was determined based on the conversion of ^{3}H-L-arginine to ^{3}H-L-citrulline, and was performed according to the protocol provided by the manufacturer (Calbiochem, Nitric Oxide Synthase Assay Kit). Data are presented as pmol ^{3}H-L-citrulline per min per protein. In some experiments, cells were treated with VEGF (10–50 ng/ml) for 30 min as a positive control for NO release, eNOS phosphorylation, and eNOS enzymatic assays.

Aortic-ring assay was used for NO-dependent vessel function studies. The thoracic aorta is dissected from both male and female mice (8–10 weeks old) and cut into cylindrical, 3-mm-long segments. The rings are suspended by two tungsten wires mounted in a vessel myograph system (Danish Myotechnologies, Aarhus, Denmark). The aortas were bathed in oxygenated Krebs buffer and submitted to a resting tension of 9.8 mN. After 60 min of equilibration with frequent washings, concentration response curves for PE were generated to determine vasoconstrictor responses. To study vasodilator and L-nitro arginine methyl ester (L-NAME) responses, the rings were preconstricted with a submaximal concentration of PE, and acetylcholine (ACh) (10^{-9}–10^{-5} M), sodium nitroprusside (SNP) (10^{-9}–3×10^{-7} M), or L-NAME (100 μM) was injected at the plateau of the PE-induced contraction.

2.3.1. Isolation of mitochondria from aortic and heart tissues

Isolation of mitochondria was performed according to the instructions of Mitochondria Isolation kit for Tissue (PIERCE, 89801). A Dounce grinder was prechilled on ice. Protease inhibitors (Roche, 11697498001) were added to Mitochondria Isolation Reagent A, C, and Wash Buffer immediately before use. The whole aorta or heart tissues were washed twice in PBS, and then were cut into small pieces. Pellets in Reagent A solution were homogenized in the Dounce grinder. After adding Mitochondria Isolation Reagent C, cover the top of the homogenizer with parafilm and mix by inverting for several times. The homogenate were centrifuged at $700 \times g$ for 10 min at 4 °C. The crude mitochondria fractions those in supernatant were collected by further centrifugation at $3000 \times g$ for 15 min at 4 °C. If desired, save the supernatant as the cytosolic fraction for analysis. Mitochondria pellets were surface washed and collected in Wash Buffer at $12,000 \times g$ for 5 min at 4 °C. Protein concentrations were determined with a protein assay reagent (Bio-Rad) using BSA (Sigma-Aldrich) as standard.

2.3.2. Mitochondria purification

Mitochondria pellets were layered onto $1–1.5\ M$ discontinuous sucrose in 10 mM Tris–HCl, 1 mM EDTA, pH 7.5, plus protease inhibitors, and centrifuged at $10,000\ g$ in Beckman SW28 rotor for 30 min at 4 °C. The mitochondria formed a layer at $1–1.5\ M$ sucrose interface. The layer were collected and washed in $1\times$ MS buffer (210 mM mannitol, 70 mM sucrose, 5 mM Tris–HCl, 1 mM EDTA, pH 7.5).

2.4. Mouse EC culture and ROS measurement

Mouse aortic EC isolation from WT and Trx2 TG mice was performed as follows. For immunoselection, 10 μl beads (per T-75 of mouse lung cells) were washed thrice with 1 ml of buffer A (PBS +2% FBS) and resuspended in 100 μl of buffer A. Ten microliters (10 μg) of anti-mouse ICAM-2 or 10 μl (10 μg) of PECAM-1 were added and rocked at 4 °C for 2 h. Beads were washed thrice and resuspended in 160 μl of buffer A. Confluent mouse aortic cells cultured in a T-75 flask were placed at 4 °C for 5 min and incubated with the beads at 4 °C for 1 h. Cells were then washed with warm PBS and treated with 3 ml of warm trypsin/EDTA. When cells were detached, 7 ml of growth media were added. An empty 15-ml tube in the magnetic was placed on the holder and the cell suspension ($\sim$10 ml) was added slowly by placing the pipette on the wall of the tube so that the cells pass through the magnetic field. Cells were incubated for 5 min, and the media are carefully aspirated. The 15-ml tube was removed from the magnetic holder and the beads/cells were resuspended in 10 ml of media. The selected cells were plated on 0.2% gelatin-coated flasks and cultured for

3–7 days. When the cells were confluent, another round of immunoselection was performed. The purity of EC population was more than 98% as determined by DiI-LDL-fluorescein isothiocyanate (FITC) staining followed by flow cytometry analyses.

For measurement of total intracellular ROS, we used 5,6-chloromethyl-2′,7′-dichlorodihydrofluorescein diacetate (CM-H$_2$DCFDA, preferentially for H$_2$O$_2$), or dihydroethidium (DHE, preferentially for superoxide). For measurement of total mitochondrial ROS, we used dihydrorhodamine 123 (DHR123, preferentially for H$_2$O$_2$), or MitoTracker Red CM-H$_2$XROS (MitoROS, preferentially for superoxide). All probes were purchased from Molecular Probes. Cells were loaded with 5 μM of a indicated probe and were incubated at 37 °C for 20 min, and were immediately detached and subjected to analytic flow cytometry on a FACSort (BD Biosciences). The fluorescence signal was recorded on the FL1 (green) or FL3 (red) channel and analyzed by using CellQuest software. As controls, cells were treated with TNF (10 ng/ml), paraquat (1 mM), or H$_2$O$_2$ (1 mM) for indicated times.

2.5. Localization of Trx2 by confocal immunofluorescence microscopy

Cells fixation, permeabilization, and staining were described previously (Zhang et al., 2004). For mitochondria visualization, cells were stained with 20 nM tetramethylrhodamine methyl ester (TMRM) (Molecular Probes, Eugene, OR) for 30 min at 37 °C and washed twice with PBS. For indirect fluorescence microscopy, the FITC-conjugated anti-IgG, and tetramethylrhodamine isothiocyanate (TRITC)-conjugated anti-IgG were purchased from Molecular Probes. Confocal immunofluorescence microscopy was performed using an Olympus confocal microscope.

2.6. Apoptosis assays

2.6.1. Cell apoptotic assay
Cells were treated with TNF (10 ng/ml) plus cycloheximide (10 μg/ml) for 6 h. Cell nuclei were stained with DAPI (2.5 mg/ml) and apoptotic cell (nucleus fragmentation) were counted under UV microscope.

2.6.2. Caspase-9 activity assay
Caspase-9 activity was measured with a Caspase-9 peptide substrate acetyl-ASP-Glu-Val-Asp-7-amido-4-methylcoumarin (Ac-DEVD-AMC) (Sigma) as described previously for Caspase 3 (Liu and Min, 2002). Briefly, BAEC were harvested in a lysis buffer (25 mM Hepes, pH 7.4, 5 mM CHAPS, 5 mM DTT) and incubated on ice for 15–20 min followed by a centrifugation at 14,000×g for 10–15 min at 4 °C. For each reaction, 5 ml (200 mg) of cell

lysate was incubated with 200 ml of 16 mM Caspase-9 peptide substrate acetyl-Val-Glu-His-Asp-7-amido-4-methylcoumarin (Ac-VEHD-AMC) in the assay buffer (25 mM Hepes, pH 7.4, 5 mM EDTA, 0.1% CHAPS, 5 mM DTT). The reaction was incubated in the dark for 1–1.5 h and fluorescence was measured in a fluorescence plate reader. The measured fluorescence was used as an arbitrary unit.

2.6.3. Cytochrome c release assay

To obtain the cytosolic fraction, cells were washed with PBS and resuspended in 50 ml of 250 mM sucrose and 70 mM Tris (pH 7.0) with protease inhibitors mixture (Roche Diagnostics Corp., Indianapolis, Indiana, USA). Ten milliliters of 4 mg/ml digitonin was added followed by incubation at room temperature for 2 min. Two milliliters of cells were stained with trypan blue followed by direct observation with light phase contrast microscopy to obtain lysis of 90–95% of cells. This approach allows preparation of cytosol that is essentially free of mitochondrial contamination (Chen *et al.*, 2002). After centrifugation at 600×g for 2 min at room temperature, the supernatant was collected as the cytosolic fraction. To prepare the mitochondrial fraction, cells were washed once with PBS, resuspended in ice-cold hypotonic buffer (10 mM NaCl, 1.5 mM CaCl$_2$, 10 mM Tris–HCl, pH 7.5, with protease inhibitors mixture, and kept on ice for 10 min. MS buffer (210 mM mannitol, 70 mM sucrose, 5 mM EDTA, 5 mM Tris, pH 7.6) was then added, and cells were homogenized using a Dounce homogenizer with 30 strokes. Disruption of plasma membrane was monitored by trypan blue staining. After removing the nuclear fraction by two successive centrifugations at 3000×g for 10 min, the supernatant was centrifuged at 12,000×g for 10 min. The pellet was collected as the mitochondrial portion and resuspended in lysis buffer (50 mM HEPES, pH 7.0, 500 mM NaCl, and 1% Nonidet P-40), supplemented with a mixture of protease inhibitors. After measuring the protein concentrations (Bio-Rad reagents), the recombinant protein contents in both fractions were analyzed by Western blot analysis for cytochrome *c* (BD PharMingen, San Diego, California, USA), Trx1, Trx2, and ASK1 as indicated.

REFERENCES

Andreyev, A. Y., Kushnareva, Y. E., and Starkov, A. A. (2005). Mitochondrial metabolism of reactive oxygen species. *Biochemistry (Mosc).* **70,** 200–214.

Chang, T. S., Cho, C. S., Park, S., Yu, S., Kang, S. W., and Rhee, S. G. (2004). Peroxiredoxin III a mitochondrion-specific peroxidase, regulates apoptotic signaling by mitochondria. *J. Biol. Chem.* **279,** 41975–41984.

Chen, Y., Cai, J., Murphy, T. J., and Jones, D. P. (2002). Overexpressed human mitochondrial thioredoxin confers resistance to oxidant-induced apoptosis in human osteosarcoma cells. *J. Biol. Chem.* **277,** 33242–33248.

Conrad, M., Jakupoglu, C., Moreno, S. G., Lippl, S., Banjac, A., Schneider, M., Beck, H., Hatzopoulos, A. K., Just, U., Sinowatz, F., Schmahl, W., Chien, K. R., *et al.* (2004). Essential role for mitochondrial thioredoxin reductase in hematopoiesis, heart development, and heart function. *Mol. Cell. Biol.* **24,** 9414–9423.

Dai, S., He, Y., Zhang, H., Yu, L., Wan, T., Xu, Z., Jones, D., Chen, H., and Min, W. (2009). Endothelial-specific expression of mitochondrial thioredoxin promotes ischemia-mediated arteriogenesis and angiogenesis. *Arterioscler. Thromb. Vasc. Biol.* **29,** 495–502.

Fulton, D., Gratton, J. P., McCabe, T. J., Fontana, J., Fujio, Y., Walsh, K., Franke, T. F., Papapetropoulos, A., and Sessa, W. C. (1999). Regulation of endothelium-derived nitric oxide production by the protein kinase Akt. *Nature* **399,** 597–601.

Hansen, J. M., Go, Y. M., and Jones, D. P. (2006). Nuclear and mitochondrial compartmentation of oxidative stress and redox signaling. *Annu. Rev. Pharmacol. Toxicol.* **46,** 215–234.

Holmgren, A. (2000). Antioxidant function of thioredoxin and glutaredoxin systems. *Antioxid. Redox Signal.* **2,** 811–820.

Landmesser, U., Dikalov, S., Price, S. R., McCann, L., Fukai, T., Holland, S. M., Mitch, W. E., and Harrison, D. G. (2003). Oxidation of tetrahydrobiopterin leads to uncoupling of endothelial cell nitric oxide synthase in hypertension. *J. Clin. Invest.* **111,** 1201–1209.

Liu, Y., and Min, W. (2002). Thioredoxin promotes ASK1 ubiquitination and degradation to inhibit ASK1-mediated apoptosis in a redox activity-independent manner. *Circ. Res.* **90,** 1259–1266.

Matsushima, S., Ide, T., Yamato, M., Matsusaka, H., Hattori, F., Ikeuchi, M., Kubota, T., Sunagawa, K., Hasegawa, Y., Kurihara, T., Oikawa, S., Kinugawa, S., *et al.* (2006). Overexpression of mitochondrial peroxiredoxin-3 prevents left ventricular remodeling and failure after myocardial infarction in mice. *Circulation* **113,** 1779–1786.

Nonn, L., Williams, R. R., Erickson, R. P., and Powis, G. (2003). The absence of mitochondrial thioredoxin 2 causes massive apoptosis, exencephaly, and early embryonic lethality in homozygous mice. *Mol. Cell. Biol.* **23,** 916–922.

Tanaka, T., Hosoi, F., Yamaguchi-Iwai, Y., Nakamura, H., Masutani, H., Ueda, S., Nishiyama, A., Takeda, S., Wada, H., Spyrou, G., and Yodoi, J. (2002). Thioredoxin-2 (TRX-2) is an essential gene regulating mitochondria-dependent apoptosis. *EMBO J.* **21,** 1695–1703.

Yamawaki, H., Haendeler, J., and Berk, B. C. (2003). Thioredoxin: A key regulator of cardiovascular homeostasis. *Circ. Res.* **93,** 1029–1033.

Zhang, R., Al-Lamki, R., Bai, L., Streb, J. W., Miano, J. M., Bradley, J., and Min, W. (2004). Thioredoxin-2 inhibits mitochondria-located ASK1-mediated apoptosis in a JNK-independent manner. *Circ. Res.* **94,** 1483–1491.

Zhang, H., Luo, Y., Zhang, W., He, Y., Dai, S., Zhang, R., Huang, Y., Bernatchez, P., Giordano, F. J., Shadel, G., Sessa, W. C., and Min, W. (2007). Endothelial-specific expression of mitochondrial thioredoxin improves endothelial cell function and reduces atherosclerotic lesions. *Am. J. Pathol.* **170,** 1108–1120.

Author Index

A

Adams, J. D. Jr., 299
Adcock, I. M., 222
Addanki, S., 150
Adenuga, D., 218, 225–228
Aesif, S. W., 160, 289–295
Ago, T., 76
Ahn, B., 230
Ahsan, M. K., 74
Aitken, C. J., 74
Akamatsu, Y., 72
Alary, J., 307
Allis, C. D., 216–217
Allison, W. S., 46
Alvarez, C. E., 74
Ames, B. N., 125
Anathy, V., 291
Andersen, H. R., 32
Anderson, C. L., 169, 177
Anderson, L. O., 182
Anderson, M., 132
Andoh, T., 198–204, 206–207
Andreyev, A. Y., 316
Andringa, K. K., 83–106
Arai, R. J., 71
Arner, E. S. J., 110, 117–118, 124, 140, 246, 248, 256–257
Åslund, F., 16–17
Atkinson, H. J., 2

B

Babbitt, P. C., 2
Bai, J., 70, 202
Bailey, S. M., 83–106
Balaban, R. S., 124
Banjac, A., 161
Banks, A. S., 218
Barford, D., 37, 39, 45
Barnes, P. J., 225
Barrett, W. C., 39
Bartone, N. A., 24–25
Baty, J. W., 52, 64, 144
Beato, M., 73
Beatty, P., 167, 177
Becker, K., 118, 257
Beckman, K. B., 125
Beech, D. J., 157
Beer, S. M., 130–132, 134

Benedetti, A., 298
Benhar, M., 293
Benitez, L. V., 46
Berlett, B. S., 230
Bernardes, G. J., 24
Berndt, C., 69, 199, 201
Bernstein, S., 149–162
Bersani, N. A., 142
Bertos, N. R., 217
Betley, J. R., 47
Bhatnagar, A., 297–311
Biaglow, J. E., 256
Bindoli, A., 109–120
Biswas, S. K., 231
Biteau, B., 52
Biterova, E. I., 111, 246
Bjornstedt, M., 117, 199, 246, 260
Blankenberg, S., 32
Blum, J., 25
Bodnar, J. S., 74
Boivin, B., 35–48
Boja, J. W., 208
Bonetto, V., 230
Bonini, M. G., 62
Brennan, J. P., 64
Brookes, P. S., 89, 97, 102, 149–162
Brown, D. I., 39
Brown, K. K., 52–53, 140
Buettner, G. R., 124
Buhrman, G., 39
Burgess, R. R., 151
Burkitt, M. J., 62
Burwell, L. S., 102, 150

C

Caito, S., 213–239
Capeillere-Blandin, C., 191
Carballal, S., 182
Carlson, B. A., 255–272
Carroll, K. S., 63
Ceccarelli, J., 156, 159, 161
Chaiswing, L., 161
Chang, E. Y., 110
Chang, T. S., 316
Chappell, J. B., 127
Charles, R. L., 46
Chen, J., 74–75
Chen, K. S., 74
Chen, L., 222

Chen, T. Y., 203
Chen, W. L., 200
Chen, Y. Y., 39, 119, 142, 317, 323
Chen, Z., 67–76
Cheng, Q., 246
Chew, E. H., 257
Chiueh, C. C., 197–209
Cho, C.-S., 23–32
Chou, 203
Chouchani, E. T., 123–144
Chow, K. L., 73
Chrestensen, C. A., 291
Chung, D. G., 40
Chutkow, W. A., 74–75
Circu, M. L., 150
Ckless, K., 294
Combs, G. F. Jr., 256
Conklin, D. J., 308–309
Conrad, M., 111, 316
Consevage, M. W., 25
Conway, M. E., 46
Cooper, J. A., 278
Coppede, F., 238
Corbett, J. A., 74
Cornish-Bowden, A., 15
Costa, N. J., 84–85, 124
Cotter, T., 232
Cox, A. G., 51–64, 119, 124, 140
Cox, D. A., 151
Curry, S., 183
Czech, M. P., 36

D

Dahm, C. C., 84–85, 135–136
Dahm, L. J., 177
Dai, S., 317–318
Dalle-Donne, I., 231, 290–291
Dansen, T. B., 69, 75
Das, D., 198, 201
Das, K. C., 201
Delaunay, A., 126
DeLuca, H. F., 74
Denu, J. M., 37, 39–40
Deshpande, T., 36
Deutscher, M. P., 151
Dianzani, M. U., 299
Dietz, K. J., 4
Dittberner, U., 299
Diwadkar-Navsariwala, V., 256
Dixon, J. E., 37
Doorn, J. A., 85
Drisdel, R. C., 45
Dröge, W., 191

E

Eaton, P., 233, 239, 298
Ellis, H. R., 4, 9, 11, 46

Ellman, G., 127–128
Ema, M., 72
England, K., 232
Era, S., 182, 191
Eriksson, S. E., 263, 271–272
Erusalimsky, J. D., 135
Essex, D. W., 167
Esterbauer, H., 298, 300, 309

F

Falk, M. H., 161
Fang, J., 159
Feelisch, M., 136
Feron, V. J., 298
Ferrari, D. M., 73
Finkel, T., 124
Fischle, W., 217
Flint, A. J., 47
Flohe, L., 25
Forrester, M. T., 295
Foster, M. W., 232
Fridovich, I., 25
Friedman, D. B., 157
Friedman, M., 25
Fu, C., 144
Fujii, S., 70
Fujino, G., 256–257, 279
Fukuda, H., 216
Fulton, D., 320
Funato, Y., 73

G

Galkin, A., 135
Gallogly, M. M., 291
Gandin, V., 257
Gao, L., 218
Garel, J. R., 40
Garland, P. B., 152
Gasdaska, P. Y., 246
Gavaret, J. M., 24–25
Gelderman, K. A., 157
Genestra, M., 278
Ghezzi, P., 230, 290–291
Gidday, J. M., 198
Gilbert, H. F., 170–171
Giles, G. I., 290
Gilmour, P. S., 217
Gladyshev, V. N., 245–253, 255–272
Glockshuber, R., 6
Go, Y. M., 165–178
Gotoh, Y., 278
Gow, A. J., 293–294
Gregory, P. D., 216–217
Griendling, K. K., 39
Grimsrud, P. A., 233, 239
Grippo, J. F., 200
Gromer, S., 117–118, 246, 248, 256

Gross, E., 24–25
Grozinger, C. M., 217
Grunstein, M., 216–217
Gundimeda, U., 257
Gutteridge, J. M. C., 188, 191

H

Hall, A., 3–4, 52
Halliwell, B., 188, 191
Hampton, M. B., 51–64, 123–144
Han, S. H., 70, 74
Hanahan, D., 271–272
Hansen, J. M., 171, 176, 316
Hansford, R. G., 127
Harada, D., 188
Harrison, J. H. Jr., 30
Hashemy, S. I., 199–200
Hashimoto, S., 71
Hatfield, D. L., 245–253, 255–272
Hayashi, S., 200
Hayashi, T., 183–186, 188–189
He, Y., 315–323
Hedley, D. W., 156, 159
Hedstrom, E., 257
Hennecke, J., 6
Hess, D. T., 232, 293
Hilger, M., 47
Hill, B. G., 85, 105, 299, 305
Hill, K. E., 117, 260
Himmelfarb, J., 182, 191
Hintze, K. J., 260
Hirata, H., 69
Hirota, K., 72–73
Hofer, S., 199, 203
Hofmann, B., 4
Hogg, N., 135
Holmgren, A., 2, 5, 16–17, 69–70, 73–74, 110, 113, 117, 124, 140, 199–200, 246, 248, 256–257, 260, 290–291, 316
Honeggar, M., 257, 263, 272
Horvath, J. J., 299
Horz, W., 216–217
Hovius, R., 111
Huang, P. L., 198
Huang, Q., 315–323
Hui, S. T., 75
Hui, T. Y., 74–75
Hurd, T. R., 123–144, 156–157, 233
Hwang, C., 169, 171

I

Ichijo, H., 277–287
Imai, H., 187–188
Imanishi, H., 30
Imhoff, B. R., 176
Ingram, D. K., 202
Ireland, R. C., 72

Irons, R., 255–272
Ischiropoulos, H., 120
Ishii, T., 233, 299
Ito, K., 218–221, 223–224, 227, 238
Iyer, S. S., 175
Izawa, S., 73

J

Jacob, C., 290
Jaffrey, S. R., 45, 232, 294
Jakupoglu, C., 256
Janatova, J., 182
Jang, H. H., 52
Janssen-Heininger, Y. M., 126–127, 135–136, 157, 278, 287, 289–295
Jeon, J. H., 74
Ji, Y., 290
Jiang, S., 169, 175–176
Johnstone, R. W., 217, 238
Jollow, D. J., 30
Jonas, C. R., 166, 168, 172–173
Jones, A. J., 24
Jones, D. P., 161, 165–178, 191–192, 200
Jönsson, T. J., 7, 11, 52
Juarez, J. C., 39
Jullien, M., 40

K

Kadokura, H., 6
Kagoshima, M., 219
Kamata, H., 40, 69
Kapinya, K. J., 198, 201
Karck, M., 201
Karplus, P. A., 1–19
Katagiri, K., 277–287
Kawai, H., 111
Keeney, P. M., 102
Kettenhofen, N. J., 85
Kiltz, H. H., 25
Kim, K. J., 111
Kim, S. O., 290
Kim, Y. C., 70
Kimura, Y., 150
Kingston, R. E., 216–217
Kinnula, V. L., 70
Kitano, H., 201
Klaidman, L. K., 299
Knight, Z. A., 25
Ko, F. C., 73
Kolle, D., 223–224
Kondo, N., 199, 202
Kouzarides, T., 216–217
Krause, G., 5
Krief, A., 24
Kuge, S., 73
Kuiper, H. C., 307
Kumar, S., 40–41, 47

Kumar, V., 53, 61–63, 140
Kuo, M. H., 216–217
Kurooka, H., 73
Kwon, J., 40, 45

L

Lam, J. B., 257
Lamas, S., 155
Lamprecht, M., 188
Landmesser, U., 317
Laragione, T., 157, 159
Larkin, M. A., 5
Lavu, S., 218
Lee, K. N., 74
Lee, S. R., 39–40, 71, 111, 116, 124, 140, 246
Lee, S. Y., 203
Lemasters, J. J., 85
Lenartowicz, E., 117
Leonard, S. E., 46
Lescure, A., 111
Leslie, N. R., 40
Levander, O. A., 260
Levengood, M. R., 24, 26
Levine, R. L., 230
Lewis, P. N., 40
Li, M., 167
Li, S., 39, 257
Liang, Y., 166–167, 169
Lilley, K. S., 157
Lillig, C. H., 126, 130, 246
Lim, J. C., 63
Lin, T. K., 85–87, 89, 155, 231
Liu, Y., 322
Liu, Z. B., 263, 272
Lou, Y. W., 39–40, 46
Low, F. M., 52, 54
Low, S. C., 263
Lowther, W. T., 52
Lu, J., 199, 203, 256
Luo, Y.,
Luthman, M., 113, 117
Lysko, H., 127–128

M

Ma, S., 24
Maalouf, M., 202
Mahadev, K., 39
Makino, Y., 73
Mallis, R. J., 290
Manevich, Y., 52
Mangelsdorf, D. J., 73
Mani, A. R., 153
Marsche, G., 181–193
Martin, J. L., 71
Martinez-Ruiz, A., 155
Maruyama, T., 73
Marwick, J. A., 225

Masri, M. S., 25
Mastroberardino, P. G., 47
Masutani, H., 67–76
Matlib, M. A., 151
Matsui, M., 70
Matsuo, Y., 67–76, 202
Matsushima, S., 316
Matsuyama, Y., 189, 191
Matsuzawa, A., 277–287
McDonagh, B., 144
McMonagle, E., 182, 191
Mega, T., 25
Meng, T. C., 39–40, 45, 290
Menger, K. E., 123–144
Merola, F., 6
Merrill, G. F., 256
Michan, S., 218
Mieyal, J. J., 291
Migliore, L., 238
Miller, R. A., 256
Min, W., 315–323
Miranda-Vizuete, A., 110–111, 124, 140
Mitsui, A., 70
Mitsumoto, A., 63
Miyazaki, K., 70
Mochizuki, M., 71
Mohanakumar, K. P., 205
Mohanty, J. G., 119
Moncada, S., 135
Montfort, W. R., 278
Montine, K. S., 299
Moodie, F. M., 218
Moon, K. H., 106
Moos, P. J., 256
Morell, J. L., 24
Morgan, M. E., 157
Moriarty-Craige, S. E., 161, 176
Mortier, E., 201
Motohashi, K., 247
Mueller, S., 30
Muller, F. L., 62
Murphy, M. P., 123–144, 155–156
Murtaza, I., 231
Mustacich, D., 110
Myers, C. R., 53
Myers, J. M., 53

N

Nakamura, H., 69–71, 161
Neel, B. G., 45
Nelson, K. J., 9
Niimura, Y., 4
Nishinaka, Y., 70, 74
Nishitoh, H., 278
Nishiyama, A., 74
Nkabyo, Y. S., 166, 173
Noda, M., 258
Noguchi, T., 278

Author Index

Nonn, L., 316
Nordberg, J., 256

O

Oettl, K., 181–193
Ogusucu, R., 52
Ohmiya, Y., 24–25
Oka, S., 74–75
Olm, E., 161
Ooi, L., 233–234
Orlando, V., 234
Orrenius, S., 278
Osada, H., 218
Ou, W., 157

P

Pacholec, M., 227
Parikh, H., 74
Paro, R., 234
Parola, M., 299
Parsonage, D., 1–19, 52
Patwari, P., 74
Perkins, D. N., 135
Persson, C., 40, 46
Peskin, A. V., 52, 54, 61
Peters, T., 182
Phillips, A. T., 25
Pigeolet, E., 25
Pineda-Molina, E., 291
Poole, L. B., 1–19, 46
Powis, G., 110, 278
Prime, T. A., 130, 135–137, 139

Q

Qin, J., 72

R

Rabilloud, T., 63
Rahman, I., 213–239
Rajendrasozhan, S., 218
Ramana, K. V., 297–311
Ramirez, A., 169, 176
Rao, A. K., 72
Rauhala, P., 199–200, 204–205, 208
Rayman, M. P., 256
Reed, D. J., 167, 177
Reeves, S. A., 1–19
Reichard, P., 73
Requejo, R., 61, 123–144
Retey, J., 24
Reynaert, N. L., 289–295
Reznick, A. Z., 231
Rhee, S. G., 23–32, 39, 52, 124, 159, 200, 203, 287
Riddles, P. W., 10
Rigobello, M. P., 109–120

Rittner, H. L., 298
Ritz, D., 73
Robino, G., 299
Rossignol, R., 151
Roth, S. Y., 216
Rouhier, N., 3, 203
Royall, J. A., 120
Ruef, J., 299
Rundlöf, A. K., 256–257

S

Sahaf, B., 157
Sahlin, L., 73
Saitoh, M., 71, 199–200, 206, 278–279, 282, 286
Salazar, M., 156
Salmeen, A., 39
Sandalova, T., 246
Sano, T., 201
Sasaki, H., 218
Sato, N., 70
Scarlett, J. L., 132
Schafer, F. Q., 124
Schagger, H., 89, 134
Schilling, B., 104
Schroder, E., 53, 61
Schuster, B., 24
Schwertassek, U., 157, 247
Sebastia, J., 156
Sedlic, F., 201
Selenius, M., 256–257
Seo, J. H., 63
Seo, M. S., 52
Seo, Y. H., 63
Sethuraman, M., 85, 89, 144, 233
Seyfried, J., 150
Shacter, E., 182
Shahid, M.,
Shan, X., 150
Sheehan, D., 239
Shelton, M. D., 291
Shen, X., 263, 272
Shepherd, D., 152
Sies, H., 256
Silver, J., 157
Sinclair, D., 218
Sinskey, A. J., 169, 171
Skalska, J., 149–162
Slaby, I., 5
Smith, A. D., 260
Smith, P. K., 128
Smith, R. A., 155
Snyder, S. H., 45, 294
Soejima, A., 191
Sogami, M., 182–183, 189
Soling, H. D., 73
Spange, S., 222
Spite, M., 299, 309

Spyrou, G., 73, 110, 124, 140
Srivastava, S., 297–311
Srivastava, S. K., 297–311
Stacey, M. M., 52
Stadtman, E. R., 230
Stadtman, T. C., 110, 117
Stamler, J. S., 293
Stauber, R. E., 182
Steinbrenner, H., 256
Stern, A., 26, 30–32
Strumeyer, D. H., 24
Sukumar, P., 157
Sullivan, S. G., 36, 39
Sumida, Y., 70
Sun, J. M., 239
Sun, Q. A., 246
Sundar, I. K., 213–239
Sundaresan, M., 36, 39
Suzuki, E., 188–189
Szabo, C., 125

T

Tagaya, Y., 69
Takagi, Y., 203
Takakura, K., 39
Takashima, Y., 291
Takeda, K., 278, 282, 286
Tamura, T., 110, 117
Tanaka, T., 73, 316–317
Tanner, K. G., 39–40
Taplick, J., 217
Tapuria, N., 198, 201
Tassi, S., 156
Taylor, E. R., 131–132
Terada, L. S., 40
Terawaki, H., 189, 191
Thomas, J. A., 290
Tobiume, K., 278–279, 282, 286–287
Tomida, M., 185, 188–189
Tomita, K., 217
Tompkins, A. J., 95, 97
Tonissen, K. F., 70
Tonks, N. K., 35–48
Torta, F., 85
Toyokuni, S., 231
Tsuji, P., 255–272
Turanov, A. A., 110, 245–253
Turell, L., 182
Turoczi, T., 198, 201

U

Uchida, K., 298
Ueno, M., 72
Umstead, T. M., 231–232
Unlu, M., 157
Urig, S., 118, 257

V

Valko, M., 39
van der Donk, W. A., 24, 26
Venkatraman, A., 85–89, 95, 97, 105–106
von Jagow, G., 89, 134

W

Wang, D., 73
Wang, H., 217
Wang, Q., 47
Wang, X., 137, 199, 207
Wardman, P., 62
Warnke, M. M., 300, 306
Watabe, S., 111–112
Watson, W. H., 263, 271–272
Weinberg, R. A., 271–272
Weinmann, A. S., 234
Wells, J. R., 70
Wendel, A., 26
Whorton, A. R., 39
Wickner, R. B., 24–25
Williams, C. H. J., 113
Wilm, M., 135
Winterbourn, C. C., 26, 30–32, 51–64, 125, 127
Witze, E. S., 85
Wolffe, A. P., 216–217
Woo, H. A., 61, 63, 233
Wood, I. C., 233–234
Wood, Z. A., 3–4, 52
Workman, J. L., 216–217
Wu, R. F., 40
Wudarczyk, J., 117
Wunderlich, M., 6

X

Xanthoudakis, S., 72
Xu, D., 39
Xu, L., 315–323
Xu, W. S., 218
Xu, X. M., 258, 263–264, 269–270, 272

Y

Yamada, S., 231
Yamamoto, Y., 9, 16
Yamawaki, H., 198, 201, 317
Yan, C., 257
Yan, Y., 234
Yang, J., 39
Yang, S. R., 217–218, 220–221, 225–228, 234
Yang, W., 222, 233
Yang, Y., 47
Yao, H., 213–239
Ying, J., 85
Yodoi, J., 67–76
Yoo, M. H., 255–272

Yoritaka, A., 299
Yoshihara, E., 67–76
Yu, Y. P., 161

Z

Zeng, H., 256
Zhang, H., 315–323
Zhang, J., 87
Zhang, R., 317, 322

Zhao, P., 198, 201
Zhe, X., 315–323
Zhong, L., 200, 203, 246, 248, 256
Zhou, F., 203
Zhou, H., 315–323
Zhu, L., 42
Zhu, X., 73
Zielonka, J., 62
Zuo, Z., 198, 201

Subject Index

A

Acrolein metabolism, 307–309
Activator protein-1 (AP-1), 72
Acute-on-chronic liver failure (ACLF), 190–191
Acute peritonitis, 309–310
Acute stimulus-induced reversible oxidation, 44–46
2', 5'-ADP-Sepharose 4B affinity chromatography, 114
Adult T-cell leukemia-derived factor (ADF), 69
Aging, albumin, 191–193
Albumin
  hemolysis, 186
  HPLC analysis
    HMA and HNA1 preparation, 184–185
    rat *vs.* bovine serum albumin, 186–187
    redox state, 184
    sample preparation, 185–186
    sample storage, 186, 187
    separation, 182–184
  nutritional supplementation, 188
  oxidation
    aging, 191–193
    aqueous humor, 189
    diabetes, 189
    exercise, 187–188
    hemodialysis, 191
    liver disease, 189–191
  plasma and serum, 185
  structure, 182, 183
Anchorage-independent growth assay, 261–262
Angiogenesis, 317–318
Antioxidant defense system, 124, 125
Antioxidant responsive element (ARE), 70
Aortic-ring assay, 320
Apoptosis, 207, 272
Apoptosis signal-regulating kinase 1 (ASK1), 71
  ASK1-induced EC apoptosis, 317
  assay materials
    antibodies, 282
    cell culture, 281
    plasmids, 282
    reagents, 282
    solutions, 282–283
  domain structure, 280–281
  endogenous ASK1 activation detection, phospho-ASK antibody, 285–286
  H_2O_2-induced activation, 280, 281
  kinase activity, 280–281
  oxidative stress, 278
  redox state, 279–280, 286–287
  ROS-dependent activation, 278–279
  and Trx interaction, 279–280
  in vitro binding assay
    ASK1 protein preparation, 284–285
    glutathione S-transferase-tagged thioredoxin (GST-Trx) preparation, 283–284
    procedure, 285
  in vitro kinase assay, 286
Aqueous humor, 189
Arteriogenesis, 318
Artifactual oxidation, 54–55, 128, 142
ASK1. *See* Apoptosis signal-regulating kinase 1 (ASK1)
Autoxidation, 26
2,2'-Azino-di-[3-ethylbenzthiazoline sulfonate] (ABTS), 319

B

Bacterial peroxiredoxin
  catalytic cycle, 3
  cellular reductant, 3–4
  rate limiting assay, 4
Bacterioferritin comigratory protein (BCP), 6, 18
β-elimination reaction, 24, 31
Bicinchoninic acid (BCA) assay, 128, 132
Biotin-conjugated cysteamine, 26–28
Biotin-switch assay
  cysteine group modifications, deacetylases, 233
  HDACs and SIRT1, 232–233
Blot-based detection, 28–29, 32
Blue native gel electrophoresis (BN-PAGE), 86
  2D BN-PAGE, protein separation
    IBTP-labeled mitochondrial protein detection, 96
    recipes, 93
    SYPRO® Ruby stain, 93
  first dimension gel electrophoresis
    Coomassie Blue R-250, 91–92
    preparation, 89, 91
    recipes, 92

C

Caspase activity assay, 207, 322–323
Cell apoptotic assay, 322

Centrifugal gel filtration, 128, 130, 142
Chemical preconditioning, 200–201
Chemotaxis and invasion assays, 262
ChIP. *See* Chromatin immunoprecipitation (ChIP) assay
Chromatin immunoprecipitation (ChIP) assay
  consideration, 237–238
  flow chart, 234
  protein–DNA interactions, 233–234
  using cells, 234–236
  using tissues, 236–237
2-Cys Prxs, redox transformations. *See* Peroxiredoxins (Prxs)
Cysteinylation, 6
Cysteinyl-labeling assay, 40–41, 44
Cytochrome c release assay, 207, 323
Cytosolic thiol status measurement, 156–157

D

DEAE-Sephacel chromatography, 113
Dehydroalanine (DHA)-containing glutathione peroxidase (DHA-GPx)
  biotin-conjugated cysteamine preparation, 27–28
  blot-based detection, RBC, 28–29
  DHA
    detection, 24–25
    residues, 24
  GPx oxidative inactivation and conversion
    biotinylation, 26–27
    GPx catalysis, 25
    GPx Sec to DHA conversion, 25–26
    hemoglobin autoxidation, 26
  oxidative damage protection, 31
  oxidative stress effect, RBCs
    aging, 29–30
    environmental chemicals, 30
    hematocrit, 30–31
    reduced and oxidized glutathione (GSH:GSSG) ratio, 30
Diabetes, 189
Dihydroxynonanol
  esterified forms, 303–305
  synthesis and characterization, 303
Dimerization, peroxiredoxins
  adherent cell alkylation, 56
  artefactual oxidation, limitation, 54–55
  Jurkat cells, 57–59
  nonreducing electrophoresis, 56–57
  Prx Western blotting, 56–57
  redox immunoblot method, 55
  suspension cell alkylation, 55–56
Disulfide bonded fluorescent reporter, 7–8
5,5'-dithiobis(2-nitrobenzoic acid) (DTNB), 199
Dithionitrobenzoic acid (DTNB) assay, 127–128
Dithiothreitol (DTT), 293
Dual-specificity phosphatases (DSPs), 36–39

E

Electron transfer detection. *See* Fluorescence-based peroxidase activity assay
Enals
  diseased tissues, 298–299
  HNE (*see* 4-Hydroxy trans-2-nonenal (HNE))
Epigenetics
  chromatin immunoprecipitation (ChIP) assay
    consideration, 237–238
    flow chart, 234
    protein-DNA interactions, 233–234
    using cells, 234–236
    using tissues, 236–237
  cytoplasmic and nuclear protein preparation
    from cells, 229
    from tissues, 230
  HAT activity assay, 222–223
  HDAC activity assay
    commercial kits, 226
    HDAC classification, 217–218
    [^{3}H]-labeled histones, 223–225
    immunoblotting, 227–228
    immunoprecipitation, 226–227
    posttranslational modifications, 228
  histone acetylation
    gene expression, 216–217
    [^{3}H]-acetate incorporation, 219–221
    immunoblotting, 221
  redox-mediated posttranslational modification assays
    biotin-switch assay, 232–233
    protein carbonylation, 230–232
  SIRT posttranslational modifications, 228
  whole cell lysate preparation, 229
Exofacial thiol status
  BIAM staining, 159
  cancer
    extracellular microenvironment, 161
    proliferation, 161
    redox state, 160–161
  fractionation, 158–159, 160
  mercury orange staining, 159
  mixed disulfides, 159–160
  in situ measurement, 157–158

F

Fluorescence-based peroxidase activity assay
  buffer, 14
  fluorescence measurement, 14–15
  Hanes plot, AhpC and t-butyl hydroperoxide, 15, 16
  reaction rate, S128W NTD, 15
  solutions, 13–14
  stopped-flow spectrophotometer, 14–15
Fluorescent redox reporters
  bacterial peroxiredoxin
    catalytic cycle, 3

Subject Index

cellular reductant, 3–4
rate limiting assay, 4
disulfide bonded reporter, electron transfer, 7–8
engineering material
  chemical modification agents, 9
  proteins, 9
  solution, 8–9
fluorescence-based peroxidase activity assay, AhpC with S128W NTD
  buffer, 14
  fluorescence measurement, 14–15
  Hanes plot, 15, 16
  reaction rate, 15
  solutions, 13–14
  stopped-flow spectrophotometer, 14–15
F6W mutant Grx1
  BCP, 18
  catalytic activities, 16–17
  expression, 15–16
  fluorescence, 16, 17
modified AhpC protein
  AhpC-FAM(165) reduction, 11–12
  fluorescein fluorescence, 10–11
  target protein modification, 10
  two-step procedure, 10
N-terminal domain (NTD), AhpF and *E. coli* Grx1
  Trp into electron transfer protein, 6–7
  Trx-fold protein sequence alignment, 4–6
reaction, 2–3
S128W NTD fluorescence and activity, 12–13
tryptophan (Trp), 2
Forkhead box class O (FoxO), 75

G

Glutaredoxins (Grx), 291
Glutathione peroxidase (GPx)
  dehydroalanine (*see* Dehydroalanine (DHA)-containing glutathione peroxidase (DHA-GPx))
Glutathione system, 246
S-Glutathionylated protein
  glutathionylation
    glutaredoxins, 291
    GSH, 290–291
  in situ detection, 291–292
Glutathionylation
  glutaredoxins, 291
  glutathionylated proteins and cysteine residues
    BN-PAGE, 134
    MALDI-TOF-TOF mass spectrometer, 135
    SDS–PAGE, 134–135
  GSH, 290–291
  protein-bound glutathione, 131–132
  recycling assay, 132–133

Glutathionyl conjugates, enals
  enals, diseased tissue, 298–299
  esterified glutathionyl conjugates
    characterization, 304–305
    synthesis, 303–304
  GS-HNE
    characterization, 301–302
    purification and quantification, 301
    synthesis, 300–301
  HNE
    acrolein metabolism, 307–309
    cellular metabolism, 305–306
    redox signaling, 310–311
    signaling properties, 309–310
    systemic metabolism, 306–307
  lipid oxidation, 298
  reagent GS-DHN synthesis and characterization, 303
Grx1-catalyzed cysteine derivatization protocol
  confocal imaging, 293
  nuclear counter stains, 292–293
  schematic representation, 291

H

HAT. *See* Histone acetyltransferase (HAT) activity assay
HDAC. *See* Histone deacetylase (HDAC) activity assay
Hemodialysis, 191
Hemolysis, 186
Histone acetylation assays. *See also* Histone acetyltransferase (HAT) activity assay
  gene expression, 216–217
  [^{3}H]-acetate incorporation
    cell culture, 219
    procedure, 220–221
    protein extraction, 219–220
    radioactive acetyl-CoA, 219
  immunoblotting, 221
Histone acetyltransferase (HAT) activity assay, 222–223
Histone deacetylase (HDAC) activity assay, 216
  biotin-switch assay, 232–233
  commercial kits, 226
  HDAC classification, 217–218
  [^{3}H]-labeled histones
    HDAC assay sample preparation, 223–224
    preparation, 223
    procedure, 224–225
  immunoblotting, 227–228
  immunoprecipitation, 226–227
  posttranslational modifications, 228
HMA. *See* Human mercaptalbumin (HMA)
HNA1. *See* Human nonmercaptalbumin1 (HNA1)
Hoechst 33258 staining, 207
Hormesis, 198–199

Human mercaptalbumin (HMA), 182
Human nonmercaptalbumin1 (HNA1), 182
Human retinal pigment epithelial (hRPE) cells, EhCySS
   age-related macular degeneration, 176
   oxidant-induced apoptosis, 175–176
Hydroxyl radical trapping, 204
Hydroxypropyl mercapturic acid (HPMA), 307–309
4-Hydroxy trans-2-nonenal (HNE)
   acrolein metabolism, 307–309
   cellular metabolism, 305–306
   glutathione conjugate-1,4-dihydroxynonanol (GS-DHN)
      esterified forms, 303–305
      synthesis and characterization, 303
   glutathionyl conjugates (GS-HNE)
      characterization, 301–302
      esterified forms, 303–305
      purification and quantification, 301
      synthesis, 300–301
   JNK activation, 299
   redox signaling, 310–311
   signaling properties, 309–310
   systemic metabolism, 306–307
   VSMC, 299–300
Hyperoxidation, peroxiredoxins (Prxs)
   general method, 60–61
   Jurkat cells, 61, 62
   limitation, 63
   redox immunoblot method, 59–60
Hypoxia-inducible factor (HIF), 72
Hypoxic glove box, 43

I

4-Iodobutyl triphenylphosphonium (IBTP) thiol tagging strategy
   BN-PAGE, 86
   immunoblotting protocol
      IBTP-labeled protein thiol detection, 94
      posttranslational modification, 97
      protein modification, 104–105
      specific protein amount, 105
   mass spectrometry, 95–97
   mitochondrial protein labeling
      pretreatment, 88–89
      protocol, 87–88
      SDS–PAGE approach, 88
   triphenylphosphonium (TPP) moiety, 85–86
Ischemic preconditioning, 201
Isoprostane, 319

K

Keratinocyte growth factor (KGF)-stimulated cell proliferation, 173

L

Lipid peroxidation, 205
Liver disease, 189–191

M

Metastatic analysis, 268–269
Michael addition reaction, 25, 27
Microarray analysis, 174–175
Mitochondrial protein thiols
   antioxidant defense system, 124, 125
   BN-PAGE, oxphos complexes, 90
      1D BN-PAGE gels, 89, 91–92
      preparation and running, 92–93
      protein complex separation, 92–93
   glutathionylation
      glutathionylated proteins and cysteine residues, 133–135
      oxidative stress, 130
      protein-bound glutathione, 131–132
      recycling assay, 132–133
   GSH/GSSG ratio, 124
   IBTP-labeled protein thiol detection, 94
   imaging and analysis, 95–97
   liver mitochondria
      IBTP labeling, 87–89
      preparation, 86–87
   mass spectrometry identification
      complex I band, 98–99
      complex III band, 100–101
      complex IV band, 101–102
      complex V band, 103–104
   peroxiredoxin 3, Western blotting, 140
   posttranslational protein thiol modifications, 85–86
   quantification
      bicinchoninic acid (BCA) assay, 128
      DTNB assay, 127–128
      experimental scheme, 129
      oxidative stress, 128, 130
      solvent-exposed thiols, 127
   reactive species, 84–85
   redox modification and reversal mechanism, 125–126
   redox signaling, 126–127
   ROS and RNS, 124–125
   S-nitrosated protein thiols, 135
      protein S-nitrosothiols, 136–137
      selective labeling, 137, 139
      visualization, 138
   thioredoxin redox, PEGylation assay, 140–143
   toxic effects, 84–85
Mitochondrial thiol status measurement
   mitochondria
      ATP synthesis, 150
      glutathione (GSH), 150

Subject Index

isolation, 151–152
pH gradient, 150
in situ, iodobutyl triphenylphosphonium (IBTP), 155–156
thiol reagents
  assay reagent, 153
  bioGEE labeling, 155
  biotin-PEO-maleimide labeling, 154
  incubation medium, 153
  SDS–PAGE, 153–154
  yield and purity, 152–153
Mitochondrial thioredoxin reductase
  activity estimation, 117–118
  2', 5'-ADP-Sepharose 4B affinity chromatography, 114
  ω−Aminohexyl-Sepharose 4B, 114
  ammonium sulfate fractionation, 113
  cell signaling, 118–120
  DEAE-Sephacel chromatography, 113
  heat treatment, 113
  inhibitor studies, 118, 119
  mitochondria
    freeze/thaw cycles and disruption, 112–113
    preparation and purification, 111–112
  splice variants, 100
  TrxR2 purification, 116
Mitogen-activated protein kinase (MAPK) pathway, 173

N

N-acetylcysteine-1,4-dihydroxynonanol (NAC-DHN), 307–308, 310
NADPH, 246, 250, 252
Nitric oxide (NO) bioavailability, 317–318
Nitric oxide synthase (NOS1) activity, 204–205
S-Nitrosated protein
  protein S-nitrosothiol quantification, 136–137
  selective labeling, 137, 139
  S-nitrosation, 135–136
  visualization, 138
S-Nitrosylated protein
  nitrosylation, 293–294
  schematic representation, 294
  in situ detection, 294–295
S-Nitrosylation, 200, 293–294
Nod-like receptor proteins (NLRP3) inflammasome complex dissociation, 76
Nuclear factor-erythroid 2-related factor 2 (Nrf2) activation, 176

P

Palmitoylation, 45
Peroxiredoxins (Prxs)
  bacterial peroxiredoxin
    catalytic cycle, 3
    cellular reductant, 3–4
    rate limiting assay, 4

catalytic cycle, 53
dimerization measurement
  adherent cell alkylation, 56
  artefactual oxidation, limitation, 54–55
  Jurkat cells, 57–59
  nonreducing electrophoresis, 56–57
  Prx Western blotting, 56–57
  redox immunoblot method, 55
  suspension cell alkylation, 55–56
hyperoxidation measurement
  acidic shift assays, 63–64
  general method, 60–61
  Jurkat cells, 61, 62
  limitation, 63
  redox immunoblot method, 59–60
  sulfinic acid, 52–53
kinetic studies, 52
mammals, 53–54
peroxiredoxin 3, Western blotting, 140
redox-sensitive probes, 62
pK_a values, 290
Polyoma virus enhancer-binding protein-2 (PEBP-2), 72
Posttranslational modification
  glutathionylation, 130–131
  HDAC activity assay, 228
  mitochondrial protein thiol, 85–86
  S-nitrosylation, 293–294
  redox-mediated assay
    biotin-switch assay (*see* Biotin-switch assay)
    protein carbonylation (*see* Protein carbonylation)
  SIRT, 228
Programmed cell death, 199
Protein carbonylation
  covalent SIRT1 modification, 231
  proteomics approach, 231–232
  reaction, 230–231
Protein thiol detection, mitochondrial oxidative phosphorylation complex
  BN-PAGE, oxphos complexes
    1D BN-PAGE gels, 89, 91–92
    preparation and running, 92–93
    protein complex separation, 92–93
  IBTP-labeled protein thiol detection, 94
  imaging and analysis, 95–97
  liver mitochondria
    IBTP labeling, 87–89
    preparation, 86–87
  mass spectrometry identification
    complex I band, 98–99
    complex III band, 100–101
    complex IV band, 101–102
    complex V band, 103–104
  posttranslational protein thiol modifications, 85–86
  reactive species, 84–85
  toxic effects, 84–85

Protein tyrosine kinase (PTK), 36
Protein tyrosine phosphatase (PTP) analysis
  cell lysate preparation, 43–44
  cysteinyl-labeling assay
    principle, 40–41
    procedure, 44
  hypoxic glove box preparation, 43
  lysis buffer preparation, 42–43
  PTP
    active-site structure, 37
    acute stimulus-induced reversible oxidation, 44–46
    catalysis, 37–39
    detection method, 39–40
    inactivation, 47–48
    member structural diversity, 36–37
    oxdiation, 38–39
    proteome characteristics, 46–47
    PTK, 36
    reactivity, 38
    reversibly oxidization, 45–46
  solutions, 42
Protein tyrosine phosphatase 1B (PTP1B), 71
PTP. *See* Protein tyrosine phosphatase (PTP) analysis
pU6-m3 knockdown construct, 258
Pyridine nucleotide-disulfide oxidoreductase, 246, 253

Q

QuikChange method, 15

R

Redox clamp model
  Cys/CySS redox potential (EhCySS)
    aortic endothelial cells and monocytes, 173–174
    cell density, 172
    cysteinylation, 170
    Eh verification, potentiometric electrode, 169–170
    glutamine (Gln), 173
    human colon carcinoma cell line (Caco-2), 172–173
    human retinal pigment epithelial (hRPE), 175–176
    keratinocyte growth factor (KGF)-stimulated cell proliferation, 173
    microarray analysis and spectrometry-based proteomics, 174–175
    mitochondrial signaling pathway, 174
    mitogenic p44/p42 mitogen-activated protein kinase (MAPK) pathway, 173
    nuclear factor-erythroid 2-related factor 2 (Nrf2) activation, 176
    pool size, 168–169, 171
    profibrotic signaling pathway activation, 176
    proinflammatory signaling, 175
    protein synthesis, 170
    semilog plots, 171
    serum, 170
    thiol and disulphide concentration, 167–168
  GSH/GSSG redox potentials (E_hGSSG)
    colon carcinoma cell line (Caco-2), 173
    human plasma, 166–167, 169
    pool size, 172
Redox factor-1 (Ref-1), 72
Redox-mediated posttranslational modification assays, 230–233
Redox signaling
  ASK1 regulation (*see* Apoptosis signal-regulating kinase 1 (ASK1))
  epigenetics (*see* Epigenetics)
  mitochondrial protein thiols (*see* Mitochondrial protein thiols)
  redox clamp model (*see* Redox clamp model)
  thioredoxin (*see* Thioredoxin)
Reversible cysteine oxidation, 295
RNAi technology, 263

S

^{75}Se-labeling, 260
Selegiline, 203
Selenocysteine insertion sequence (SECIS), 263, 264
Selenoenzyme. *See* Thioredoxin reductase 1 (TR1)
Selenoproteins, 256
Sirtuins (SIRT1)
  activity assay, 226–227
  biotin-switch assay, 232–233
  posttranslational modifications, 228
  protein carbonylation, 231–232
S-Nitrosylated glutathione (GSNO). *See* S-Nitrosylated protein
Sodium dodecyl sulfate–polyacrylamide gel electrophoresis (SDS–PAGE), 250, 253
SpeedVac system, 28
SYPRO® Ruby stain, 93

T

Thiol redox transitions
  reactive oxygen species (ROS), 68
  redox-thiol modification, 68–69
  thiol-redox regulation
    kinase-mediated cellular signal transduction, 71
    nuclear receptors and transcription factors-mediated cellular signal transduction, 72–73
  thioredoxin
    adult T-cell leukemia-derived factor (ADF), 69

Subject Index

mice experiments, 70
stress-responsive elements, 70
structure, 69
TRX expression, 69–70
thioredoxin-binding protein-2 (TBP-2)
binding motif, 74
biological functions, 74
FoxO4, p300/CBP-mediated acetylation, 75–76
insulin sensitivity and secretion, 75
NLRP3 inflammasome complex dissociation, 76
thioredoxin superfamily, 73–74
Thiol status measurement
cytosol, 156–157
exofacial thiol status
BIAM staining, 159
cancer, 160–161
fractionation, 158–159, 160
mitochondria
ATP synthesis, 150
glutathione (GSH), 150
isolation, 151–152
pH gradient, 150
in situ, IBTP, 155–156
thiol reagents, 153–155
yield and purity, 152–153
Thioredoxin (Trx). *See also* Apoptosis signal-regulating kinase 1 (ASK1); Thioredoxin reductase 1 (TR1)
apoptosis, 207
endogenous Trx redox function, 206
ex vivo autoradiographic brain image, 207–208
hormetic mechanism, 198–199
materials, 208
mimetics, 202–203
molecular biology method, 206–207
PEGylation assay, 140–143
preconditioning protection
preclinical and clinical studies, 200–202
and signaling, cells, 202–203
redox function, 199–200
redox protein thioredoxin-2 (Trx2), vasculature
angiogenesis, 317–318
arteriogenesis, 318
ASK1-induced EC apoptosis, 317
caspase-9 activity assay, 322–323
cell apoptotic assay, 322
cellular ROS and inflammation, 316
cytochrome *c* release assay, 323
EC culture, 321–322
echocardiograms, 320
generation and characterization, 319
hemodynamic studies, 319–320
isoprostanes, 319
localization, 322
mitochondria, 316

NO bioavailability, 317–318
ROS measurement, 321–322
serum NO level assessment, 320–321
total antioxidant studies, 319
short-lived free radicals
hydroxyl radical trapping, 204
lipid peroxidation, 205
NOS1 activity, 204–205
statistical analysis, 209
in vitro preconditioning human cell model, 203–204
Thioredoxin reductase 1 (TR1)
apoptosis, 272
cancer
prevention, 256
promotion, 257
cells and tissues, 271
expression monitoring
Northern blot analysis, 259
^{75}Se-labeling, 260
thioredoxin reductase activity assay, 260–261
Western blot analysis, 260
knockdown, 264
level assessment, 263–264
mammalian cell lines and mice, 257–258
removal, RNAi technology, 263
selenocysteine insertion sequence (SECIS), 263
selenoproteins, 256
targeted removal and reexpression, 258–259
TR1-deficient cells
phenotypic changes, 264–266
TR1 reexpression, 269–270
tumorigenesis, 266–268
in vitro and *in vivo* metastatic analysis, 268–269
tumor formation activity
anchorage-independent growth assay, 261–262
tumor formation assays, 262
in vitro chemotaxis and invasion assays, 262
Thioredoxin reductase activity assay, 260–261
Thioredoxin reductases (TrxRs)
accessibility and high-reactivity selenocysteine, 246–247
cancer etiology (*see* Thioredoxin reductase 1 (TR1))
cellular pathways, 246
forms, 246
isoforms, 100
mitochondria
2', 5'-ADP-Sepharose 4B affinity chromatography, 114
ω–Aminohexyl-Sepharose 4B, 114
ammonium sulfate fractionation, 113
cell signaling, 118–120
DEAE-Sephacel chromatography, 113
freeze/thaw cycles and disruption, 112–113

Thioredoxin reductases (TrxRs) (cont.)
  heat treatment, 113
  inhibitor studies, 118, 119
  preparation and purification, 111–112
  TrxR2 purification, 116
 protein affinity resin preparation
  CNBr-activated matrices, 247
  intermediate selenenylsulfide formation, 247–248
  materials, 249
  mechanism-based method, 248
  method, 249–250
  recombinant mammalian forms, 247
 splice variants, 100
 substrate identification, 247
 target identification, cell lysate
  C and SS forms, 251
  materials, 250, 252
  method, 252
  scheme, 251
  SDS-PAGE, 250
  TR-affinity columns, 250
  TrxR2, 100–111
Transmembrane thioredoxin-related protein (TMX), 73–74
Triphenylphosphonium (TPP) moiety, 85–86
Trx. See Thioredoxin (Trx)
Tumor formation assays, 262
Tumorigenesis, 266–268

V

Vascular smooth muscle cells (VSMC), 299–300

W

Whole cell lysates, 229